Wood Adhesives

Wood Adhesives

Edited by

A. Pizzi and K. L. Mittal

CRC Press
Taylor & Francis Group
Boca Raton London New York

CRC Press is an imprint of the
Taylor & Francis Group, an **informa** business

CRC Press
Taylor & Francis Group
6000 Broken Sound Parkway NW, Suite 300
Boca Raton, FL 33487-2742

© 2011 by Taylor & Francis Group, LLC
CRC Press is an imprint of Taylor & Francis Group, an Informa business

First issued in paperback 2019

No claim to original U.S. Government works

ISBN 13: 978-0-367-44596-6 (pbk)
ISBN 13: 978-90-04-19093-1 (hbk)

Visit the Taylor & Francis Web site at
http://www.taylorandfrancis.com

and the CRC Press Web site at
http://www.crcpress.com

Contents

Preface ix

Part 1: Fundamental Adhesion Aspects in Wood Bonding

Natural Lignans as Adhesives for Cellulose: Computational Interaction Energy *vs* Experimental Results
M. Sedano-Mendoza, P. Lopez-Albarran and A. Pizzi 3

Evaluation of Some Synthetic Oligolignols as Adhesives: A Molecular Docking Study
C. Martínez, M. Sedano, A. Munro, P. López and A. Pizzi 21

Modification of Sugar Maple (*Acer saccharum*) and Black Spruce (*Picea mariana*) Wood Surfaces in a Dielectric Barrier Discharge (DBD) at Atmospheric Pressure
F. Busnel, V. Blanchard, J. Prégent, L. Stafford, B. Riedl, P. Blanchet and A. Sarkissian 35

Determination of the Microstructure of an Adhesive-Bonded Medium Density Fiberboard (MDF) using 3-D Sub-micrometer Computer Tomography
G. Standfest, S. Kranzer, A. Petutschnigg and M. Dunky 49

Influence of the Degree of Condensation on the Radial Penetration of Urea-Formaldehyde Adhesives into Silver Fir (*Abies alba*, Mill.) Wood Tissue
I. Gavrilović-Grmuša, J. Miljković and M. Điporović-Momčilović 63

Radial Penetration of Urea-Formaldehyde Adhesive Resins into Beech (*Fagus Moesiaca*)
I. Gavrilović-Grmuša, M. Dunky, J. Miljković and M. Djiporović-Momčilović 81

A Flexible Adhesive Layer to Strengthen Glulam Beams
M. Brunner, M. Lehmann, S. Kraft, U. Fankhauser, K. Richter and J. Conzett 97

Properties Enhancement of Oil Palm Plywood through Veneer Pretreatment with Low Molecular Weight Phenol-Formaldehyde Resin
Y. F. Loh, M. T. Paridah, Y. B. Hoong, E. S. Bakar, H. Hamdan and M. Anis 135

Reaction Mechanism of Hydroxymethylated Resorcinol Adhesion
Promoter in Polyurethane Adhesives for Wood Bonding
A. Szczurek, A. Pizzi, L. Delmotte and A. Celzard   145

Part 2: Synthetic Adhesives

Optimization of the Synthesis of Urea-Formaldehyde Resins using
Response Surface Methodology
*J. M. Ferra, P. C. Mena, J. Martins, A. M. Mendes, M. R. N. Costa,
F. D. Magalhães and L. H. Carvalho*   153

Characterization of Urea-Formaldehyde Resins by GPC/SEC and HPLC
Techniques: Effect of Ageing
*J. M. Ferra, A. M. Mendes, M. R. N. Costa, F. D. Magalhães and
L. H. Carvalho*   171

Formaldehyde-Free Dimethoxyethanal-Derived Resins for Wood-Based
Panels
M. Properzi, S. Wieland, F. Pichelin, A. Pizzi and A. Despres   189

Melamine–Formaldehyde Resins without Urea for Wood Panels
E. Pendlebury, H. Lei, M.-L. Antoine and A. Pizzi   203

Bonding of Heat-Treated Spruce with Phenol-Formaldehyde Adhesive
M. Kariz and M. Sernek   211

Influence of Nanoclay on Phenol-Formaldehyde and
Phenol-Urea-Formaldehyde Resins for Wood Adhesives
H. Lei, G. Du, A. Pizzi, A. Celzard and Q. Fang   225

Emulsion Polymer Isocyanates as Wood Adhesive: A Review
K. Grøstad and A. Pedersen   235

Adhesives for On-Site Rehabilitation of Timber Structures
H. Cruz and J. Custódio   261

Part 3: Environment-Friendly Adhesives

Thermal Characterization of Kraft Lignin Phenol-Formaldehyde Resin
for Paper Impregnation
A. R. Mahendran, G. Wuzella and A. Kandelbauer   291

Acacia mangium Tannin as Formaldehyde Scavenger for Low Molecular
Weight Phenol-Formaldehyde Resin in Bonding Tropical Plywood
*Y. B. Hoong, M. T. Paridah, Y. F. Loh, M. P. Koh, C. A. Luqman and
A. Zaidon*   305

Synthesis of Modified Poly(vinyl acetate) Adhesives
A. Salvini, L. M. Saija, M. Lugli, G. Cipriani and C. Giannelli   317

Acrylated Epoxidized Soy Oil as an Alternative to Urea-Formaldehyde
in Making Wheat Straw Particleboards
M. Tasooji, T. Tabarsa, A. Khazaeian and R. P. Wool   341

Gluten Protein Adhesives for Wood Panels
H. Lei, A. Pizzi, P. Navarrete, S. Rigolet, A. Redl and A. Wagner   353

Wood Panel Adhesives from Low Molecular Mass Lignin and Tannin
without Synthetic Resins
P. Navarrete, H. R. Mansouri, A. Pizzi, S. Tapin-Lingua,
B. Benjelloun-Mlayah, H. Pasch and S. Rigolet   367

Part 4: Wood Welding and General Paper

Wood-Dowel Bonding by High-Speed Rotation Welding — Application
to Two Canadian Hardwood Species
G. Rodriguez, P. Diouf, P. Blanchet and T. Stevanovic   383

Chemical Changes Induced by High-Speed Rotation Welding of Wood
— Application to Two Canadian Hardwood Species
Y. Sun, M. Royer, P. N. Diouf and T. Stevanovic   397

Moisture Sensitivity of Scots Pine Joints Produced by Linear Frictional
Welding
M. Vaziri, O. Lindgren, A. Pizzi and H. R. Mansouri   415

High Density Panels Obtained by Welding of Wood Veneers without any
Adhesives
H. R. Mansouri, J.-M. Leban and A. Pizzi   429

Overview of European Standards for Adhesives Used in Wood-Based
Products
F. Simon and G. Legrand   435

Preface

Wood adhesives are of tremendous industrial importance as more than two-thirds of wood products today in the world are totally, or at least partially, bonded together using a variety of adhesives. The reason being that adhesive bonding offers many advantages *vis-à-vis* other joining methods for wood components.

Even a cursory look at the literature will evince that currently there is brisk R&D activity in devising new wood adhesives or ameliorating the existing ones. The modern mantra in all industrial sectors is: 'Think green, go green' and the wood industry is no exception. This new refrain has spurred much research activity in synthesizing environmentally-benign and human-friendly wood adhesives. One specific example is the elimination of formaldehyde emissions from wood adhesives and many alternate avenues have been explored in this vein.

Considering the industrial and commercial importance of wood adhesives and the high tempo of research in understanding and improving adhesion strength of wood adhesives we decided to bring out this special volume which reflects the collective wisdom of a contingent of world-class researchers in this technologically highly important field.

This book is based on the Special Issue of the *Journal of Adhesion Science and Technology* (*JAST*) which was published as Vol. 24, Nos 8–10 (2010). Based on the widespread interest and tremendous importance of wood adhesives in construction and other industries, we decided to bring out this book as a single and easily available source of information. The papers as published in the above-mentioned Special Issue have been rearranged in a more logical fashion in this book.

This book contains a total of 28 papers (reflecting overviews and original research) covering many ramifications of wood adhesives and is divided into four parts as follows: Part 1: Fundamental Adhesion Aspects in Wood Bonding; Part 2: Synthetic Adhesives; Part 3: Environment-Friendly Adhesives; and Part 4: Wood Welding and General Paper. The topics covered include: computation of interfacial interactions between resin oligomers and cellulose substrates by theoretical molecular mechanics means; treatment of wood surfaces by atmospheric pressure plasma; application of computer tomography in determination of microstructures in bonded fiberboards; penetration of adhesives in wood species; veneer pretreatment; strengthening of glulam beams; resorcinol adhesion promoter; synthesis and characterization of a variety of adhesives for wood bonding; influence of fillers (e.g., nanoclay) on resins for wood bonding; adhesives for on-site rehabilitation of timber structures; development of adhesives to eliminate or reduce formaldehyde emission; gluten protein adhesives for wood panels; lignin- and tannin-based adhesives; wood welding; wood connections without use of adhesives; and overview of European standards for adhesives used in wood-based products.

We sincerely hope this book containing bountiful up-to-date information, including some novel approaches to wood bonding/wood connections, will be of great value to anyone interested in this highly industrially important topic. This book should serve as repository of information and a commentary on current R&D activity dealing with wood adhesives. We further hope this book will serve as a fountainhead for new research ideas to further enhance the performance and durability of wood-based products.

Acknowledgements

Now it is our great pleasure to thank all those who made this book possible. First and foremost our thanks are extended to the authors for their interest, enthusiasm, cooperation and contribution without which this book could not be materialized. We profusely thank the unsung heroes (reviewers) for their time and efforts in providing many valuable comments as comments from peers are *sine qua non* to maintain the highest standard of publication. We would be remiss if we do not extend our appreciation to the appropriate individuals at Brill (publisher) for giving this book a body form.

A. PIZZI
ENSTIB-LERMAB
University Henri Poincare — Nancy 1
BP 1041
88051 Epinal cedex 9, France
E-mail: antonio.pizzi@enstib.uhp-nancy.fr

K. L. MITTAL
P.O. Box 1280
Hopewell Jct., NY 12533, USA

Part 1

Fundamental Adhesion Aspects in Wood Bonding

Natural Lignans as Adhesives for Cellulose: Computational Interaction Energy *vs* Experimental Results

M. Sedano-Mendoza [a,b], **P. Lopez-Albarran** [b,*] **and A. Pizzi** [a]

[a] ENSTIB-LERMAB, Nancy University, University Henri Poincaré, 27 rue du Merle Blanc, BP 1041, 88051 Epinal, France

[b] Facultad de Ingenieria en Tecnologia de la Madera, Universidad Michoacana de San Nicols de Hidalgo, Campus Morelia, Av. Fco. J. Muciga S/N, CP: 58030, Morelia, Michoacan, Mexico

Abstract
Comparison between the molecular mechanics calculated energy of interaction of lignan dimers and trimers with a cellulose I crystallite and the experimental values of Young's modulus obtained by thermomechanical analysis (TMA) of cellulose paper impregnated with low molecular weight lignins showed good correlation between calculated and experimental results. The oligomer composition of the four low molecular weight lignins tested was obtained by MALDI-TOF mass spectrometry. This showed that these lignins were predominantly composed of dimers and trimers rendering them ideal for correlation testing. The lignan/cellulose crystallite interaction energy is determined by the oligomer molecular weight as well as the type of linkage within the lignan oligomers. Lignans with higher molecular weight in which the units are linked as β–O–4 give interaction energy values indicating stronger attraction with cellulose.

Keywords
Adhesives, molecular mechanics, lignans, lignins, thermomechanical analysis, interaction energy, cellulose

1. Introduction

Lignin is the second most abundant renewable biopolymer in the world, exceeded in quantity only by cellulose, comprising 30% of all non-fossil organic carbon [1] and constituting from a quarter to a third of the dry mass of wood. Its precursors are three monolignol monomers, methoxylated to various degrees by enzymatic reaction *via* a free radical route: *p*-coumaryl alcohol, coniferyl alcohol, and sinapyl alcohol [2].

The amounts and types of monolignols produced or present in lignin depend on the treatment that the raw lignin has undergone. Guaiacyl units are usually more abundant in softwood than hardwood, the latter one presenting a similar proportion of guaiacyl and syringyl units. Lignin in solution has a composition which differs

---

* To whom correspondence should be addressed. E-mail: plopez@umich.mx

Wood Adhesives

from protolignin (native lignin as present in wood before extraction) due to the extraction process used. However, in general, the most abundant linkages between lignin units are the β–O–4, β5 and $\beta\beta$ [3].

The need to substitute synthetic thermosetting wood adhesives with more environmentally acceptable resins has led to intense research on adhesives derived from natural, non-toxic materials. Extensive reviews on the subject exist [4, 5]. It is sufficient here to state that lignin is one of the materials at the forefront of these studies. Numerous wood adhesive fomulations based on lignin have been published over the years [6].

In the study presented in this article several types of industrial lignins have been investigated: (a) a low molecular weight organosolv grass lignin from India, (b) an organosolv lignin from miscanthus grass, (c) a kraft lignin depolymerized according to a published procedure, and (d) a depolymerized form of the lignin in (a) above. The molecular mechanics adhesion calculations of all the secondary forces interactions with crystalline cellulose have been limited so far to the short oligomers found to compose these lignins.

To determine the main dimers and trimers in such lignins, MALDI-TOF (matrix assisted laser desorption ionisation time-of-flight) mass spectrometry was used. In principle MALDI-TOF is matrix-assisted, thus, the sample under test is dried together with a light absorbing compound (matrix). A pulse of laser light is used to force molecules into gas phase and ionize them. The ions are then accelerated in an electrical field. They are then allowed to drift in the mass spectrometer and are picked up by a detector. The drift time is measured electronically. The drift time thus measured allows determination of the molecular weight because drift time is proportional to velocity and the acceleration is proportional to mass. Calibration is done by simultaneous analysis of standards of known masses. The result is a mass spectrum where the peaks indicate clearly the oligomers present.

However, fundamental molecular mechanics studies on the interaction of the lower molecular weight oligomers of lignin with cellulose have been carried out but without checking if the values obtained corresponded or could be correlated to Young's modulus values obtained experimentally. Such an approach has already proven its worth in the correlation found between adhesive/substrate interactions calculated by molecular mechanics and experimental bond strength obtained for phenol-formaldehyde and urea-formaldehyde thermosetting adhesives to cellulose [7–10]. The same approach but using considerably more modern molecular mechanics algorithm has been adopted in this article for the interaction of short lignin oligomers with cellulose.

2. Experimental

The types of lignins used were: (a) an industrial organosolv grass lignin (Ln_India) Protobind 100SA.140 India Lignin, provided by Granit®, Switzerland [11]; (b) a laboratory organosolv lignin from miscanthus (*Miscanthus giganteus*) grass

(Ln_miscanthus) [12]; (c) a depolymerized kraft lignin (Ln_CO2) Ln-T-CO2-1 [13], and (d) (Ln_India_depo) depolymerised Indian lignin from sample (a) [14].

AutoDock [15] is a tool for molecular modelling that provides a procedure for predicting the interaction of small molecules with macromolecular targets. The ideal procedure would find the global minimum in the interaction energy between the substrate and the target molecule, exploring all available degrees of freedom for the system. In order to represent our specific interest the ligand is represented by the lignan (oligolignols) target and the substrate by crystalline celulose. AutoDock has a free energy scoring function that is based on a linear regression analysis, the AMBER [16] force field.

The MALDI-TOF spectra were recorded on a KRATOS Kompact MALDI AX-IMA TOF 2 instrument. The irradiation source was a pulsed nitrogen laser with a wavelength of 337 nm. The length of one laser pulse was 3 ns. The measurements were carried out using the following conditions: polarity-positive, flight path-linear, mass-high (20 kV acceleration voltage), and 100–150 pulses per spectrum. The delayed extraction technique was used by applying delay times of 200–800 ns.

The samples were prepared by mixing the lignin water solutions with acetone (4 mg/ml, 50/50 water/acetone by volume). The sample solutions so prepared were mixed with an acetone solution of the matrix (10 mg matrix solution per ml acetone). As the matrix 2,5-dihydroxy benzoic acid was used. For enhancement of ion formation, NaCl was added to the matrix (10 mg/ml in water). The solutions of the sample and the matrix were mixed in proportions 3 parts matrix solution + 3 parts lignin solution + 1 part NaCl solution, and 0.5 to 1 μl of the resulting solution mix were placed on the MALDI target. After evaporation of the solvent the MALDI target was introduced into the spectrometer. The dry droplet sample preparation method was used.

Commercial α-cellulose sheets (14.5 mm × 5.5 mm × 0.17 mm) were used as sample with 89.62 g/m^2 of lignin, hexamine hardener (5% on lignan grammage) and with 1.53% paper moisture content. The lignins were dissolved at 42% solids content at pH 12. The paper samples were then impregnated, and dried in an oven at 36°C.

TMA tests were carried out on the dried, lignin impregnated papers. Tensile tests on the dry specimens were conducted on a Mettler 40 TMA apparatus. The specimen was subjected to an oscillating force. The Young's modulus was obtained as a function of temperature by applying the formula:

$$E = \Delta F L_0 / (A \Delta L), \tag{1}$$

where:

E: Young's modulus of lignin impregnated paper sheet (N/mm^2),

ΔF: difference between maximum and minimum values of the oscillating force applied (N),

L_0: sample length (mm),

A: sample area (thickness) (width) (mm^2),

ΔL: length change (mm).

The weight percentage of each oligomer in the MALDI-TOF spectrum of each lignin was determined. These weight percentages were multiplied by the maximum value of Young's modulus obtained by TMA for the lignin sample being tested and thus an approximation of the contribution of each oligomer to the strength of the lignin/cellulose composite was determined.

The theoretical molecular mechanics calculation of the highest energy of interaction of individual oligolignans dimers and trimers with model cellulose I crystallite composed of three layers each of 6 by 12 glucose residues has been carried out with AutoDock Version 4.2 [17, 18]. Visualization of the minimum energy conformation for the interaction is made with MOLDEN 3.9 [19], gOpenMol 2.32 [20, 21] and Molekel 4.3 [22, 23] for Linux computer operating system. AutoDock calculations are performed in several steps: (1) preparation of the coordinates files of a model substrate and of the lignan including its hydrogen atoms, atomic partial charges and atoms types, and also information on the torsional degrees of freedom; (2) recalculation of atomic affinities, which is a rapid energy evaluation that is achieved by calculating atomic affinity potential for each atom type in the lignan, and (3) Docking simulation carried out using a semi-empirical free energy force field to evaluate conformations during docking simulations (Fig. 1).

The force field calculates the interaction energies in two steps. The lignan and the substrate start in an unbound conformation. In the first step, the intermolecular energetics is estimated for the transition from this unbound state to the conformation of the lignan and substrate in the bound state. The second step then evaluates the intermolecular energetics of combining the lignan and the substrate in their bound conformation.

The force field includes six pair-wise evaluations (V) and estimate of the conformational entropy lost upon binding (ΔS_{conf}):

$$\Delta G = (V^{L-L}_{bound} - V^{L-L}_{unbound}) + (V^{P-P}_{bound} - V^{P-P}_{bound})$$
$$+ (V^{P-L}_{bound} - V^{P-L}_{unbound} + \Delta S_{conf}), \tag{2}$$

where L refers to the "lignan" and P refers to the "substrate" in a lignan–substrate docking calculation. Each interaction includes evaluations for dispersion/repulsion van der Waals (vdW), hydrogen bonding (hbond), electrostatic (elec), and desolvation (sol) energies between a pair of atoms i and j:

$$V = W_{vdW} \sum_{i,j} [(A_{ij}/r_{ij}^{12}) - (B_{ij}/r_{ij}^{6})] + W_{hbond} \sum_{i,j} E(t)[(C_{ij}/r_{ij}^{12}) - (D_{ij}/r_{ij}^{10})]$$

$$+ W_{elec} \sum_{i,j} [q_i q_j / \varepsilon(r_{ij}) r_{ij}] + W_{sol} \sum_{ij} (S_i V_j + S_j V_i) e^{(-r_{ij}^2 / 2\sigma^2)}, \tag{3}$$

where q_i and q_j are, respectively, the electrostatic charges of atoms i and j; r_{ij} is the distance between atoms i and j.

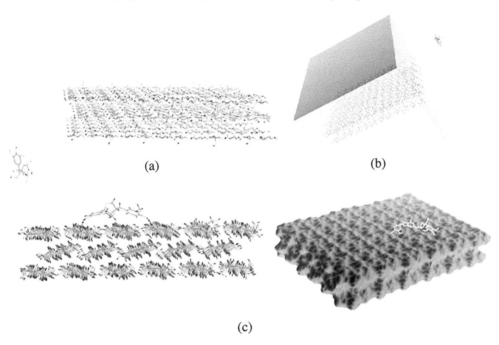

(a)   (b)

(c)

Figure 1. Examples of molecular mechanics interaractive docking of a lignan trimer on a cellulose I crystallite. (a) and (b) different positions of the oligomer in relation to the cellulose crystallite at the start of the docking simulation, as the oligomer progressively approaches the cellulose crystallite. (c) Different views of the position, conformation and site of minimum energy of the lignan/cellulose system.

The weighting constants W have been optimized to standerdize the empirical free energy based on a set of experimentally-determined binding constants. The first term is a typical 6/12 potential function of the Lennard–Jones type [24, 25] for dispersion/repulsion interactions. The parameters are based on the Amber force field. The second term is a directional H-bond term based on a 10/12 potential function. The parameters C and D are assigned to give a maximun energy well depth of 5 kcal/mol at a 1.9 Å distance for hydrogen bonds between oxygen and nitrogen (C), and an energy well depth of 1 kcal/mol (D). The function $E(t)$ provides directionality based on the angle t from the ideal H-bonding geometry. The third term is a screened Coulomb potential for electrostatic non-directional interactions. The final term is a desolvation potential based on the volume of atoms (V) that surround a given atom and shelter it from the solvent, weighted by solvation parameter (S) and an exponential term with distance weighting factor $\sigma = 3.5$ Å [26, 27].

It must be clearly pointed out that in molecular mechanics, by convention, an energy with a negative sign indicates that two molecules attract each other, hence, their interaction is attractive. A positive sign energy value indicates instead that the two molecules repel each other. Thus, the more negative is the energy of interaction, the more strongly attracted to each other are the two molecules.

3. Results and Discussion

In Fig. 1 is shown the example of a lignan trimer docking onto the surface of a cellulose crystallite. Fig. 1(a) and (b) show the initial positions at which the lignan is placed in relation to the cellulose crystallite, at a distance where interaction between the two is too small to be of any significance. As the lignan approaches the surface its conformation changes to adapt itself to the surface of the cellulose crystallite to minimize the total energy of the system until the lignan conformation of minimum energy is reached (Fig. 1(c)). In Fig. 1(c) the position of the trimer in relation to the section of the cellulose crystallite is shown.

The experimental TMA tensile test results in Table 1 show that the interaction with cellulose of the (Ln_India) lignin yields a Young's modulus higher than that of the three other lignins used independently of solids content of the impregnating lignin solution as shown from the higher Young's modulus obtained. Fig. 2 shows the elongation curves as a function of temperature obtained in the TMA tensile test that have allowed to calculate the modulus values in Table 1. MALDI TOF analysis of the four lignins tested indicated that these were very low molecular weight lignins composed almost exclusively of dimers and trimers. The MALDI-TOF spectra of these 4 lignins are shown in Figs 3–6 for the interval 200–750 Da. The MALDI-TOFspectra between 60 and 200 Da were recorded but are not re-

Table 1.
Results of: (1) Young's modulus of cellulose paper sheets impregnated with solutions of different lignins (Ln-CO2; Ln-depol; Ln-miscanthus; Ln-India) at different lignin solids concentrations of the impregnating aqueous solution (42%, 22%, 15%, 11%), and (2) the percentage modulus increase over non-impregnated paper control

Non-impregnated control		Young's modulus (MPa)	Young's modulus increase (%)
		0.80	–
42% solids	Ln_CO2	2.01	151
	Ln_depol	2.63	228
	Ln-miscanthus	3.33	317
	Ln_India	**4.50**	**463**
22% solids	Ln_CO2	1.22	53
	Ln_depol	1.71	114
	Ln-miscanthus	1.84	130
	Ln_India	**3.80**	**375**
15% solids	Ln_CO2	1.76	120
	Ln_depol	1.84	129
	Ln-miscanthus	1.92	140
	Ln_India	**2.60**	**225**
11% solids	Ln_CO2	1.00	25
	Ln_depol	1.48	85
	Ln-miscanthus	1.73	117
	Ln_India	**2.48**	**210**

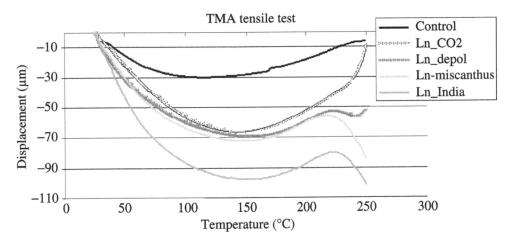

Figure 2. Thermomechanical analysis of tensile test on paper inmpregnated with different types of low molecular weight lignins: displacement as a function temperature. Standard is the displacement obtained with non-impregnated paper used as a control.

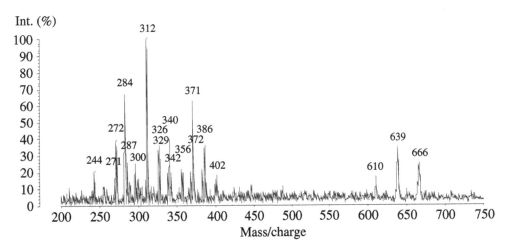

Figure 3. MALDI-TOF spectrum of original lignin Ln_India.

ported here. However, the relevant assignment of the MALDI-TOF peaks and their distribution are reported in Tables 2 and 3. In Table 2 are shown all the relevant peaks in the spectra in Figs 3–6 with the calculated molecular mass + 23 Da (for the Na^+ matrix) for the relevant corresponding oligomers and their relative intensity expressed as a percentage of the highest peak.

To explain the shortened nomenclature used in Table 2, lignin units are of the H-type, guayacil-type (G-type) and syringyl-type (GS-type) [28]. As examples, for Lignin Ln_CO$_2$ the peaks at 113 and 181 indicate that $H_{FRACTION}$ and $GS_{FRACTION}$ correspond to fragments, respectively, of H-type and GS-type monomers the calculated peak values of which are given by the molecular weight of the oligomer

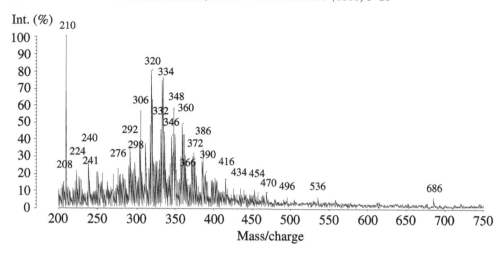

Figure 4. MALDI-TOF spectrum of lignin Ln_India depolymerized.

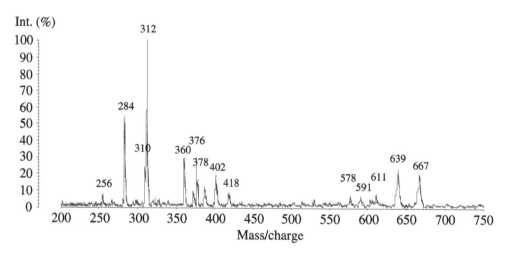

Figure 5. MALDI-TOF spectrum of lignin Ln_CO$_2$.

plus that of the Na matrix, thus peaks at $93 + 23(\text{Na}^+)$ Da $= 116$ Da (117 Da is protonated) and $154 + 23(\text{Na}^+)$ Da $= 177$ Da are obtained. These correspond to structures (I) and (II), respectively.

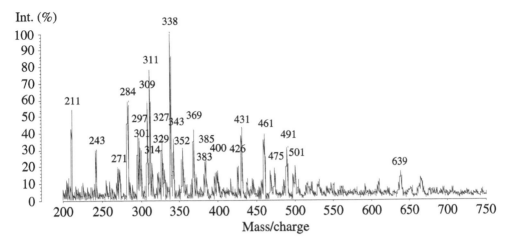

Figure 6. MALDI-TOF spectrum of depolymerized miscanthus lignin, Ln_miscanthus.

Table 2.
Interpretation of MALDI TOF peaks in Figs 3–6

Dimer/trimer	Molecular weight +Na	Peak (Da)	Relative peak height* (%)	Young's modulus related to relative peak height (MPa)
Lignin Ln_CO2-1				
H$_{FRACTION}$	117	113	65	1.03
GS$_{FRACTION}$	177	181	40	0.84
H + GS	383	378	30.5	0.61
G + GS$_{FRACTION}$	356	360	30	0.6
GS + G$_{FRACTION}$				
G + G + GS	593	591	8	0.16
H + GS + GS				
Lignin miscanthus				
H$_{FRACTION}$	117	121	18	0.6
G$_{FRACTION}$	147	147	25	0.83
GS$_{FRACTION}$	177	177	18	0.6
G	203	198	10.5	0.35
H + H$_{FRACTION}$	266	271	20	0.66
H + G$_{FRACTION}$	296	297	38	1.26
G + H$_{FRACTION}$				
GS (–CH$_3$) + H$_{FRACTION}$	337	338	100	
H + H	323	322	17	0.56
H + GS$_{FRACTION}$	326	327	38	1.26
G + G$_{FRACTION}$				
GS + H$_{FRACTION}$				
G + GS$_{FRACTION}$	356	352	24.5	0.82

Table 2.

(Continued.)

Dimer/trimer	Molecular weight +Na	Peak (Da)	Relative peak height* (%)	Young's modulus related to relative peak height (MPa)
GS + G$_{FRACTION}$				
G + GS$_{FRACTION}$	356	355	32	1.07
GS + G$_{FRACTION}$				
H + GS	383	383	15.5	0.52
G + G				
GS + GS$_{FRACTION}$	386	385	23.6	0.79
H + H + H	473	475	19	0.82
G + G + H$_{FRACTION}$	477			
H + H + G	503	501	21	0.7
Lignin India depolymerized				
H$_{FRACTION}$	117	117	13	0.34
G$_{FRACTION}$	147	149	32	0.84
GS$_{FRACTION}$	177	177	23	0.61
G	203	208	16	0.42
G + H$_{FRACTION}$	296	298	38	0.1
H + H	323	322	43.2	1.14
H + GS$_{FRACTION}$	326	326	34.7	0.91
G + GS$_{FRACTION}$	356	352	20.5	0.54
GS + G$_{FRACTION}$				
G + GS$_{FRACTION}$	357	360	52	1.37
GS + G$_{FRACTION}$				
GS + GS$_{FRACTION}$	387	386	36	0.95
G + GS	413	416	18	0.47
H + H + H$_{FRACTION}$	417			
H + H + H	473	470	10	0.26
G + GS + G$_{FRACTION}$	537	536	7	0.18
GS + GS + H$_{FRACTION}$				
H + GS + GS				
Lignin India original				
G + H$_{FRACTION}$	297	300	17.3	0.78
H + GS$_{FRACTION}$	327	326	38.2	1.72
G + G$_{FRACTION}$				
GS + H$_{FRACTION}$				
G + GS$_{FRACTION}$	356	356	22.7	1.02
GS + G$_{FRACTION}$				
GS + GS$_{FRACTION}$	386	386	37.3	1.70

Equally, as an example, the dimer H + GS at an actual peak 385 Da corresponds to an oligomer of molecular weight 360 this plus the 23 Da of the Na$^+$ gives 360 + 23(Na$^+$) Da = 383 Da (structure III)

Table 3.
MALDI-TOF peaks with specific links to lignan oligomers of which the interaction energy with cellulose has been calculated

Dimer/trimer	Molecular weight +Na	Peak (Da)	Relative peak height* (%)	Young's modulus related to relative peak height (MPa)
Lignin Ln_CO2				
G_$\gamma\beta$_G	383	378	16	0.32
G_βO4_G	399	402	20	0.40
G_αO4_G				
G_$\alpha\beta$_G	417	418	10	0.20
G_βO4_G_$\beta\beta$_G	577.6	578	7.7	0.16
G_βO4_G_β5_G				
G_β5_G_$\gamma\beta$_GS	591.6	591	7.7	0.16
G_βO4_G_$\beta\beta$_GS	607.6	611	8.8	0.18
G_βO4_GS_β5_G				
G_βO4_GS_$\beta\beta$_GS	637.6	639	24.4	0.49
GS_βO4_GS_$\beta\beta$_GS	667.6	667	20	0.40
Lignin miscanthus				
G_$\gamma\beta$_G	383	383	14.5	0.48
G_βO4_G	399	400	17.3	0.58
G_αO4_G				
G_βO4_GS	429.4	431	43.6	1.45
GS_βO4_GS	459.4	461	39.1	1.3
GS_αO4_GS				
GS_$\alpha\beta$_GS	477.5	475	18.2	0.61
G_βO4_GS_$\beta\beta$_GS	637.6	639	16.4	0.55
Lignin India depolymerized				
G_$\gamma\beta$_G	383	386	36	0.95
G_$\alpha\beta$_G	417	416	18	0.47
G_βO4_G	429.4	434	12	0.32
GS_βO4_GS	459.4	454	10	0.26
GS_αO4_GS				

(III)

Where the H- and GS-type units can either be linked with the most common β–O–4 linkage or with other linkages which are allowed between lignin units [28]. Equally the peak at 360 Da corresponds to a dimer G + GS$_{\text{FRACTION}}$ the calculated molecular weight of which is 337, thus giving $337 + 23(Na^+)$ Da = 360 Da which corresponds to structure (IV)

Thus, while it is quite possible to identify which compounds are dimers and which are trimers, fragments or monomers, and which lignin units are involved, it is not strictly possible to define which types of linkages tie the units together, although these types are relatively few in lignin.

Table 3 shows the dimers and trimers that could be exactly identified in the MALDI-TOF spectra and for which the minimized interaction energies between the oligomer and the cellulose surface were calculated exactly. For each oligomer the calculated and experimental molecular weight + 23 (Na^+) Da peaks are shown, as well as their interaction energy calculated by molecular mechanics and their percentages relative to the highest peak. From these the percentage by weight of each oligomer in the sample can be calculated. To explain the nomenclature used again for Lignin Ln_CO$_2$, G_βO4_G indictes a dimer composed of two guayacil units linked by a β–O–4 linkage [28]. For example, in the G_βO4_G_$\beta\beta$_G trimer two of the guayacil units are linked by a β–O–4 linkage and the third is linked to one of the other two by $\beta\beta$ linkage corresponding to structure (V).

As only lignins of low molecular weight are studied, dimers and trimers have exclusively been found in the composition of the four lignins studied. However, it is not necessary to calculate the theoretical interaction energy with cellulose for

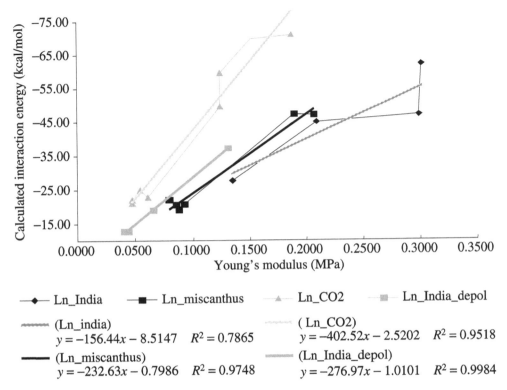

Figure 7. Correlation of experimental Young's modulus and molecular mechanics calculated (kcal/mol) interaction energies for low molecular weight lignins with cellulose.

all the dimers that are shown to exist in different percentages in the MALDI-TOF spectra. The results in Table 6 and in Fig. 7 indicate that if the calculation of the energy of interaction with the substrate of 20% or more by mass of the dimers and trimers is taken into account the trends correspond well to the Young's modulus obtained experimentally. That this is the case is evident from the results of the non-depolymerized (Ln_India) where the interaction with the substrate of only 10% of the weight of the oligomers present in the mix could be calculated. In this case the coefficient of correlation (Fig. 7) is poor. This is shown in Table 4 presenting, for the different oligomers considered, predicted energies of interaction obtained by molecular mechanics calculation and experimental Youngs modulus values obtained by TMA (Table 4). The Young's modulus values obtained by TMA have already been shown to correlate with interaction energies obtained by molecular mechanics [29, 30]. In order to compare the two parameters Fig. 7 is presented. It shows that the trend in Young's modulus for each lignan is maintained for both calculated energy and Young's modulus. The correlation coefficients R^2 between the experimental values for the interaction energy with cellulose represented by Young's modulus and the theoretically calculated values for the different lignans are excellent at 0.95, 0.97 and 0.998 for (Ln_CO2), (Ln_miscanthus) and (Ln_India_depol), respectively.

Table 4.

Calculated interaction energy of lignan components with crystalline cellulose

Dimer/trimer	Composition (%)	ADT 42 (kcal/mol) per lignan	ADT 42 (kcal/mol) per lignin mix type	Young's modulus (MPa)
Ln_India				
G_βO4_GS_$\beta\beta$_GS	6.69	−9.28	−62.08	0.3011
G_$\beta\beta$_G	6.64	−7.08	−47.01	0.2988
GS_βO4_GS_$\beta\beta$_GS	4.65	−9.67	−44.97	0.2093
G_βO4_GS_β5_G	2.99	−9.24	−27.63	0.1346
TOTAL	**20.97**	−35.27	**−181.69**	**0.9437**
Ln_miscanthus				
G_βO4_GS	6.21	−7.58	−47.07	0.2068
GS_βO4_GS	5.68	−8.32	−47.26	0.1891
G_$\gamma\beta$_G	2.77	−7.50	−20.78	0.0922
GS_$\alpha\beta$_GS	2.64	−7.30	−19.27	0.0879
G_βO4_G	2.57	−8.05	−20.69	0.0856
G_βO4_GS_$\beta\beta$_GS	2.38	−9.28	−22.09	0.0793
TOTAL	**22.25**	−48.03	**−177.15**	**0.7409**
Ln_CO2				
G_$\alpha\beta$_G	9.30	−7.63	−70.96	0.1869
G_βO4_GS_$\beta\beta$_GS	7.52	−9.28	−69.79	0.1512
GS_βO4_GS_$\beta\beta$_GS	6.16	−9.67	−59.57	0.1238
G_βO4_G	6.16	−8.05	−49.59	0.1238
G_$\gamma\beta$_G	3.05	−7.50	−22.88	0.0613
G_βO4_GS_b5_G	2.71	−9.24	−25.04	0.0545
G_β5_G_$\gamma\beta$_GS	2.37	−9.32	−22.09	0.0476
G_βO4_G_β5_G	2.37	−8.85	−20.97	0.0476
TOTAL	**24.92**	−69.54	**−340.88**	**0.7968**
Ln_India_depolymerized				
G_$\gamma\beta$_G	4.97	−7.50	−37.28	0.1307
G_$\alpha\beta$_G	2.50	−7.63	−19.08	0.0658
G_βO4_GS	1.69	−7.58	−12.81	0.0444
GS_βO4_GS	1.55	−8.32	−12.90	0.0408
TOTAL	**10.71**	−31.03	**−82.06**	**0.2817**

Such high coefficients of correlation indicate that the energy of interaction that could be calculated for the oligomers for which theortical interaction energy values are not known must present the same trend as those of the oligomers that constitute between 20% and 25% of the sample. These are the oligomers which were effectively used for the correlation between calculated and experimental results.

The coefficient of correlation is too low for the (Ln_India) non-depolymerized, for the reasons given above. This is expected if one considers that this is the raw product obtained industrially. A more negative value of the theoretically calcu-

Table 5.
Calculated interaction energies of lignins with cellulose and experimental Young's modulus of lignins/cellulose composites

Designation	Dimers		Trimers	
	AutoDock (kcal/mol)	Young's modulus (MPa)	AutoDock (kcal/mol)	Young's modulus (MPa)
Ln_India	−47.01	0.2988	−134.68	0.6449
Ln_miscanthus	−155.06	0.6617	−22.09	0.0793
Ln_CO2	−143.42	0.3721	−197.46	0.4247
Ln_India_depolymerized	−82.06	0.2817	–	–

lated interaction energy means that the system of lignan and cellulose substrate is more stable and that the secondary forces attraction between lignan and substrate is stronger. Although from the interaction energy results in Table 5 and Fig. 8(a), (b) the trimers appear, in general, to be more strongly attracted to cellulose than dimers, this is not always the case (see Ln_miscanthus, Table 5, Fig. 8(a), (b)). The molecular weight of the lignan thus does not appear to be the predominant factor in its interaction energy with the substrate. Higher molecular weight lignans such as trimers are sometimes linked more strongly to the substrate than lower molecular weight lignans such as dimers, and *vice versa*. Therefore, the important point is the ability of the oligomer to adapt itself to the surface of the substrate in such a manner that the interaction energy yields the strongest attraction of lignan and cellulose: this can happen for a dimer as well as for a trimer. This trend is evident for both theoretically calculated and for experimental results, as shown in Table 3 for the total interaction energy of lignan dimers and trimers with the cellulose substrate.

As the (Ln_India_depolym) has been depolymerised, it is exclusively composed of dimers. According to Fig. 5, the interaction energy with cellulose of (Ln_India) and (Ln_CO2) is determined mainly by trimers while for (Ln_miscanthus) and (Ln_India_depol) lignins mainly by dimers.

4. Conclusion

The low molecular weight lignins used are mostly composed of dimers and trimers. Lignins with high amounts of trimers appear to present a stronger interaction with the cellulose substrate. Due to the different compositions in oligomers of the four lignins it is difficult to make a comparison among them. Referring to Table 4, Ln_India and Ln_miscanthus had almost the same percentages of components and presented also similar interactions with the substrate both by molecular mechanics calculation as well as experimentally in terms of Young's modulus by TMA. Ln_India showed highest interaction energy both by molecular mechanics calculation and experimental Young's modulus as a result of the highest proportion of trimers being present. The Ln_India_depolymerized lignin presented the lowest cal-

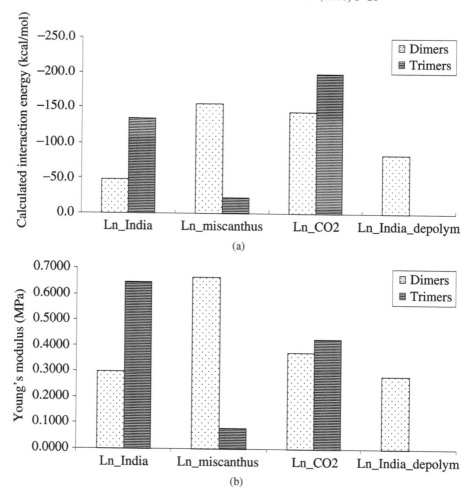

Figure 8. Strength of paper strip impregnated with 4 different types of lignins calculated on the basis of the trimers and dimers present from (a) molecular mechanics calculation, and (b) experimental thermomechanical analysis results.

culated interaction energy to which corresponded the lowest experimental value of Young's modulus too. This is expected as only 10% of the oligomers were used in the calculation and these were all dimers, hence, in general, yielding weaker interactions with the substrate. The conclusion appears to be that at least 20% of the oligomers need to be considered to expect some form of correlation between calculated and experimental results.

Ln_CO_2 lignin presented a somewhat anomalous behaviour: it presented a stronger calculated attraction but the experimental Young's modulus value, although the second best of the four lignins tested, did not maintain the same expected trend and was lower than expected.

Usually trimers appear to present higher interaction energy than dimers although this depends on the type of linkage among their units. The molecular mechanics calculations indicated that a β–O–4-linked dimer has stronger attraction for the substrate than $\beta\beta$ or $\beta5$-linked dimers. Consequently, the interaction energy is dictated by molecular weight and type of linkage within the lignan oligomers. Lignans with higher molecular weight in which the lignin units are linked by β–O–4 bonds [28] give interaction energy values indicating that the attraction between lignan and cellulose is stronger.

Acknowledgements

The French author gratefully acknowledges the financial support of the CPER 2007-2013 "Structuration du Pôle de Compétitivité Fibres Grand'Est" (Competitiveness Fibre Cluster), through local (Conseil Général des Vosges), regional (Région Lorraine), national (DRRT and FNADT) and European (FEDER) funds.

References

1. W. Boerjan, J. Ralph and M. Baucher, *Ann. Rev. Plant Biol.* **54**, 519–549 (2003).
2. T. Higuchi, in: *Biosynthesis and Biodegradation of Wood Components*, T. Higuchi (Ed.), pp. 141–160. Academic Press, Orlando, FL (1985).
3. E. Sjöström and A. Raimo, *Analytical Methods in Wood Chemistry, Pulping and Papermaking.* Springer, New York (1998).
4. A. Pizzi, *J. Adhesion Sci. Technol.* **20**, 829–846 (2006).
5. A. Pizzi, *Advanced Wood Adhesives Technology.* Marcel Dekker, New York (1994).
6. H. Nimz, in: *Wood Adhesives: Chemistry and Technology*, A. Pizzi (Ed.), Chapter 5. Marcel Dekker, New York (1983).
7. A. Pizzi and N. J. Eaton, *J. Adhesion Sci. Technol.* **1**, 191–200 (1987).
8. A. Pizzi, *J. Adhesion Sci. Technol.* **4**, 573–588 (1990).
9. A. Pizzi, *J. Adhesion Sci. Technol.* **4**, 589–595 (1990).
10. A. Pizzi and G. De Sousa, *Chem. Phys.* **164**, 203–216 (1992).
11. Granit® Lignin, Granit Recherche Développement SA, Lausanne, Switzerland.
12. R. El Hage, L. Chrusciel, N. Brosse, T. Pizzi, P. Navarrete, P. Sannigrahi and A. Ragauskas, in: *Proc. 2nd Nordic Wood Biorefinery Conference*, Helsinki, Finland, September 2–4, 2009.
13. S. Tapin-Lingua, unpublished results (2009).
14. N. Brosse, unpublished results (2009).
15. G. M. Morris, D. S. Goodsell, R. S. Halliday, R. Huey, W. E. Hart, R. K. Belew and A. Olson, *J. Comput. Chem.* **19**, 1639–1662 (1998).
16. D. A. Case, T. A. Darden, T. E. Cheatham, III, C. L. Simmerling, J. Wang, R. E. Duke, R. Luo, K. M. Merz, D. A. Pearlman, M. Crowley, R. C. Walker, W. Zhang, B. Wang, S. Hayik, A. Roitberg, G. Seabra, K. F. Wong, F. Paesani, X. Wu, S. Brozell, V. Tsui, H. Gohlke, L. Yang, C. Tan, J. Mongan, V. Hornak, G. Cui, P. Beroza, D. H. Mathews, C. Schafmeister, W. S. Ross and P. A. Kollman, AMBER 9 program, University of California, San Francisco (2006).
17. G. M. Morris, D. Goodsell, M. Pique, W. Lindstrom, R. Huey, S. Forly, W. Hart, S. Halliday, R. Belew and A. J. Olson, AutoDock Version 4.2. Automated Docking of Flexible Ligands to

Flexible Receptors. User Guide, Copyright © The Scripps Research Institute, La Jolla, CA, 1991–2009.

18. G. M. Morris, R. Huey, W. Lindstrom, M. F. Sanner, R. Belew, D. Goodsell and A. Olson, *J. Comput. Chem.* **28**, 1145–1152 (2009).

19. G. Schaftenaar and J. H. Noordik, *J. Comput.-Aided Mol. Design* **14**, 123–134 (2000).

20. L. Laaksonen, *J. Mol. Graph.* **10**, 33–34 (1992).

21. D. L. Bergman, L. Laaksonen and A. Laaksonen, *J. Mol. Graph.* **15**, 301–306 (1997).

22. P. Flükiger, H. P. Lüthi, S. Portmann and J. Weber, MOLEKEL 4.3 program, Swiss Center for Scientific Computing, Manno, Switzerland, 2000–2002.

23. S. Portmann and H. P. Lüthi, *Chimia* **54**, 766–769 (2000).

24. J. E. Lennard-Jones, *Proc. Roy. Soc.* **A 106**, 463 (1924).

25. D. Frenkel and B. Smit, *Understanding Molecular Simulation*, 2nd edn. Academic Press, San Diego, CA (2002).

26. R. Huey, G. M. Morris, A. Olson and D. S. Goodsell, *J. Comput. Chem.* **28**, 1145–1152 (2007).

27. R. Huey, D. S. Goodsell, G. M. Morris and A. J. Olson, *Lett. Drug Design Discov.* **1**, 178 (2004).

28. D. Fengel and G. Wegener, *Wood: Chemistry, Ultrastructure, Reactions*, p. 613. Walter de Gruyter, Berlin (1984).

29. A. Pizzi, F. Probst and X. Deglise, *J. Adhesion Sci. Technol.* **11**, 573–590 (1997).

30. A. Pizzi, *J. Appl. Polym. Sci.* **63**, 603–617 (1997).

Evaluation of Some Synthetic Oligolignols as Adhesives: A Molecular Docking Study

Carmen Martínez [a], **Miriam Sedano** [a], **Abril Munro** [a], **Pablo López** [a,*] and **Antonio Pizzi** [b]

[a] Facultad de Ingeniería en Tecnología de la Madera, Universidad Michoacana de San Nicolás de Hidalgo, Morelia, México

[b] ENSTIB-LERMAB, University of Nancy 1, B. P. 1041, F-88051 Epinal cedex, France

Abstract

We have investigated alternative adhesives based on some oligomers that have structural units similar to lignins. The favorable interactions for a set of synthetic oligolignols were determined through a computational docking scheme, considering their conformational changes to adapt themselves on the surface of a three-dimensional model of cellulose I-β, to describe and elucidate the interactions that drove their adhesion ability based on the minimum interaction energy calculated for each of the oligolignols conformations. The selected oligolignols included nine dimers, five trimers and one tetramer all of these were built from two lignin precursors, coniferol and sinapol, bonded in four different linkages: β–O4′, β–β′, β–5′ and γ–β′. Our results showed that trimers were the most favorable oligolignols to dock over the cellulose model, considering their individual docking energy. We hope that this work contributes to the field of wood adhesives design and helps in understanding the interaction between the cellulose and structural components of lignin.

Keywords

Lignin, adhesive, docking, lignin–cellulose interaction, molecular modeling

1. Introduction

Chemists and engineers have made several studies about the interaction between cellulose and lignin in the field of wood adhesives to improve their adhesion characteristics [1–5], all of these on an empirical basis [6, 7]. Considering the fact that lignin is the second major component of wood and its main biological function is as a cementing agent for wood cells [8], several studies have attempted to develop adhesives that reproduce or mimic these lignin properties [9, 10]. Furthermore, due to

---

* To whom correspondence should be addressed. Facultad de Ingeniería en Tecnología de la Madera, Universidad Michoacana de San Nicolás de Hidalgo, Melon 398, Fracc. La Huerta, CP: 58080, Morelia, Michoacan, México. e-mail: plopez@umich.mx

Wood Adhesives
© Koninklijke Brill NV, Leiden, 2010

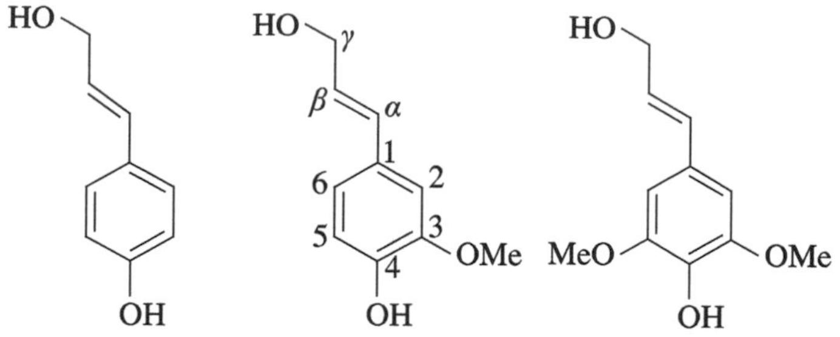

Figure 1. Lignin precursors: p-coumarol (left), coniferol (middle) and sinapol (right). Typical numbering is shown for the coniferol structure.

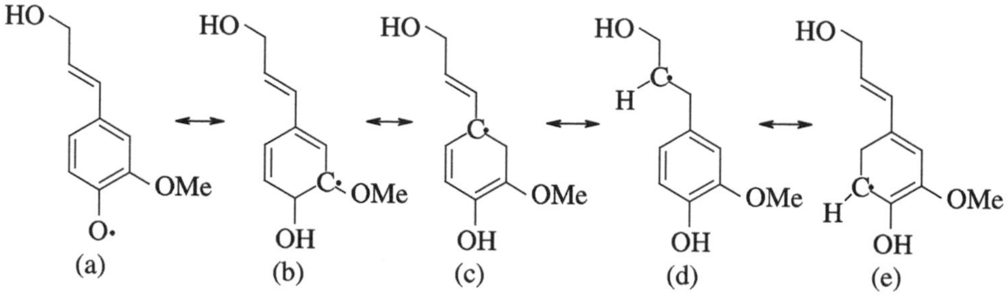

Figure 2. Free radical resonance structures of monolignols according to the resonance theory. Only coniferol is illustrated.

the fact that lignin occurs as a waste product in pulp processing in several countries (particularly in emerging economies) it becomes a very useful raw material.

Recently, adhesives that include lignin in their formulation have shown several properties similar to phenol-formaldehyde resins [11, 12] and are extensively utilized in the plywood, particleboard, fiberboard and laminated wood industries. The chemistry of the lignin-based adhesives has been studied from an empirical point of view [13]; however, the molecular mechanism and the interactions of the structural cross-linked 4-hydroxyphenylpropanoids of lignin that drive the adhesion phenomenon remain unclear.

Lignin structure is based on the assumption that it is derived from simple units of 4-hydroxyphenylpropane, and typically only three hydroxycinnamyl alcohols (p-coumarol, coniferol and sinapol) [14], these alcohols are named monolignols [15] and are shown in Fig. 1.

Lignin polymerization is initiated by an oxidative enzymatic dehydrogenation of monolignols [6], which form five different resonance structures [16] shown in Fig. 2. This leads to a large number of probable coupling reactions, yielding a disordered polymer.

The macromolecular structure of lignins could be produced by different combinations of interatomic linkages between two monolignols [17], but their relative abundance has not been established clearly [18]. The different lignin linkages are not clearly determined, because the structural complexity of lignin makes it difficult to dissolve the whole polymer and thus hampers chemical and physical characterization [19]. Furthermore, inside wood fibers the coupling reactions between monolignol units are influenced by the environment of the cell wall [20], leading to a racemic polymer. Lignin is often called the "cementing agent" of wood cells [21], thus the characterization of its bonds is the first step to understand its cementing behavior.

Several attempts to obtain lignins *in vitro* have been made to synthesize oligolignols that reproduce some of the lignin linkages [22]; however, their exploitation in an adhesive formulation has not been reported. Moreover, in the computational chemistry framework some studies have explored the molecular interactions between monolignols or dilignols and polysaccharides surfaces [23]. Both procedures, experimental and computational, have provided some valuable insights into the nature of the interactions of lignin substructures and woody tissues; however, the complete set of the most abundant oligolignols in lignins has not been included in these studies.

The objective of the present computational chemistry study was to explore and to elucidate the most favorable interactions between a set of 15 oligolignols and a three-dimensional model of cellulose I-β, evaluating their interaction energies in a molecular docking study. A set of synthetic oligolignols were selected to interact through a docking analysis with the cellulose model, because in this manner the most abundant linkages that occur in lignin are included.

2. Theoretical Background

In computational chemistry the procedure for predicting the interactions between a substrate and macromolecular targets in a molecular mechanics framework has been developed as *molecular docking scheme* [24]. Docking can be used to predict where and in which relative orientation a substrate binds to a macromolecule (also referred to as the binding mode or pose). This information may, in turn, be used to design more potent and selective analogs of the substrate [25]. Docking methods would find the global minimum in the interaction energy between the substrate and the target by exploring all available degrees of freedom for the substrate. Molecular docking is basically a conformational sampling procedure in which various docked conformations are explored to identify the most favorable substrate conformer that binds to a macromolecule. Rapid evaluation of the interaction energy between a substrate and a macromolecular target is achieved by precalculating atomic affinity potential for each atom type in the substrate [26]. In this procedure, the macromolecule is embedded in a three-dimensional grid and a probe atom is placed at each grid point. The interaction energy of this single atom with the macromolecule is assigned to a point inside the grid, yielding an affinity grid for each type of atom in the substrate, as well as a grid of electrostatic potential [27, 28]. Each point within the grid stores the potential energy of each atom type in the substrate that is due to all the atoms in the macromolecule. The energy of a particular substrate configuration is then found by tri-linear interpolation of affinity values of the eight grid points surrounding each of the atoms in the substrate. Once the atomic affinity potential is precalculated, the docking simulation is carried out with the macromolecular target stationary throughout the simulation, i.e., its molecular geometry is frozen, and the substrate molecule performs a random walk in the space around the macromolecular target, applying a small random displacement to each degree of freedom of the substrate. These displacements result in a new macromolecule-substrate complex, whose energy is evaluated using the grid interpolation procedure described above. This new energy is compared to the energy of the previous step. If the new energy is lower, the new complex is immediately accepted. If the new energy is higher, the complex is accepted or rejected based on a Boltzmann distribution [29]. This search algorithm is known as Monte Carlo simulated annealing [30]. Simulated annealing allows an efficient exploration of the space of configurations generated by the macromolecule-substrate complexes, with multiple minima, which is typical of a docking problem. The separation of the calculation of the molecular affinity grids from the docking simulation provides modularity to the procedure, and allows exploring the molecular interactions in several substrate-macromolecule complexes, from constant dielectrics to finite difference methods and from standard 12-6 potential functions to distributions based on observed binding sites [31].

3. Lignin Precursors (Synthetic Oligolignols)

The selection of the synthetic oligolignols studied in this work was based on the fact that they reproduced many of the cross-linkages most frequent in lignins native structures [32]. The dominant moieties in lignins structures are the guaiacyl and the syringyl types. From a modeling perspective, we considered the major features of lignin to be aromatic rings, hydroxyl groups and methoxy groups and coniferol and sinapol exhibit all of these features.

We use the following nomenclature to identify the oligolignol structure: CA for the guaiacyl moiety and SA for the syringyl moiety and for the combinations of interatomic linkages between two moieties the nomenclature was assigned according to the typical numbering of monolignols. Dilignols are displayed in Fig. 3, trilignols in Fig. 4 and a tetralignol in Fig. 5.

4. Three-Dimensional Model of Cellulose I-β

The model of a wood cell includes cellulose, hemicelluloses, metal ions, pectin analogues, lignin precursors and water; however inclusion of all of these components is computationally not feasible these days. Thus in order to delimit our study we have chosen the secondary wood cell wall, in which lignin and cellulose are dominant.

Cellulose is a homopolymer of β-(1-4) D-glucose molecules linked in a linear chain, with alternating sub-units in the crystalline structure being rotated through 180°. Native cellulose I microfibrils are highly ordered crystals and evidence from

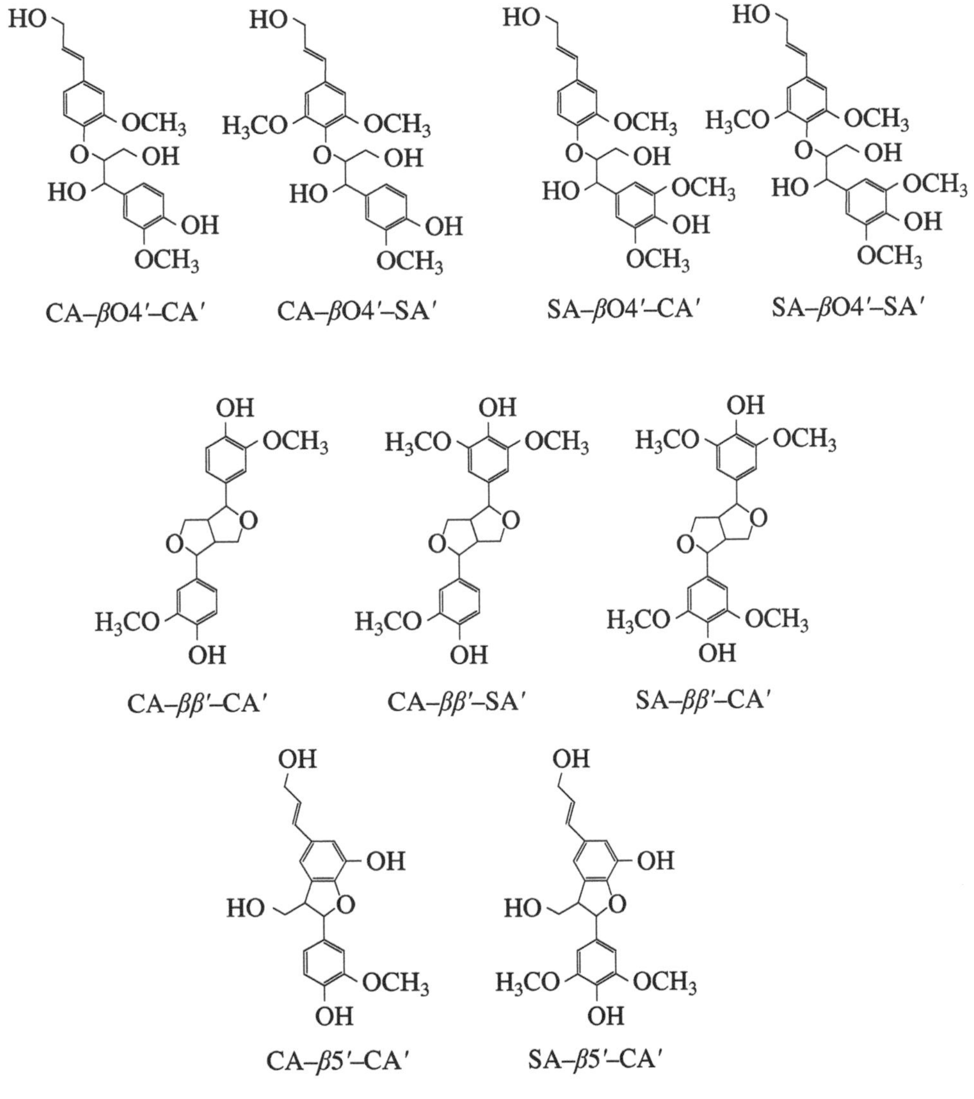

Figure 3. Dilignol structures used as the basis for studied synthetic oligolignols. The CA means guaiacyl moiety (coniferyl alcohol) and SA means syringyl moiety (sinapyl alcohol).

^{13}C-NMR spectroscopy and electron diffraction suggests that these consist of both triclinic (I-α) and monoclinic (I-β) crystalline forms [33–36]. Contrary to lignin, the macromolecular structure of cellulose is relatively well elucidated. Our study is limited to the I-β phase since it is reported to be dominant over the I-α phase.

The three-dimensional model of cellulose I-β was built based on the crystal structure of native cellulose from X-ray measurements [37]. Our model has 12 D-glucose molecules along the cellulosic axle, 6 D-glucose molecules transverse to cellulosic axle and 3 superimposed lattices ($6 \times 6 \times 3$ model) in a parallel arrangement and its unit cell is presented in Fig. 6.

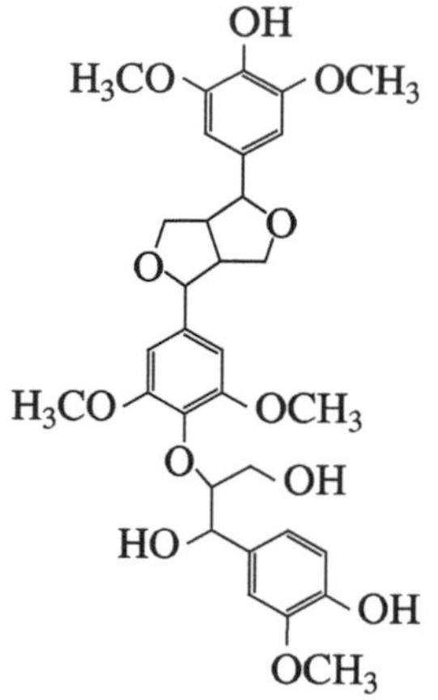

CA–βO4′–SA′–β′β″–CA″

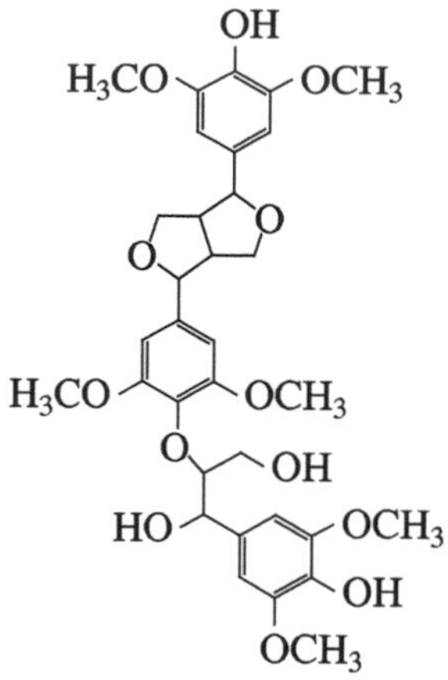

CA–βO4′–SA′–β′β″–SA″

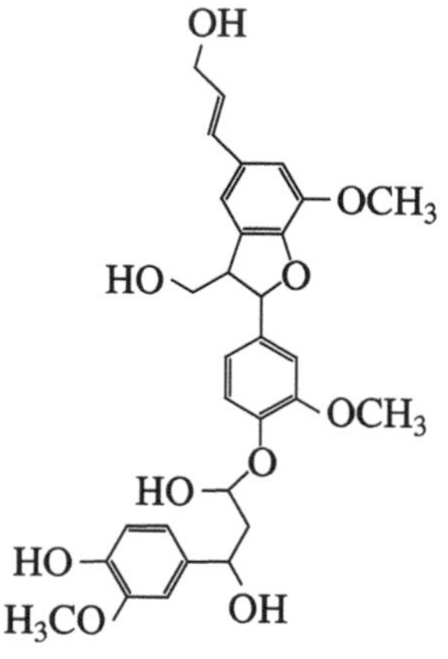

CA–βO4′–CA′–β′5″–CA″

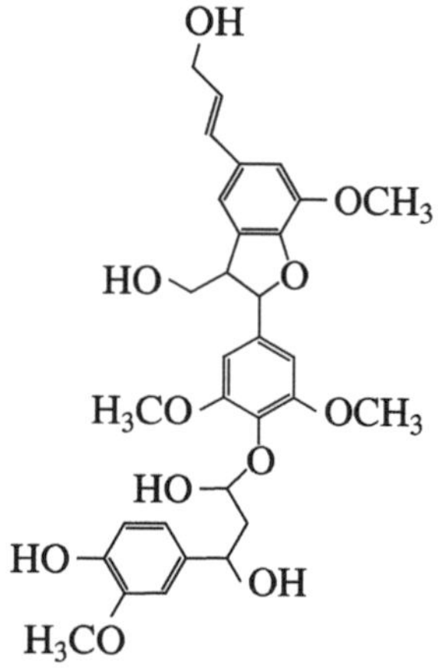

CA–βO4′–SA′–β′5″–CA″

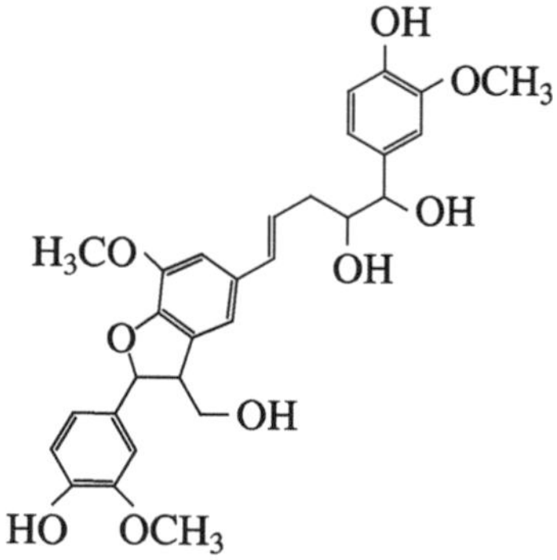

CA–β5′–CA′–γ′β″–CA″

Figure 4. Trilignol structures used as the basis for studied synthetic oligolignols. The CA means guaiacyl moiety (coniferyl alcohol) and SA means syringyl moiety (sinapyl alcohol).

This model contains enough information to simulate a crystallite of the cellulose I-β native structure, which emulates the interaction in the docking process with the oligolignols.

5. Methodology

All molecular mechanics calculations were done using the program AutoDock, version 4.2.1. [38], which was developed to provide an automated procedure for predicting the interaction energy of substrates with macromolecular targets in a docking scheme. In our study, the ligands were the 15 synthetic oligolignols and the macromolecular target was the $6 \times 6 \times 3$ model to simulate a cellulose I-β crystallite described above. ChemBioOffice ultra, version 11.0.1, [39] was selected to

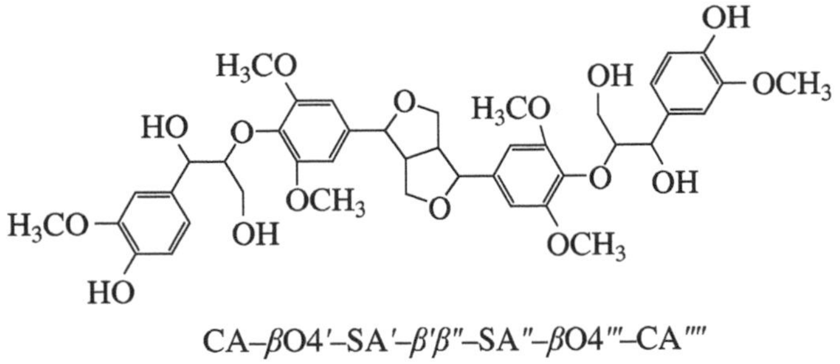

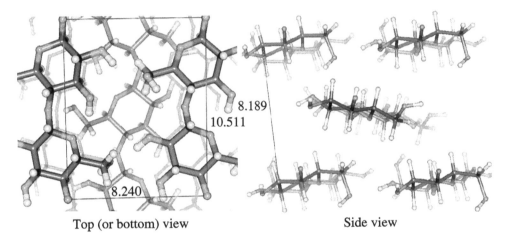

CA–βO4′–SA′–β′β″–SA″–βO4‴–CA‴′

Figure 5. Tetralignol structure used as the basis for studied synthetic oligolignols. The CA means guaiacyl moiety (coniferyl alcohol) and SA means syringyl moiety (sinapyl alcohol).

Top (or bottom) view   Side view

Figure 6. Schematic representation of the cellulose I-β model, units are in angstroms. The dimensions of the cellulose unit cell are displayed in two views: top (or bottom) view (left) show the distance along the cellulosic axle (10.511 Å) and the distance between cellulose chains at the same lattice (8.240 Å), meanwhile the side view displays the thickness of all three lattices (8.189 Å) of the cellulose unit cell.

sketch and build the molecular structures of the 15 oligolignols and the cellulose I-β three-dimensional model. Also the MGL-tools, version 1.5.4, [40] was utilized to set up, run and analyze AutoDock dockings and to visualize molecular structures.

The automated docking procedure for predicting the interaction energy was carried out independently for each oligolignol with the three-dimensional model of the cellulose I-β crystallite, and the initial configuration to start the docking calculations was built considering the substrates relatively far away from the macromolecule. We propose a three-dimensional arrangement similar to a plane (oligolignol) landing over an airstrip (macromolecule) as displayed in Fig. 7.

AutoDock provides several methods for doing the conformation search; however, the Lamarckian Genetic Algorithm (LGA) [41] provides the most efficient search for general applications and was our choice for the docking analysis. LGA begins with a population of random substrate conformations in random orientations and we

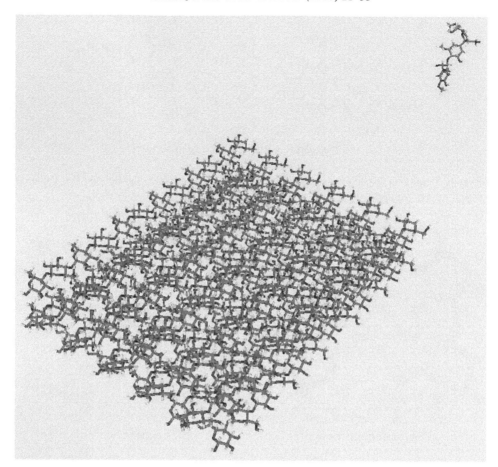

Figure 7. Model of the initial configuration to start the docking calculations built considering the oligolignols (substrates) relatively far away from the macromolecule cellulose I-β crystallite (macromolecules). In our model, each oligolignol starts its docking as a plane landing on the cellulose I-β crystallite (the airstrip).

found that setting up 50 orientations in the population of conformers covered the highest number of rotamers in all the oligolignols. The space where the oligolignols moved over the cellulose I-β model was fixed considering the macromolecule to be inside and centered in a grid-box, whose dimensions were 126 grid-points in the long axis, 126 grid-points transversally to long axis and 90 grid-points in the layer axis, all with a spacing of 0.485 Å. Figure 8 displays the grid-box for the cellulose I-β three-dimensional model.

The selected dimensions assumed that the hydroxyl groups at the edges of the cellulose I-β did not generate any interaction with the oligolignols when docking process occurred, signifying that oligolignols interacted over only the surface of the three-dimensional model.

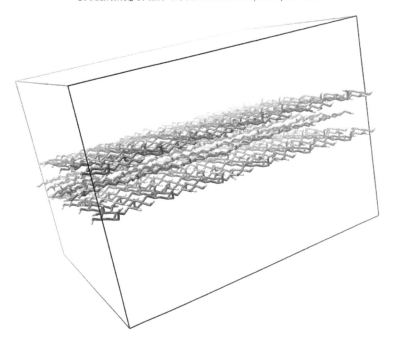

Figure 8. The grid-box surrounding the cellulose I-β three-dimensional model, where the oligolignols moved in the docking simulation.

6. Results and Discussion

Study of chemical reactions is a simple matter, i.e., to determine whether a bond exists or not, since covalent bonds have rather fixed geometries. Adhesion begins with physical adsorption, but the process becomes complicated by ever-changing molecular geometries. Nevertheless, after a molecular mechanics calculation, we have the conformers of oligolignols in energetically favorable positions to interact with the cellulose I-β macromolecular model.

The results of interaction energy and the number of H-bonds between each oligolignol and the cellulose I-β macromolecular model are presented in Table 1.

The computational docking results indicate that the most favorable interaction was found when the tetralignol binds to the cellulose I-β macromolecular model. In fact Table 1 has sorted the values for interaction energy from the most negative, i.e., the minimum, when the oligolignol–cellulose pair is very stable, to the least negative, i.e., the maximum which means poor stability of the oligolignol–cellulose pair. The first observation in the interaction energies is that the largest oligolignol presents the highest interaction energy, although the gap between the values for the trilignols and the dilignols was very close. Furthermore, the number of H-bonds was counted considering a binding distance less than 2.5 Å showing that the dilignols present similar results as trilignols and even as tetralignol.

The most favorable docked structures for the tetralignol, a trilignol and a dilignol are presented in Fig. 9, including the H-bonds formed.

Table 1.

Interaction energies obtained after the docking process in AutoDock, values are sorted from the minimum of interaction energy (most favorable interactions) to the maximum of interaction energy (least favorable interactions)

Oligolignols	Interaction (binding) energy in kcal/mol	Number of H-bonds
Tetralignol		
CA–βO4–SA$'$–$\beta\beta$–SA$''$–βO4–CA$'''$	−10.21	3
Trilignols		
SA–βO4–SA$'$–$\beta\beta$–SA$''$	−9.67	6
CA–βO4–SA$'$–$\beta\beta$–SA$''$	−9.28	4
CA–βO4–SA$'$–β5–CA$''$	−9.24	3
CA–β5–CA$'$–$\gamma\beta$–CA$''$	−8.91	3
CA–βO4–CA$'$–β5–CA$''$	−8.85	3
Dilignols		
SA–βO4–CA$'$	−8.71	5
SA–βO4–SA$'$	−8.32	4
CA–βO4–CA$'$	−8.05	3
CA–$\beta\beta$–SA$'$	−7.69	3
CA–βO4–SA$'$	−7.58	3
SA–β5–CA$'$	−7.55	3
CA–β5–CA$'$	−7.10	3
CA–$\beta\beta$–CA$'$	−7.08	3
SA–$\beta\beta$–SA$'$	−5.70	2

Considering the fact that the adhesion phenomenon begins with physical interactions, it is necessary to consider that the coupling of oligolignols with the cellulose I-β macromolecular model implies that favorable docked conformations and the H-bonds are generated. Observing Table 1 it appears that the longest oligolignol achieves the highest adhesion; however the interaction energy was obtained considering a single molecule of each oligolignol on the cellulose I-β macromolecular model. Thus while tetralignol can achieve the minimum interaction energy, i.e., most favorable docking, the trilignols and the dilignols achieve more H-bonds. These results indicate that the number of dilignol or trilignol docked molecules on the cellulose I-β macromolecular model can be higher than the tetralignol molecules.

An interesting observation for all the monolignols when they were docked on the cellulose I-β macromolecular model was that all of these present their phenyl moiety almost parallel with the surface of the cellulose model. The tetralignol CA–βO4–SA$'$–$\beta\beta$–SA$''$–βO4–CA$'''$ docked conformer presents a diagonal coupling over the surface of the cellulose I-β crystallite model. Although the tetralignol can be considered to be energetically favorable, its configuration over the cellulose I-β crystallite model hampers the coupling of another tetralignol in its vicinity, yielding a poor adhesion interaction. From trilignols group in Table 1, the configuration SA–βO4–SA$'$–$\beta\beta$–SA$''$ was the most favorable docked structure, which was observed

Oligolignol	Docked complex	H-bonds
Tetralignol CA–βO4– SA'–ββ– SA"–βO4– CA'''		
Trilignol SA–βO4– SA'–ββ–SA"		
Dilignol SA–βO4– CA'		

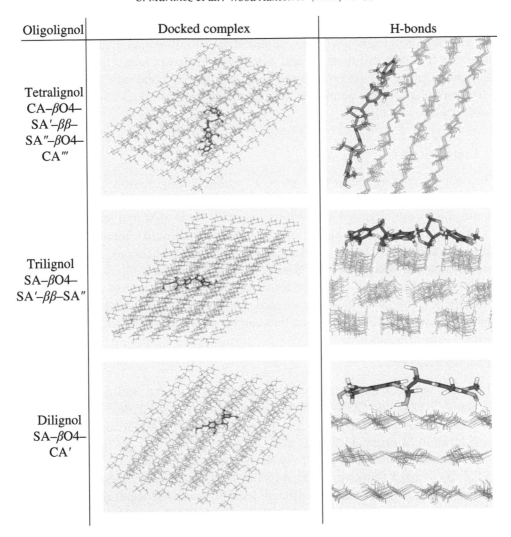

Figure 9. Most favorable docked conformations obtained for each synthetic oligolignol over the cellulose I-β macromolecular model. The H-bonds are represented with blue dashed lines. Pictures in the first column display the oligolignol–cellulose complex after the docking calculation, whereas pictures in the second column display a close-up view of the H-bonds formed.

perpendicular to the cellulosic axle of the crystallite model. Its interaction energy was slightly lower than that of tetralignol (but trilignols permit that other similar molecules can dock or couple in their vicinity, yielding a set of trilignols that can maximize the adhesion interaction). The most favorable docked conformer of dilignols was the SA–βO4–CA' which was observed parallel to the cellulosic axle of the crystallite model, thus it permits the maximum number of similar molecules to dock or couple over the cellulose I-β crystallite model. Nevertheless, higher calculated interaction energy does not guarantee a good adhesion.

The hydroxymethyl groups in lignin precursors, the monolignols, appear to drive the adhesion phenomenon with the cellulose I-β crystallite model, thus the syringyl moiety in the most favorable docked oligolignol promotes better interactions than the guaiacyl moiety. Furthermore, the $\beta O4'$ and the $\beta\beta'$ bonds are the linkages between monolignols that permit the favorable conformers to dock over the cellulose I-β crystallite model.

7. Conclusions

The coupled conformations for a set of 15 synthetic oligolignols over a cellulose I-β macromolecular model were studied and characterized. The interaction energy obtained for each oligolignol–cellulose pair provides insight into its adhesion ability. Although the longest oligolignol presented the minimum interaction energy, its size can hamper the docking of another tetralignol. The trilignols showed results that promote the adhesion interactions of the spatial kind as well as energetic ones. Furthermore these results show the importance of methoxy groups in the adhesion interactions. The methodology presented can be useful to describe the adhesive ability of synthetic oligolignols and thus can be a helpful tool in adhesives design.

Acknowledgements

This work has been supported by a research grant from the National Science and Technology Council (CONACYT-México, Ciencia Básica 2006-57669), the Michoacan State Science and Technology Council (COECYT-Michoacan, 2007-2) and the Education Ministry (PROMEP-SEP).

The French author also gratefully acknowledges the financial support of the CPER 2007-2013 "Strucuturation dú Pôle de Compétiticité Fibres Grand'Est" (Competitiveness Fibre Cluster), through local (Conseil Géneéral des Vosges), regional (Région Lorraine), national (DDRT and FNADT) and European (FEDER) funds.

References

1. H. Nimz, in: *Wood Adhesives Chemistry and Technology*, A. Pizzi (Ed.), chapter 5. Marcel Dekker, New York (1983).
2. A. Pizzi and N. J. Eaton, *J. Adhesion Sci. Technol.* **1**, 191–200 (1987).
3. A. Pizzi, *J. Adhesion Sci. Technol.* **4**, 573–588 (1990).
4. A. Pizzi, *J. Adhesion Sci. Technol.* **4**, 589–595 (1990).
5. A. Pizzi and G. de Sousa, *Chem. Phys.* **164**, 203–216 (1992).
6. A. Pizzi (Ed.), *Advanced Wood Adhesives Technology*. Marcel Dekker, New York (1994).
7. A. Pizzi and K. L. Mittal (Eds), *Handbook of Adhesive Technology*, 2nd edn. Marcel Dekker, New York (2003).
8. D. Fengel and G. Wegener, *Wood Chemistry, Ultrastructure, Reactions*, chapter 6, pp. 132–181. De Gruyter, Berlin (1989).

9. R. A. Young, M. Fujita and B. H. River, *Wood Sci. Technol.* **19**, 363–381 (1985).

10. N. E. El Mansouri, A. Pizzi and J. Salvado, *J. Appl. Polym. Sci.* **103**, 1690–1699 (2006).

11. M. A. Khana, S. Marghoob and V. P. Malhotra, *Int. J. Adhesion Adhesives* **24**, 485–493 (2004).

12. A. Pizzi, *J. Adhesion Sci. Technol.* **20**, 829–846 (2006).

13. A. Krzysik and A. R Young, *Forest Products T.* **36**, 39–44 (1986).

14. T. Higuchi, in: *Biosynthesis and Biodegradation of Wood Components*, T. Higuchi (Ed.), pp. 141–160. Academic Press, Orlando, FL (1985).

15. K. Freudenberg, *Science* **148**, 595–600 (1965).

16. S. Barsberg, P. Matousek, M. Towrie, H. Jorgensen and C. Felby, *Biophys. J.* **90**, 2978–2986 (2006).

17. C. Martinez, J. L. Rivera, R. Herrera, J. L. Rico, N. Flores, J. G. Rutiaga and P. López, *J. Mol. Model.* **14**, 77–81 (2008).

18. R. S. Ward, *Chem. Soc. Rev.* **11**, 75–125 (1982).

19. E. Sjöstrom, *Wood Chemistry — Fundamentals and Application*, 2nd edn. Academic Press, San Diego, CA (1993).

20. C. Martínez, M. Sedano, J. Mendoza, R. Herrera, J. G. Rutiaga and P. Lopez, *J. Mol. Graph. Model.* **28**, 196–201 (2009).

21. A. Sakakibara and Y. Sano, in: *Wood and Cellulosic Chemistry*, D. N. S. Hon and N. Shirasi (Eds), chapter 4, pp. 109–174. CRC Press, Boca Raton, FL (2007).

22. J. Ralph and Y. Zhang, *Tetrahedron* **58**, 1349–1354 (1998).

23. C. J. Houtman and R. H. Atalla, *Plant Physiol.* **107**, 977–984 (1995).

24. W. L. Jorgensen, *Science* **254**, 954–955 (1991).

25. P. López-Albarran, Theoretical study of the drug-receptor interactions in the histamine H2-receptor, *PhD Thesis*, Departament of Chemistry, Universidad Autonoma Metropolitana, México (2005).

26. P. J. Goodford, *J. Med. Chem.* **28**, 849–857 (1985).

27. K. Sharp, R. Fine and B. Honig, *Science* **236**, 1460–1463 (1987).

28. S. A. Allison, R. J. Bacquet and J. McCammon, *Biopolymers* **27**, 251–269 (1988).

29. T. J. A. Ewing and I. D. Kuntz, *J. Comput. Chem.* **18**, 1175–1189 (1997).

30. D. Williamson and G. Jackson, *Mol. Phys.* **83**, 603–611 (1994).

31. B. Q. Wei, L. H. Weaver, A. M. Ferrari, B. W. Matthews and B. K. Shoichet, *J. Mol. Biol.* **337**, 1161–1182 (2004).

32. L. J. Landucci, *Wood. Chem. Technol.* **15**, 349–368 (1995).

33. R. H. Atalla and D. L. VanderHart, *Science* **223**, 283 (1984).

34. D. L. VanderHart and R. H. Atalla, *Macromolecules* **17**, 1465 (1984).

35. J. Sugiyama, T. Okano, H. Yamamoto and F. Horii, *Macromolecules* **23**, 3196 (1990).

36. J. Sugiyama, R. Vuong and H. Chanzy, *Macromolecules* **24**, 4168 (1991).

37. Y. Takabashi and H. Matsunaga, *Macromolecules* **24**, 3968 (1991).

38. R. Huey, G. M. Morris, A. J. Olson and D. S. A. Goodsell, *J. Comput. Chem.* **28**, 1145–1152 (2007).

39. CambridgeSoft, http://www.cambridgesoft.com

40. M. F. Sanner, *J. Mol. Graphics Mod.* **17**, 57–61 (1999).

41. G. M. Morris, D. S. Goodsell, R. S. Halliday, R. Huey, W. E. Hart, R. K. Belew and A. J. Olson, *J. Comput. Chem.* **19**, 1639–1662 (1998).

Modification of Sugar Maple (*Acer saccharum*) and Black Spruce (*Picea mariana*) Wood Surfaces in a Dielectric Barrier Discharge (DBD) at Atmospheric Pressure

Frédéric Busnel [a], **Vincent Blanchard** [b,*], **Julien Prégent** [c], **Luc Stafford** [c],
Bernard Riedl [a], **Pierre Blanchet** [b] **and Andranik Sarkissian** [d]

[a] Centre de Recherche sur le Bois, Faculté de Foresterie et de Géomatique, Université Laval,
Québec (Québec) G1V 0A6, Canada
[b] FPInnovations–Division Forintek, 319 rue Franquet, Québec (Québec) G1P 4R4, Canada
[c] Département de Physique, Université de Montréal, Montréal (Québec) H3C 3J7, Canada
[d] Plasmionique, Montréal (Québec) J3X 1S2, Canada

Abstract

This work examines the adhesion properties of sugar maple (*Acer saccharum*) and black spruce (*Picea mariana*) wood surfaces following their exposure to a dielectric barrier discharge at atmospheric pressure. Freshly sanded wood samples were treated in Ar, O_2, N_2 and CO_2-containing plasmas and then coated with a waterborne urethane/acrylate coating. In the case of black spruce wood, pull-off tests showed adhesion improvement up to 35% after exposure to a N_2/O_2 (1:2) plasma for 1 s. For the same exposure time, adhesion improvements on sugar maple wood up to ∼25% were obtained in Ar/O_2 (1:1) and CO_2/N_2 (1:1) plasma mixtures. Analysis of the wettability with water contact angle measurements indicate that the experimental conditions leading to adhesion improvement are those producing more hydrophobic wood surfaces. In the case of sugar maple samples, X-ray photoelectron spectroscopy investigations of the near-surface chemical composition indicate an increase of the O/C ratio due to the formation of functional groups after exposure to oxygen-containing plasmas. It is believed that a combination of structural change (induced by UV radiation, metastable particles impingement, or both) and chemical change due to surface oxidation is responsible for the observed surface modification of black spruce and sugar maple wood samples.

Keywords
DBD, wood, adhesion, atmospheric plasma, coating, urethane

1. Introduction

Over the last decades, wood industry has faced several challenges including economic crisis, emerging economies, and the appearance of substitution products. One

---

* To whom correspondence should be addressed. E-mail: vincent.blanchard@fpinnovations.ca

Wood Adhesives
© Koninklijke Brill NV, Leiden, 2010

major problem of wood products for some applications is their relatively short durability and the fast deterioration of their appearance. Several waterborne coatings
have been developed to circumvent these limitations but these coatings are often
characterized by poor adhesion to wood surfaces. The most common method to
improve coating/wood adhesion is sanding. Freshly sanded wooden surfaces show
distinctly higher work of adhesion between water and wood as compared to an aged
wood surface [1]. This is because during aging, hydrophobic wood extractives migrate to the surface and, thus, decrease the surface energy [2]. Another approach to
improve adhesion is the use of cold plasmas [3]. One particular advantage of such
plasmas *versus* other alternatives such as heat treatment is their ability to uniformly
modify the near-surface region without affecting the bulk properties [4]. Although
low-temperature plasmas are already used in many technological fields such as mi-
croelectronics, packaging, biomaterials, decorative and functional coatings [5], it is
a relatively recent technology for the wood industry.

In the case of common polymers such as polypropylene, polyethylene, polystyrene and poly(methyl methacrylate), many studies have shown that substantial
changes in the chemical functionality, surface texture, wettability and bondability
to other materials can be achieved by plasma treatment or ion irradiation [6–11].
Depending on the nature of the polymer being processed, several mechanisms have
been invoked to explain the observed enhancement of the polymer properties *via*
ion irradiation, including chain scission [12], cross-linking [13] and carbonization
[14]. In the presence of chemical reactants, additional mechanisms need to be considered such as the adsorption or grafting of new reactive species, the elimination
of weak boundary layers, the actual chemical changes such as oxidation, and the
increase of surface roughness due to pitting [15].

While plasma-induced modification of synthetic polymers such as those listed
above is well documented in the literature and, thus, relatively well understood, the
results for wood surfaces are scarce and the level of knowledge remains at an em-
bryonic state [16–19]. Podgorski and coworkers [20, 21] studied the influence of
plasma and corona discharge treatments on fir species. They showed that wettability could be improved under specific conditions. Rehn *et al.* [22] showed that the
fracture strength of glued black locust (*Robinia pseudoacacia*) increased and coating delamination reduced after exposure to a dielectric barrier discharge (DBD) in
air. These results were attributed to the removal of the weak chemical and mechanical boundary layer on the wood surface. Lecoq *et al.* [23] exposed *Pinus pinaster*
samples to a nitrogen DBD afterglow. This treatment made the wood surface either hydrophilic or hydrophibic depending on electrical parameters. These authors
also observed an increase of the O/C ratio and the presence of carboxyl groups
on the surface after exposure. Evans *et al.* [24] investigated the impact of a glow-
discharge plasma derived from water on wettability and glue bond strength of four
eucalyptus wood species. Blantocas *et al.* [25] showed that the fire and moisture
resistance of different Philippine wood species was improved after treatment by
low energy hydrogen ion showers. Recent studies also investigated plasma poly-

merization on wood surfaces and showed enhanced hydrophobicity under certain conditions [26–28]. Finally, Wolkenhauer and coworkers [29, 30] studied the influence of a DBD treatment in air on the surface properties of wood–plastic composites (polyethylene-based) particleboards and fiberboards. Both the surface energy (primarily the polar component) and the poly(vinyl acetate) glue adhesion were improved by the plasma treatment. Most recently, these authors further demonstrated for several wood species (i.e., beech, oak, spruce and Oregon pine) that a plasma treatment was superior than sanding alone for increasing surface energy [31].

The survey presented above clearly indicates that plasma processing is a promising approach for treating wood surfaces. However, despite its great potential, more work is needed to better understand the influence of the plasma conditions on the surface modification dynamics. In this work, we examine the evolution of the surface properties of black spruce (*Picea mariana*) and sugar maple (*Acer saccharum*) wood samples following their interaction with an atmospheric pressure DBD. Such sources are particularly well-suited for the wood industry because they allow the generation of cold plasmas (with the neutrals being close to room temperature) under atmospheric pressure conditions. Particular emphasis is placed on the influence of the nature of the gas, i.e., Ar, O_2, CO_2, N_2 and their mixtures. Plasma-induced surface modification is characterized by both macroscopic and microscopic diagnostics, including coating/wood adhesion, wettability with water and X-ray photoelectron spectroscopy.

2. Experimental Setup and Diagnostics

A schematic drawing of the linear dielectric barrier discharge ATMOS (Plasmionique, Canada) used for the treatment of wood samples is shown in Fig. 1. The quartz-covered electrodes have an area of $\sim$285 cm^2 and are separated by 1.4 mm. The discharge is driven by a variable alternating current generator connected to an amplifier. In the present work, the frequency was set to 9 kHz. The output of the amplifier is connected to a step-up pulse transformer. In this setup, the rms voltage can be varied continuously between 0 and 8.4 kV. For example, the voltage used to sustain the Ar plasma was set to 2.2 kV. Higher voltages ($\sim$8 kV) were used for molecular gases (O_2, CO_2 and N_2). The total flow rate for each experiment was set to 50 l/min. This flow rate was sufficient to obtain a good plasma jet. As shown in Fig. 1, the substrates were moved towards the plasma region with a conveyer, the speed of which was set to 1.5 cm/s. The samples were exposed 3 times to the plasma, for a total exposure or treatment time of about 1 s. In contrast to other DBD setups used for modification of wood surfaces (see, for example, Ref. [31]), the substrates in our setups do not move between the electrodes but along the sides. This is particularly interesting from an industrial point of view because it allows the treatment of thick wood samples without modifying the discharge gap and, thus, the plasma characteristics.

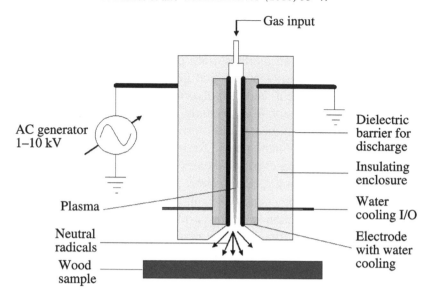

Figure 1. Schematic view of dielectric barrier atmospheric pressure linear plasma source used.

As mentioned above, the wood substrates investigated were black spruce (*Picea mariana*) and sugar maple (*Acer saccharum*). For each experimental condition investigated, all specimens came from the same board and cut in the longitudinal direction. Before each treatment, the samples were conditioned at 20°C and 65% RH for 1 week and sanded with 150 grit. Each series consisted of 10 specimens, each 4 × 85 × 100 mm in thickness, length and width, respectively.

For pull-off tests, a waterborne ultraviolet (UV) curable polyurethane/polyacrylate resin layer was deposited on the substrate after each plasma treatment. The components and their amounts were supplied by an industrial partner. The coating was applied with a square applicator to a thickness of 60 μm and then dried at 60°C for 10 min for water evaporation and polyacrylate cross-linking. In the second step, it was exposed to UV radiation (Sunkist mercury lamp/UVA = 53 J/m^2) to cross-link the polyurethane and complete the curing process. Coating adhesion was measured with a protocol similar to the ASTM D4541-02 test method. The technique involves gluing aluminum dollies using 24 h curable epoxy resin. The load required to pull-off coatings was measured with a 500 kN MTS Hydraulic Testing Machine with a load cell equal to 5 kN. The speed was 7 mm/min. For each experimental condition investigated, the adhesion was determined from the mean value of 28 pull-off tests. Wettability with water was analyzed by contact angle measurements which is a well-established technique for wood samples [32, 33]. Measurements were performed before and after each plasma treatment with a First Ten Ångstroms 200 Dynamic Contact Angle Analyzer. Ten sessile drops of water were placed on each sample. Contact angles were recorded by a CCD camera as a function of time. As shown in Fig. 2, the drop observation was done in the longitudinal direction of wood fibers.

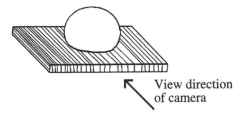

View direction
of camera

Figure 2. Camera view of sessile drop according to wood fiber orientation.

The chemical composition of selected wood substrates was investigated by X-ray Photoelectron Spectroscopy (XPS) using a PHI 5600-ci spectrometer (Physical Electronics, Eden Prairie, MN). A monochromatic aluminum X-ray source (1486.6 eV) at 300 W with a neutralizer was used to record the survey spectra while high-resolution spectra were obtained with the monochromatic magnesium X-ray source (1253.6 eV) at 300 W without charge neutralization. Spectra were recorded at 45° with respect to the sample surface.

3. Results

3.1. Pull-Off Tests

The pull-off strength of waterborne polyurethane/acrylate coating from wood following different plasma treatments are presented in Fig. 3 for black spruce and in Fig. 4 for sugar maple. These figures show the coating/wood adhesion improvement in percentage as referred to the untreated control surface. For instance, a negative value means deterioration of coating/wood adhesion with respect to control and *vice versa*. One can see in Fig. 3 that nominally pure Ar, N_2 and CO_2 plasmas tend to have a negative impact on coating/black spruce adhesion. In contrast, a 16% increase is observed after exposure to the nominally pure O_2 plasma. Based on these results, the influence of O_2 addition in N_2, CO_2 and Ar plasmas was also investigated. In Fig. 3, it can be observed that treatments in N_2/O_2, CO_2/O_2 and Ar/O_2 discharges have a positive impact on the coating/black spruce adhesion. Among all mixtures, the one producing the most improvement of 36% is N_2/O_2 (1:2).

In the case of the sugar maple, Fig. 4 shows that nominally pure gases such as N_2 and O_2 decrease adhesion, whereas Ar and CO_2 plasmas increase it up to 4% and 10%, respectively. Accordingly, the influence of Ar and CO_2 addition to N_2 and O_2 discharges was also examined. Figure 4 reveals that coating/sugar maple adhesion improvement up to 25% can be achieved after exposure to Ar/O_2 (1:1) and CO_2/N_2 (1:1) plasmas. Overall, from the results presented in Figs 3 and 4, it can be deduced that for a given wood species, the coating/wood adhesion strongly depends on the gas mixture used for the treatment. In addition, the conditions leading to the maximal coating/wood adhesion improvement differs from one wood species to another.

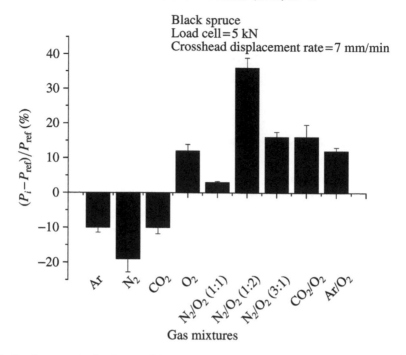

Figure 3. Coating–wood adhesion modifications as a function of plasma treatment for black spruce. Values were calculated with reference to an untreated control specimen (P_{ref}). P_i represents the strength needed to pull-off the coating after plasma exposure.

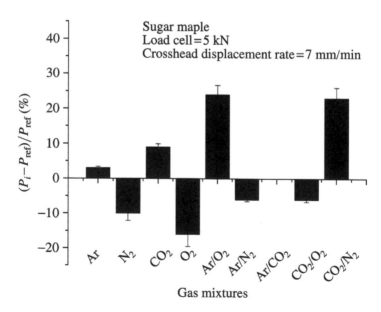

Figure 4. Coating–wood adhesion modifications as a function of plasma treatment for sugar maple. Values were calculated with reference to an untreated control specimen (P_{ref}). P_i represents the strength needed to pull-off the coating after plasma exposure.

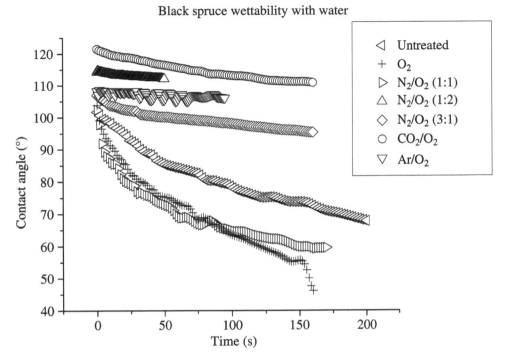

Figure 5. Evolution with time of contact angle of a drop of water according to plasma treatment of black spruce.

3.2. Wettability with Water

To gain insight into the adhesion results presented above, the wettability of the wood surface, which is a necessary condition to ensure good adhesion, was determined by contact angle measurements. The wettability results with water following different plasma treatments are presented in Fig. 5 for black spruce and in Fig. 6 for sugar maple. In both figures, the results are presented in terms of mean contact angle as a function of time. For each treatment, 10 curves were interpolated/extrapolated on the time-axis with cubic B-spline method to have the same x-axis range for all input curves and then averaged. In Fig. 5, the contact angle for the untreated spruce sample decreases from 100° to 70° in about 200 s. In comparison, treatments in N_2/O_2 (1:2), N_2/O_2 (3:1), Ar/O_2 (1:1) and CO_2/O_2 (1:1) plasmas significantly increase the contact angles up to 120°, while considerably reducing the time dependence ($d\theta/dt \rightarrow 0$). These results reveal that the black spruce surface becomes highly hydrophobic after such treatments. In the case of sugar maple, Fig. 6 shows that the contact angle varies from 70° to 10° in 40 s for the untreated sample. Treatments in CO_2/N_2 (1:1), Ar/O_2 (1:1), CO_2 and nominally pure Ar plasmas decrease the decay rate, meaning again that the surface becomes more hydrophobic. Combining the results of Figs 3–6, a tendency is revealed: wood surfaces that become more hydrophobic after plasma treatment also present improved coating/wood adhesion.

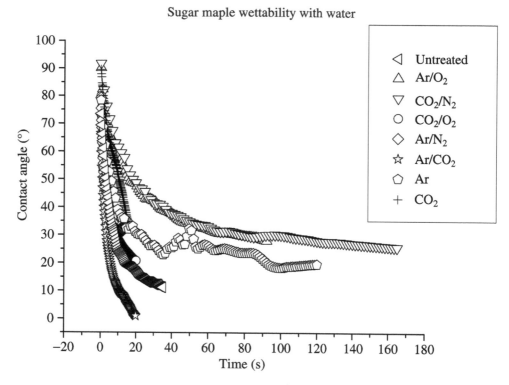

Sugar maple wettability with water

Figure 6. Evolution with time of contact angle of a drop of water according to plasma treatment of sugar maple.

In many studies reported in literature, plasma treatment has been used to increase the wetting properties of the wood surface with the objective of enhancing the bond strength between adhesives and wood [34, 35]. A reduced wettability is expected to prevent the adhesives from easily penetrating into wood and, thus, to reduce their cross-linking efficiency. In the present study, however, the waterborne urethane/acrylate coating used is an amphipantic compound, i.e., it is an emulsion with both hydrophilic and hydrophobic components. After water evaporation, the coating becomes mostly hydrophobic and its compatibility is, thus, improved with hydrophobic surfaces. Therefore, the improved adhesion observed after exposure to the N_2/O_2 plasma for black spruce and after exposure to Ar/O_2 and CO_2/N_2 plasmas for sugar maple can be explained by the less hydrophilic character of both surfaces after treatment. A similar correlation was deduced by Petric et al. [36] who investigated the influence of heat treatment on the adhesion properties of Scots pine wood (Pinus sylvestris L.) with polyacrylate waterborne coatings. An improved coating wetting was observed due to an enhanced surface hydrophobicity. This result is also consistent with the measurements of Podgorski et al. [20] who showed that both plasma and corona treatments improved the wettability of wood but no improvement in the coating adhesion was observed.

Table 1.
C/O ratio and C1 to C4 percentages according to wood species and plasma treatment

Chemical components	Black spruce		Sugar maple		
	Untreated	N_2/O_2	Untreated	Ar/O_2	CO_2/N_2
C/O	4.1	4.1	2.6	2.6	2.2
C1	67.7	66.7	50.6	46	35.7
C2	25.5	25.9	39.7	42.6	51.4
C3	3.7	3.9	6.5	7.7	9.3
C4	3.1	3.5	3.2	3.7	3.6

3.3. X-Ray Photoelectron Spectroscopy Characterization

The near-surface chemical composition of selected black spruce and sugar maple wood samples was characterized before and after plasma treatments by X-ray photoelectron spectroscopy. These measurements were performed for experimental conditions producing the maximum improvement in coating adhesion, i.e., N_2/O_2 (1:2) for black spruce and Ar/O_2 (1:1) and CO_2/N_2 (1:1) for sugar maple. Surface composition results deduced from the high-resolution C_{1s} spectra are shown in Table 1. The carbon (1s) peak was also resolved in details using the classification of Dorris and Gray [37]. For wood samples, the C_{1s} peak is generally separated into four components expressed as C1–C4. The four components are listed in order of their increasing range of binding energies. In class 1 (C1), carbon atoms are bonded only to carbon and/or hydrogen atoms (C–C/C–H, 284.6–289.18 eV). Class 2 (C2) refers to carbon atoms bonded to a single oxygen, except for carboxyl oxygen (C–O, 286.3–289.04 eV). In class 3 (C3), carbon atoms are bonded to two non-carboxyl oxygen or to a single carboxyl oxygen (O–C–O or C=O, 287.9–290.75 eV) while in class IV they are bonded to a carboxyl and a non-carboxyl oxygen (O=C–O, 289–291.78 eV). Proportions of these contributions are also shown in Table 1.

In the case of sugar maple samples, XPS results show that the C/O ratio decreases from 2.6 before treatment to 2.2 after plasma exposure. Such decrease was also observed for polyethylene and poly(ethylene terephthalate) samples treated in oxygen-containing plasmas due to the formation of carboxyl groups [38, 39]. Table 1 also shows that the proportions of C2–C4 increase after plasma treatment: for C2, the increase was >10% while for C3 and C4 it was much less (~3% for C3 and <1% for C4). A similar but much less increase of the C2–C4 proportions was observed for sugar maple samples exposed to Ar/O_2 (1:1) plasma and for black spruce specimens treated in N_2/O_2 (1:2) discharge.

In our previous work with inductively and capacitively coupled plasmas at low pressures [40], it was shown that N_2-containing plasmas produced considerable adhesion improvement (>50%) between waterborne urethane/acrylate coating and sugar maple. XPS studies further indicated a considerable amount of grafted nitrogen atoms which are likely to allow covalent bonding between urethane-based coating and wood surface. In the present study, the same coating was used but N_2-based plasmas had a negative impact on coating–wood adhesion. In addition, no nitrogen concentration was observed by XPS. Under the DBD conditions investigated, it seems that N-atoms are either not grafted or grafted to such a low extent that these were undetectable by our XPS experiments ($<1\%$).

4. Discussion

It is well established in literature that wood is a complex material, both in the near-surface region and the bulk. It is mainly composed of cellulose, hemicelluloses, lignin and extractives. These extractives are likely to interfere with coating adhesion by reducing wettability, preventing penetration into the wood, and interfering with cross-linking [41]. The high density of several wood species can also restrict coatings from penetrating into the wood and forming strong bonds [42]. Black spruce sample investigated in the present study is softwood and, therefore, has a high amount of extractives on its surface. On the other hand, sugar maple is hardwood and the amount of surface extractives is relatively low. As mentioned above, both samples were subjected to sanding before plasma treatment. Since sanding is known to remove hydrophobic extractives and, thus, to produce hydrophilic surfaces [1, 2] the more hydrophobic character observed after selective plasma treatments cannot be attributed to a better removal of surface extractives.

Literature shows that hemicellulose is, in great part, responsible for the wood hygroscopic and hydrophilic behaviour because of its free hydroxyl groups. Some authors have reported that when wood is heated, ether links can be formed by splitting of two adjacent hydroxyl groups, leading to a reduction of wood hygroscopic and hydrophilic behaviour [23]. As emphasized in [23], this cannot be understood by XPS analysis because both carbons bonded to hydroxyl groups or present in ether groups correspond to the C2 band. In the case of sugar maple, a decrease of the C/O ratio along with an increase of C2–C4 proportions was observed after exposure to oxygen-containing plasmas. In [38, 39], such decrease of the C/O ratio with the increase of the C2–C4 proportions was found to increase the surface polarity of plasma-treated polyethylene and poly(ethylene terephthalate). This suggests that the observed chemical reorganization observed by XPS is not the dominant mechanism leading to the more hydrophobic character displayed in Figs 5 and 6 and, thus, to the observed improvement in coating/wood adhesion.

Another mechanism proposed by some authors for the enhanced hydrophobicity evokes a closure of large micropores in cell walls [43]. Indeed, change in crystallinity of cellulose or conformational reorganization of polymeric components of wood could lead to more perfectly aligned bonds which may no longer be separable by intrusion of water. This would explain why water drops are not absorbed in wood for certain experimental conditions as seen in wettability with water measurements. Blantocas et al. [44] and Ramos et al. [45] observed partial closure of surface pores in wood following ion irradiation. For low ion energies, ion bombardment made the surface hydrophobic. In our experimental conditions, the samples are exposed to the spatial afterglow of the DBD such that significant ion irradiation is not expected. However, a number of other plasma species can play a role analogous to low energy ion bombardment, including UV photons and metastable atoms. Wertheimer and coworkers [46, 47] reported on UV modification of polymers, with direct breaking of carbon–oxygen bonds with radiation wavelengths of >160 nm and breaking of carbon–carbon bonds by radiation wavelengths of <160 nm, each followed by secondary reactions. Metastable atoms such as $N_2(A)$, $N_2(a')$ and Ar in the $^3P_0$ and $^3P_2$ states can also transfer their internal energy and induce structural modification by quenching with the wood surface atoms. Additional analyses by plasma emission spectroscopy will, however, be required to determine the respective role of each of these reaction pathways in the surface modification dynamics.

5. Conclusion

In summary, this work examined the influence of plasma treatments on the adhesion properties of black spruce (*Picea mariana*) and sugar maple (*Acer saccharum*) wood surfaces. The samples were treated in a dielectric barrier discharge at atmospheric pressure in Ar, O_2, N_2 and CO_2-containing plasmas and then coated with a waterborne urethane/acrylate coating. In the case of black spruce wood, adhesion improvement up to 35% was obtained after exposure to a N_2/O_2 (1:2) plasma for 6 s. For the same exposure time, adhesion improvements on sugar maple wood up to ~25% were obtained in Ar/O_2 (1:1) and CO_2/N_2 (1:1) plasma mixtures. Wettability with water pointed out that the experimental conditions leading to adhesion improvement are those producing more hydrophobic wood surfaces. In the case of sugar maple samples, the near-surface chemical composition determined by XPS indicated an increase of the O/C ratio due to the oxidation of wood surfaces. It is believed that a combination of structural change (induced by UV radiation or metastable particles impingement, or both) and chemical reorganization due to surface oxidation is responsible for the observed surface modification of black spruce and sugar maple wood samples.

Acknowledgements

The authors would like to acknowledge FPInnovations–Division Forintek, Plasmionique, and the National Science and Engineering Research Council of Canada (NSERC) for their financial support. Dr. Fabienne Poncin-Epaillard from CNRS, France is also acknowledged for her valuable input.

References

1. M. Gindl, A. Reiterer, G. Sinn and S. E. Stanzl-Tschegg, *Holz Roh Werkst.* **62**, 273 (2004).
2. M. Gindl, G. Sinn, W. Gindl, A. Reiterer and S. Tschegg, *Colloids Surfaces A* **181**, 279 (2001).
3. H. Conrads and M. Schmidt, *Plasma Sources Sci. Technol.* **9**, 441 (2000).

4. E. M. Liston, *J. Adhesion* **30**, 199 (1989).
5. C. Tendero, C. Tixier, P. Tristant, J. Demaison and P. Leprince, *Spectrochimica Acta B* **61**, 2 (2006).
6. K. L. Mittal (Ed.), *Polymer Surface Modification: Relevance to Adhesion*, vol. 5. VSP/Brill, Lieden (2009).
7. S. Nowak and O. M. Küttel, *Mater. Sci. Forum* **140–142**, 705 (1993).
8. U. Kogelschatz, *Plasma Chem. Plasma Process.* **23**, 1 (2003).
9. R. Cueff, G. Baud, J. P. Besse and M. Jacquet, *J. Adhesion* **42**, 249 (1993).
10. F. Dreux, S. Marais, F. Poncin-Epaillard, M. Metayer and M. Labbé, *Langmuir* **18**, 10411 (2002).
11. F. Massines and G. Gouda, *J. Phys. D: Appl. Phys.* **31**, 3411 (1998).
12. M. E. Fragala, G. Compagnini, L. Torrisi and O. Puglisi, *Nucl. Instrum. Meth. Phys. Res. B* **141**, 169 (1998).
13. E. H. Lee, G. R. Rao and L. K. Mansur, *Trends. Polym. Sci.* **4**, 229 (1996).
14. V. Svorcik, E. Arenholz, V. Rybka and V. Hnatowicz, *Nucl. Instrum. Meth. Phys. Res. B* **122**, 663 (1997).
15. P. Grönig, O. M. Küttel, M. Collaud-Coen, G. Dietler and L. Schlapbach, *Appl. Surface Sci.* **89**, 83 (1995).
16. M. Strobel, C. S. Lyons and K. L. Mittal (Eds), *Plasma Surface Modification of Polymers: Relevance to Adhesion.* VSP, Utrecht (1994).
17. H. Chen and E. Zavarin, *J. Wood Chem. Technol.* **10**, 387 (1990).
18. A. E. Lepska-Quinn, The effect of RF plasma on the chemical and physical properties of wood, *Wood Science Thesis*, University of California, Berkeley, CA (1994).
19. F. S. Denez, L. E. Cruz-Barba and S. Manolache, in: *Handbook of Wood Chemistry and Wood Composites*, R. M. Rowell (Ed.), pp. 447–474. Taylor and Francis, Boca Raton, FL (2005).
20. L. Podgorski, B. Chevet, L. Onic and A. Merlin, *Int. J. Adhesion Adhesives* **20**, 102 (2000).
21. L. Podgorski, C. Boustas, F. Schambourg, J. Maguin and B. Chevet, *Pigment Resin Technol.* **31**, 33 (2002).
22. P. Rehn, A. Wolkenhauer, M. Bente, S. Forster and W. Viol, *Surf. Coat. Technol.* **174–175**, 515 (2003).
23. E. Lecoq, F. Clément, E. Panousis, J.-F. Loiseau, B. Held, A. Castetbon and C. Guimon, *Europ. Phys. J. Appl. Phys.* **42**, 47 (2008).
24. P. D. Evans, M. Ramos and T. Senden, in: *Proc. European Conference on Wood Modification*, C. A. S. Hill, D. Jones, H. Militz and G. A. Ormondroyd (Eds), Wales, UK, pp. 123–133 (2007).
25. G. Q. Blantocas, P. E. R. Manteum, R. W. M. Odille, R. J. U. Ramos, J. L. C. Monasterial, H. J. Ramos and L. M. Boot, *Nucl. Instrum. Meth. Phys. Res. B* **259**, 875 (2007).
26. W. L. F. Magalhães and M. F. Souza, *Surf. Coat. Technol.* **155**, 11 (2002).
27. S. Manolache, H. Jiang, R. M. Rowell and F. S. Denes, *Mol. Cryst. Liq. Cryst.* **483**, 348 (2008).
28. B. S. Kim, B. H. Chun, W. I. Lee and B. S. Hwang, *J. Thermoplastic Composite Mater.* **22**, 21 (2009).
29. A. Wolkenhauer, G. Avramidis, Y. Cai, H. Militz and W. Viöl, *Plasma Proc. Polym.* **4**, 470 (2007).
30. A. Wolkenhauer, G. Avramidis, H. Militz and W. Viöl, *Holzforschung* **62**, 472 (2008).
31. A. Wolkenhauer, G. Avramidis, E. Hauswald, H. Militz and W. Viöl, *Int. J. Adhesion Adhesives* **29**, 18 (2009).
32. B. M. Collett, *Wood Sci. Technol.* **6**, 1 (1972).
33. M. de Meijer, S. Haemers, W. Cobben and H. Militz, *Langmuir* **16**, 9352 (2000).
34. T. Uehara and S. Jodai, *Mokuzai Gakkaishi* **33**, 777 (1987).
35. I. Sakata, M. Morita, N. Tsuruta and K. Morita, *J. Appl. Polym. Sci.* **49**, 1251 (1993).

36. M. Petrič, B. Knehtl, A. Krause, H. Militz, M. Pavlič, M. Pétrissans, A. Rapp, M. Tomažič, C. Welzbacher and P. Gérardin, *J. Coat. Technol. Res.* **4**, 203 (2007).

37. G. M. Dorris and D. G. Gray, *Cellulose Chem. Technol.* **12**, 721 (1978).

38. C. Jie-Rong, W. Xue-Yan and W. Tomiji, *J. Appl. Polym. Sci.* **72**, 1327 (1999).

39. H. Drnovska, L. Lapcik, V. Bursikova, J. Zemek and A. M. Barros-Timmons, *Colloid Polym. Sci.* **281**, 1025 (2003).

40. V. Blanchard, P. Blanchet and B. Riedl, *Wood Fiber Sci.* **41**, 245 (2009).

41. K. F. Plomley, W. E. Hillis and K. Hirst, *Holzforschung* **30**, 14 (1976).

42. G. K. Brennan and P. Newby, *Australian Forestry* **55**, 74 (1993).

43. A. W. Christiansen, *Wood Fiber Sci.* **22**, 441 (1990).

44. G. Q. Blantocas, H. J. Ramos and M. Wada, *Jpn J. Appl. Phys.* **45**, 8498 (2006).

45. H. J. Ramos, J. L. C. Monasterial and G. Q. Blantocas, *Nucl. Instrum. Meth. Phys. Res. B* **242**, 41 (2006).

46. M. R. Wertheimer, A. C. Fozza and A. Hollander, *Nucl. Instrum. Meth. Phys Res. B* **151**, 65 (1999).

47. F. E. Truica-Marasescu and M. R. Wertheimer, *Macromol. Chem. Phys.* **206**, 744 (2005).

Determination of the Microstructure of an Adhesive-Bonded Medium Density Fiberboard (MDF) using 3-D Sub-micrometer Computer Tomography

Gernot Standfest [a,*], **Simon Kranzer** [b], **Alexander Petutschnigg** [a] **and Manfred Dunky** [c]

[a] Salzburg University of Applied Sciences, Forest Products Technology and Timber Construction, Markt 136 a, A-5431 Kuchl, Austria

[b] Salzburg University of Applied Sciences, Information Technology and Systems Management, Urstein Süd 1, A-5412 Puch, Austria

[c] Kronospan GmbH Lampertswalde, Mühlbacher Strasse 1, D-01561 Lampertswalde, Germany

Abstract

The microstructure of a medium density fiberboard (MDF) has attracted much interest, because it determines most of the properties during production as well as during use of this type of board. In this paper the sub-micrometer computer tomography (sub-μm-CT) with very high resolution was used in order to investigate an industrially produced MDF. The data obtained were analyzed using mathematical morphology and image analysis in order to obtain information on the distribution of voids and the fiber material. Resulting distributions were found to follow Γ-law which was confirmed using the maximum likelihood estimation. Eventually, good correlations between void distribution and local density as well as between the proportion of cell wall material and local density were found in this investigation.

Keywords

MDF, pore size distribution, sub-μm-CT, image analysis

1. Introduction

The microstructure of fiber-based panels is one of the most important parameters concerning their performance. A detailed description of various methods for the evaluation of the 3-D microstructure in fiber-based panels was compiled by Bucur [1]. Serial microtome sections were used in combination with light microscopy to assess 3-D models of wood-fiber-based panels, where the inter-fiber voids were investigated at different depths [2]. The most promising technique for microstruc-

---

* To whom correspondence should be addressed: Tel.: (43) 5022112406; Fax: (43) 5022112099; e-mail: gernot.standfest@fh-salzburg.ac.at

tural characterization of wood-fiber-based composites is the X-ray micro computer tomography (μCT). This method was successfully used in some of the first investigations concerning material characterization and microstructure related properties, like fiber properties and fiber orientation [3, 4]. The vertical density profile, usually determined by X-ray or isotope radiation methods, was also established using μCT images and a good correlation between the results of both methods was found [5].

X-ray source from synchrotron radiation can be used as the basis for microstructural characterization of wood-fiber-based panels. Fiber-based composites as produced by different manufacturing processes have been characterized concerning fibers and fiber network in order to develop 3-D fiber network models and to characterize their thermal conductivity [6]. Morphological image analysis tools were used for fiber network characterization, in order to describe the morphology of various wood-fiber-based low-density insulation materials [7] and their thermal conductivity [8]. Badel *et al.* [9] also used synchrotron micro computer tomography (SRμCT) for the *in situ* investigation of the different compression states of low-density fiberboard. Influences of density variation and changes in microstructure during the compaction of the composite were evaluated.

SRμCT as the basis for CT image acquisition can be combined with different image analysis tools. The structure of laboratory-made wood and natural fiber-based panels was investigated using 3-D image analysis to obtain information on the fiber network, size and shape of single fibers, and fiber contact areas and fiber orientation [10–13]. Additionally, the resin distribution on the fiber surfaces was assessed. Based on these data also the permeability of the wood-fiber network was determined experimentally and calculated based on the CT data [14].

In this study, 3-D evaluation of data obtained from sub-μm-CT measurements was performed for characterizing industrially produced MDF; the main focus of this investigation was to achieve a high resolution. For such evaluation, morphological operations were used in order to determine the proportion and the distribution of the sum of inter- and intra-particle pores on the one side and of fiber cell wall material on the other side in the panel. Additionally, it is of great interest to see if and how different sub-volume sizes influence the resulting pore and cell wall material size distribution and which sub-volume size is representative for the whole board. Finally, the correlations between pore size and cell wall material size proportion and the local density at a certain position in the cross-section were investigated.

2. Material and Methods

2.1. Material and Sample Preparation

A commercial MDF board of 19 mm thickness was used for the investigations reported here. Three samples (A, B and C) with a size of approximately $50 \times 50 \times 19$ mm^3, hence including the whole thickness of the board, were cut from this MDF board and conditioned at standard conditions (20°C/65% RH) until equilibrium moisture content was reached. Average sample density was determined stereometrically according to EN 323 [15]. The densities were found to be 714 kg/m^3 for samples A and C and 710 kg/m^3 for sample B. The cross-sectional density profile was measured using a vertical density measuring device (Dense-Lab X, Electronic Wood Systems, Hameln, Germany) using the smallest possible step of sample movement (10 μm). For the tomographic scanning, small samples with dimensions approximately 4 × 4 × 19 mm^3 were prepared from the same samples as used for the vertical density profile measurements. Again the samples were conditioned (20°C/65% RH) to equilibrium moisture content. The small samples were then wrapped in a plastic foil under vacuum in order to avoid changes in the moisture content. Unpacking was done just before the samples were scanned.

2.2. Sub-micrometer Computer Tomography (Sub-μm-CT)

The three-dimensional (3-D) characterizations for wood-based panels can be performed on the basis of data obtained with the sub-micrometer computer tomography (sub-μm-CT) used for the investigation reported here. All measurements were carried out at the Upper Austrian University of Applied Sciences, Wels, Austria, using a Phoenix X-ray Nanotom CT system (Phoenix X-ray Systems + Services GmbH, Wunstorf, Germany). The resolution achieved for the above-mentioned sample size was 3.2 μm per voxel (= volumetric pixel element with 3.2 × 3.2 × 3.2 μm^3). Scanning height was approximately 10.5 mm (overall sample height was approximately 19 mm). Based on the almost symmetrical density profile (see Fig. 2) it was possible to restrict the CT analysis to only one of the two halves of thickness of the sample, thus, the measurement time could be reduced by a factor of two without losing significant information.

2.3. 3-D Image Processing and Pore Size Distribution

The data sets from the sub-μm-CT measurements were virtually divided into sub-volumes with various sizes, and the pore size distribution was determined for these various sub-volumes.

The first step in this work included a qualitative evaluation of the sub-μm-CT images. It was crucial that the sub-volumes consisted of the sample without surrounding air (region outside the sample). Selecting sub-volumes in the way described above also avoids edge effects due to mechanical treatment during sample preparation [13]. For each sample, four sub-volumes were defined with lengths and widths of 0.25 mm, 0.5 mm, 1 mm and 2 mm. This corresponds to edge lengths of 78, 156, 312 and 625 voxels, respectively. The height of the sub-volumes was limited to 9.6 mm, which corresponds to 3000 voxels and is slightly more than the half the height of the original sample. The CT data hence yield information about the full height of the board. The purpose of these sub-volumes with different heights was to clarify if and how the determined pore-size distribution might be influenced by the size of the sub-volume and what size of sub-volume was still representative of the whole board sample. It was found that the achieved solution itself is not influenced by the sizes of the selected sub-volumes.

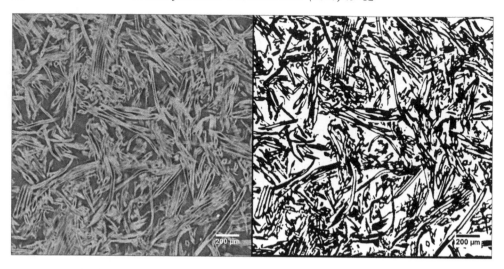

Figure 1. Left: Original image after CT data reconstruction. Right: Binary (black and white) image after filtering and thresholding.

The noise in the CT images was reduced using a despeckle filter which is based on a $3 \times 3 \times 3$ voxels median filter, which considers all adjacent voxels in x-, y- and z-direction. With this arrangement the edges and fiber cell walls are preserved and not destroyed. After noise reduction the threshold for each image in the image stack was calculated using the Otsu method [16] (see Fig. 1) generating regions of connected voxels.

The next step performed was the 3-D analysis using MATLAB [17]. When evaluating the size of a pore at a certain position in the sample the diameter of the largest spherical structural element is taken, which is completely within the pore space [18]; the method based on this definition is called 'opening' size distribution (or granulometry). With basic morphological operations (erosion and dilation) the pore size and the pore size distribution from binary images (2-D and 3-D) can be determined, a method originally introduced by Matheron [19] and Serra [20].

Erosion (equation (1)) of a set I of foreground voxels (u', v') in a matrix Q_I with a certain shaped (e.g., circular in two dimensions, spherical shaped in three dimensions) structuring element H in a matrix Q_H causes the set of foreground voxels to shrink according to pixel values (i, j). This means that the foreground voxels are converted to background voxels (if the background is eroded, the background voxels are converted to foreground voxels). Structuring elements have a 'hot spot' and a certain neighborhood (certain number of voxels around the hot spot) of voxels. The 'hot spot' is a particular point of reference (here the centre voxel of the 3-D structuring element) that defines where the placing of the mask at a certain voxel position is done. During erosion the structuring element and each foreground voxel are compared step by step. Voxels are eroded if they are not contained completely within the structuring element. Only those connected voxels where the structuring element completely fits into are not converted to background. Erosion is denoted by $\ominus$.

$$I \ominus H = \{(u', v') | (u' + i, v' + j) \in Q_I, \forall (i, j) \in Q_H\}. \tag{1}$$

A dilation operation (equation (2)), denoted by $\oplus$, results in an increase in the number of voxels in a set of voxels (I). Again it is performed with a specifically shaped structuring element H. At each point of the set of foreground voxels, where the mask of the structuring element fits completely, dilation is performed. This operation reconstructs the initial shape of the set of eroded pixels, but without the regions where the structuring element did not fit completely.

$$I \oplus H = \{(u', v') = (u + i, v + j) | (u', v') \in Q_I (i, j) \in Q_H\}. \tag{2}$$

In the so-called 'opening' process (equation (3)), denoted by, erosion is followed by dilation with the same structuring element.

$$I \circ H = (I \ominus H) \oplus H. \tag{3}$$

The pore size distribution can be obtained by performing a series of these 'opening' processes with a family of growing structuring elements [20]. For this particular purpose, spherical shaped structuring elements were used which grow by 2 voxels in diameter. So the first structuring element has a diameter of 3 voxels, the second one of 5 and so on. With each opening, the intra- (= fiber lumen) and the inter-particle background (white area in Fig. 1) shrink until only foreground pixels (black area in Fig. 1) are left. The pore size distribution is then the result of a density function $D(i)$:

$$D(i) = \frac{N_i - N_{i+1}}{N_0}, \tag{4}$$

where the difference in the numbers of voxels of the initial image stack before (N_i) and after (N_{i+1}) each morphological opening procedure with every structuring element of growing size is divided by the sum of all pore voxels (N_0, white voxels) in the image stack.

3. Results and Discussion

3.1. Density Profile

Figure 2 shows the vertical density profiles of the investigated samples. As mentioned above the vertical density profile was nearly symmetrical which justifies scanning only one half of each sample.

3.2. Sub-μm-CT

Figure 3 shows the 3-D reconstructions of 2 sub-volumes of MDF sample A, in particular one high-density surface sample and low-density core sample. The sample sizes of both were $0.5 \times 0.5 \times 4.8 \text{ mm}^3$. Density of the surface is around 1000 kg/m^3 and of the core 700 kg/m^3.

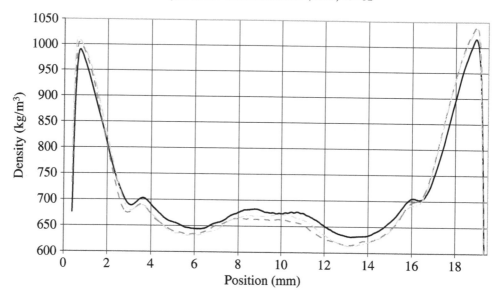

Figure 2. Symmetrical vertical density profiles of the investigated samples. The lines show the average mean density values of the samples, A (continuous line), B (dashed line) and C (dotted line).

In Fig. 3 also the cross-sections of fibers and fiber bundles as well as of small residual wooden pieces can be discerned as the consequence of often not completely achieved defibrillation in the refiner on industrial scale.

With the resolution achieved it was possible to distinguish between voids, single fibers and fiber bundles. With the sample size selected for the CT measurements the resolution achieved is 3.2 µm per voxel; this number is between the resolution of 2.3 µm per voxel for MDF [10–14] and the resolution of 4.9 µm per voxel reported for low-density fiber-based insulation materials [6–9], both values obtained with synchrotron micro computer tomography (SRµCT). It is also important to mention that the reconstructed and analyzed sub-volumes are larger than those reported in the literature up to now [10–14, 21]. This helps to avoid the risk that the sub-volumes evaluated are not representative of the whole sample investigated.

Some tiny particles in Fig. 3 show very high attenuation, appearing as very light spots. Wernersson *et al.* [22] attributed such artifacts to metallic impurities.

3.3. Pore Size Distribution

Pore size distributions in most wood fiber networks show Γ-distributions. Such behavior was also found for paper materials [23] and for low-density wood fiber insulation materials [7, 8]. The maximum likelihood estimation using the *gamfit* function in MATLAB [17] was used in order to prove that the assumption of a Γ-type pore size distribution was correct.

Figure 4 shows the cumulative density function (CDF) of the pore size distribution of sample A ($1 \times 1 \times 9.6$ mm^3). Similar curves were obtained for the other samples investigated, but are not shown here.

Figure 3. 3-D reconstructions of a part of sub-volume of MDF sample A. Left: high-density surface on top of the picture, low-density core at the bottom. Right: low-density core. Both sub-volumes have a size of $0.5 \times 0.5 \times 4.8$ mm^3.

Lux *et al.* [7] argued that the results of the granulometric analysis rather reflect the distance between fibers than the pore size distribution of a certain shaped pore. The determined pore size distributions can be significantly influenced by the choice of the shape of the structuring element. For example, if spherical structuring element is used the number of adjacent voxels increases and affects the frequency of pores detected by this structuring element.

3.4. Closing the Fiber Lumen

Fiberboards consist, in principle, of three phases: inter-particle pores, fiber lumens, and cell wall material. In the CDF plot (Fig. 4) the sum of the voids (inter-particle pores and fiber lumens) is shown. Lux *et al.* [7] used dilation in their evaluation of

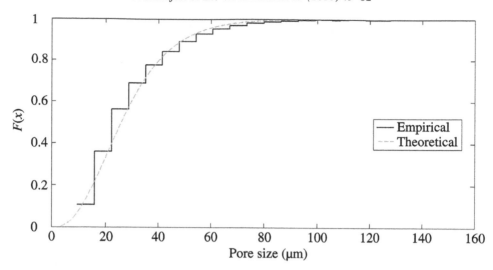

Figure 4. CDF plot of the pore size distribution of sample A.

the CT data followed by closing using reconstruction, which preserved edges and closed the fiber lumens. The segmented fiber material was then analyzed and the diameter of the connected fiber material (single fibers or fiber bundles) was determined, yielding two phases: inter-particle voids (voids in a fiberboard between the fibers) and fiber material (but including the lumens within the fibers). This method succeeded in analyzing the pore size distribution and the size of this connected fiber cell wall material with morphological operations. Another more complex method was presented by Walther and Thoemen [13], where inter-particle voids, fiber cell wall material and also fiber lumens could be distinguished. According to Walther [10] dilation was used to expand the cell wall in order to close the holes in the cell wall due to the presence of pits. Then the fiber lumen was identified by its thin elongated form. This procedure may not work with bent fibers, where the fiber lumen does not appear elongated.

In the work presented here still another approach was attempted. By morphological closing it is possible to fill the lumens of the fibers. The resulting image is then subtracted from the image obtained by the Otsu segmentation [16]. An exemplary result of this approach is shown in Fig. 5. The proportions of the various phases are 30% for the inter-particle voids, 52% for the cell wall material (fibers and fiber bundles) and 19% for the lumens within fibers. A disadvantage of this method still is the fact that areas between fibers, which are very close to each other, are wrongly interpreted as fiber lumens. Therefore, further analysis was performed on two phases: (i) on fibers with all voxels belonging to the cell wall phase and (ii) on voids with all voxels belonging to the inter-particle and lumen void phase (see following sections).

Figure 5. Segmentation into the three phases: void, fiber cell wall and fiber lumen. Left: Original gray level image. Middle: Black and white image after filtering and thresholding with two phases segmented (void = white, cell wall material = black). Right: Segmentation into three phases (void = white, cell wall material = black, lumen within the fibers = gray).

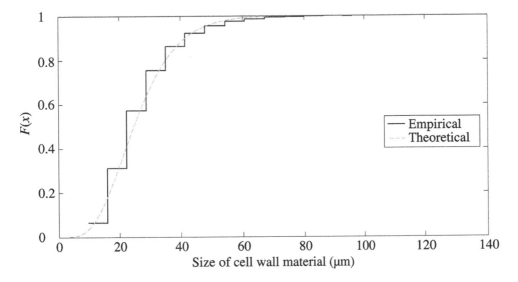

Figure 6. CDF plot of the size of connected cell wall material distribution (sample A).

3.5. Evaluation of the Cell Wall Material

Figure 6 shows the CDF plot of the analyzed fiber phase. All voxels belonging to the cell wall material phase were analyzed by means of granulometry. Hence the term 'size of connected cell wall material' here is defined as a measure describing the dimensions of configurations of the solid phase (cell wall material phase): this can be the diameter of a single fiber cell wall or the cross-fiber dimension of fiber bundles or the size of an aggregation of cell wall material due to the compression process.

3.6. Analysis of Different Sub-volumes

Figure 7 shows the average pore size distributions of the analyzed sub-volumes. The mean pore size of all investigated sub-volumes was approximately 29 µm; however, the high standard deviation of approximately 17 to 18 µm reflects a broad pore size distribution. The maximal pore size diameter was found in the sub-volumes of $1 \times 1 \times 9.6$ mm^3 and $2 \times 2 \times 9.6$ mm^3 as 221 µm. The average pore fraction (sum of all voxels that were found to be in the pore phase, independent of whether as inter-fibril (inter-particle) pore or as lumens of the fibers) in the analyzed sub-volumes was 45%. This number is very close to the range determined by Walther and Thoemen [13] who also investigated MDF with an average density of around 700 kg/m^3. An overview of the results for analyzed sub-volumes is given in Table 1.

The size distribution of the connected cell wall material as defined above is shown in Fig. 8 with the averages summarized in Table 2. The largest size of connected cell wall material detected was 138 µm. The phase fraction of fibers was around 55%.

These results show that the size distributions of pores or of the connected cell wall material are not affected by different sizes of the sub-volumes investigated. For the sake of short analysis and calculation times small sub-volumes might be preferred, but this attempt will involve risk of insufficient representation of the sub-volume compared to the total sample investigated.

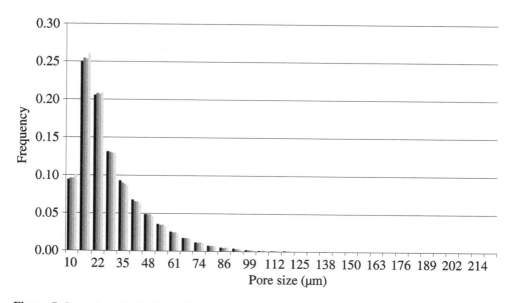

Figure 7. Pore size distribution of the analyzed sub-volumes. Bars from black to light gray indicate the analyzed sub-volumes $0.25 \times 0.25 \times 9.6$ mm^3, $0.5 \times 0.5 \times 9.6$ mm^3, $1 \times 1 \times 9.6$ mm^3 and $2 \times 2 \times 9.6$ mm^3, respectively.

Table 1.
Mean, median and standard deviation (in µm) of the pore size distribution of the analyzed sub-volumes

Size of sub-volume	0.25 × 0.25 × 9.6 mm³	0.5 × 0.5 × 9.6 mm³	1 × 1 × 9.6 mm³	2 × 2 × 9.6 mm³
Mean	29	29	29	29
Median	22	22	22	22
Standard deviation	17	17	19	18

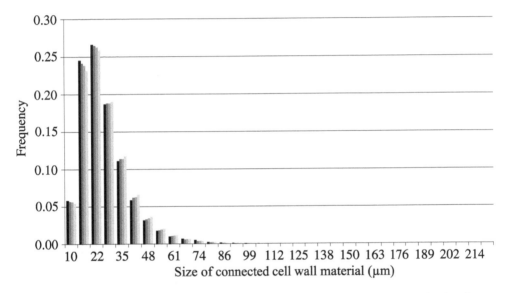

Figure 8. Size distribution of the connected fiber call wall material of the investigated sub-volumes. Bars from black to light gray indicate the analyzed sub-volumes 0.25 × 0.25 × 9.6 mm³, 0.5 × 0.5 × 9.6 mm³, 1 × 1 × 9.6 mm³ and 2 × 2 × 9.6 mm³, respectively.

Table 2.
Mean and standard deviation (in µm) of the size of the connected cell wall material of the investigated sub-volumes

Size of sub-volume	0.25 × 0.25 × 9.6 mm³	0.5 × 0.5 × 9.6 mm³	1 × 1 × 9.6 mm³	2 × 2 × 9.6 mm³
Mean	26	26	27	27
Median	22	22	22	22
Standard deviation	12	12	12	12

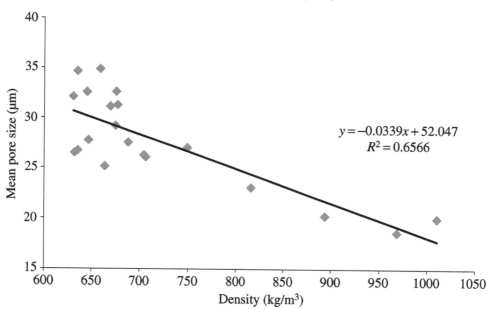

Figure 9. Influence of the local densities in various layers of the MDF sample on mean pore size. The equation describes the linear regression line.

3.7. Pore Size and Local Density

The sub-volume with approximately $1 \times 1 \times 9.6$ mm^3 size of sample A was chosen to investigate the correlation between (i) the size of connected cell wall material (fiber) and (ii) the pore size using the local density of the sample. For this evaluation, 20 image stacks out of this sub-volume of sample A with a sample height of 150 voxels (0.48 mm) were created, representing the whole height of the half sample as investigated by CT.

Figure 9 shows the correlation between mean pore size and the local board density. Looking at the whole density range investigated a fairly good correlation is given; only in the low density range (630–670 kg/m^3) a high scatter in the individual values of the pore sizes occurs. The maximal void size was found to be 176 µm at a mean density of 675 kg/m^3.

3.8. Size of Connected Cell Wall Material and Local Density

Figure 10 shows an excellent correlation between the size of the connected cell wall material as defined above and the local density ($R^2 = 0.95$). No excessive scatter in individual calculated size of connected cell wall material occurs in the density range between ~625 and 750 kg/m^3 (size range of ~20 to 27 µm). The maximum size of connected cell wall material was determined to be 138 µm as already mentioned above; this value corresponds well with the largest sizes detected in the evaluation of the sub-volume at a local density of 1010 kg/m^3.

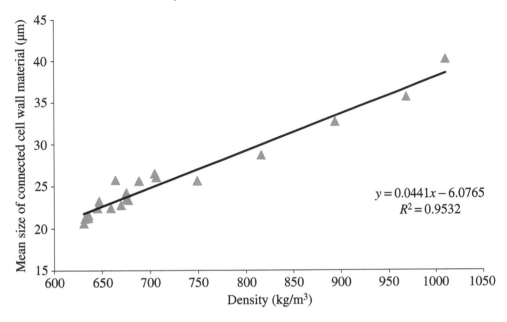

Figure 10. Influence of the local density in various layers of the MDF sample on the size of the connected cell wall material. The equation describes the linear regression line.

4. Conclusion

In this paper sub-µm-CT was used to characterize the microstructure of an MDF. In particular, the inter-fiber distances (voids) and the so-called size of connected cell wall material as a measure for connected parts of solid material (cell wall material, fibers, fiber bundles) was investigated. The results show that connected parts of both, fibers and voids (sum of voids (i) between fibers and (ii) within the fibers = lumen) are Γ-distributed, and this was also confirmed by the maximum likelihood estimation.

Sub-µm-CT was found to be a promising method for future investigations of wood-fiber-based panels and is more or less equivalent in its performance to SRµCT.

The data evaluation time (calculation time) is one of the most crucial and limiting factors for this kind of analysis using MATLAB. For very small sub-volumes the granulometric analysis can be achieved within a few minutes; sub-volumes of approximately $2 \times 2 \times 9.6$ mm^3, as used in the work here and corresponding to 4.5×10^9 voxels, however, require several hours of calculation time using a standard consumer computer.

Good correlations were found between the local density on the one side and (i) the size of connected cell wall material (as a measure for connected cell wall material in various configurations) and (ii) and the mean pore size.

Acknowledgement

This project was gratefully supported by the 'FHplus in COIN' Programme of the Austrian Research Promotion Agency (FFG).

References

1. V. Bucur, *Measurement Sci. Technol.* **14** (12), 91–98 (2003).
2. R. Ramli, S. Shaler and M. A. Jamaludin, *J. Oil Palm Res.* **14** (1), 35–44 (2002).
3. S. M. Shaler, in: *Proc. 1st European Panel Products Symposium*, Llandudno (North Wales), p. 28 (1997).
4. L. Groom, L. Mott and S. Shaler, in: *Proc. 33rd Int. Particleboard/Composite Materials Symposium*, Pullman, WA, USA, pp. 89–100 (1999).
5. G. Standfest, A. Petutschnigg, M. Dunky and B. Zimmer, *Eur. J. Wood Prod.* **67**, 83–87 (2009).
6. M. Faessel, C. Delisée, F. Bos and P. Castéra, *Composites Sci. Technol.* **65**, 1931–1940 (2005).
7. J. Lux, C. Delisée and X. Thibault, *Image Anal. Stereol.* **25**, 25–35 (2006).
8. J. Lux, A. Ahmadi, C. Gobbé and C. Delisée, *Int. J. Heat Mass Transfer* **49**, 1958–1973 (2006).
9. E. Badel, C. Delisee and J. Lux, *Composites Sci. Technol.* **68**, 1654–1663 (2008).
10. T. Walther, Methoden zur qualitativen und quantitativen Analyse der Mikrostruktur von Naturfaserwerkstoffen, *PhD Thesis*, University Hamburg, Hamburg (Germany) (2006).
11. T. Walther, K. Terzic, T. Donath, H. Meine, F. Beckmann and H. Thoemen, *Prog. Biomedical Optics Imaging* **7**, 631812.1–631812.10 (2006).
12. T. Walther, H. Thoemen, K. Terzic and H. Meine, in: *Proc. 10th European Panel Products Symposium (EPPS) 2006*, Llandudno (North Wales), pp. 23–32 (2006).
13. T. Walther and H. Thoemen, *Holzforschung* **63**, 581–587 (2009).
14. H. Thoemen, T. Walther and A. Wiegmann, *Composites Sci. Technol.* **68**, 608–616 (2008).
15. EN 323, *Wood Based Panels; Determination of Density* (1993).
16. N. Otsu, *IEEE Trans. Systems, Man and Cybernetics* **9**, 62–66 (1979).
17. The MathWorks Inc., *MATLAB.* www.mathworks.com (08.10.2009).
18. H. J. Vogel, *Eur. J. Soil Sci.* **48**, 365–377 (1997).
19. G. Matheron, *Random Sets and Integral Geometry.* Wiley, New York, NY (1975).
20. J. Serra, *Image Analysis and Mathematical Morphology.* Academic Press, London (1982).
21. S. Rolland du Roscoat, M. Decain, X. Thibault, C. Geindreau and J. F. Bloch, *Acta Materialia* **55**, 2841–2850 (2007).
22. E. L. G. Wernersson, C. L. L. Hendriks and A. Brun, in: *Proc. IEEE 6th Int. Symposium on Image and Signal Processing and Analysis*, Salzburg, Austria, pp. 365–370 (2009).
23. C. T. J. Dodson and W. W. Sampson, *Appl. Math. Letters* **10**, 87–89 (1997).

Influence of the Degree of Condensation on the Radial Penetration of Urea-Formaldehyde Adhesives into Silver Fir (*Abies alba*, Mill.) Wood Tissue

Ivana Gavrilović-Grmuša *, Jovan Miljković and Milanka Điporović-Momčilović

Faculty of Forestry, University of Belgrade, RS-11030 Belgrade, Serbia

Abstract

Penetration is the ability of an adhesive to move into the voids on the surface of a substrate or into the substrate itself. The cellular nature of wood can cause significant penetration of an adhesive into the substrate.

The objective of this work was to evaluate the influence of the degree of condensation of urea-formaldehyde (UF) resins on the radial penetration into silver fir (*Abies alba*, Mill.) and hence on the distribution of resin in the wood tissue by means of microscopic determination.

Three UF resins with different degrees of condensation together with extender and hardener were used as adhesive mixes and were applied onto one of the surfaces to be bonded by hot pressing parallel to the grain direction, with Safranin added as a coloring agent. Microtome slides (20 μm thick) were cut from each joint sample, showing the bondline and the two adherends with the resin penetrated in radial direction. The depth of adhesive penetration was determined by epi-fluorescence microscopy.

The results show a significant correlation between the penetration behavior and the degree of condensation (molecular size, viscosity) of the resins. The higher the degree of condensation, the lower is the penetration possibility, expressed as 'Average penetration' depth (AP). The portion of filled tracheids in the radial direction on both sides of the bondline ('Filled interphase region', FIR) however, did not depend statistically on the viscosity of the resin mix.

Keywords

Penetration, silver fir (*Abies alba*, Mill.), urea-formaldehyde (UF) adhesive, degree of condensation, epi-fluorescence microscopy, fluorescence confocal laser scanning microscopy (CLSM), scanning electron microscopy (SEM)

1. Introduction

Penetration is the ability of an adhesive to enter into the lumens and into the cell walls as a process of fluid movement [1]. The interphase region of the adhesive bond is defined as the volume containing both wood cells and adhesive. It is created by penetration of the adhesive into the wood surface and partly filling of the

---

* To whom correspondence should be addressed. E-mail: ivana.grmusa@sfb.rs

Wood Adhesives

lumens determined by (i) wood related parameters (such as diameter of the lumen and portion of cut open lumens on the wood surface due to the grain slope), (ii) the properties of the resin and the adhesive mix (such as molar mass distribution, composition of the adhesive mix, viscosity [2] and surface energy, amount of adhesive spread, hardening time and rate of resin curing [3]) and (iii) bonding processing parameters (such as assembly time, press temperature and pressure, or moisture level). The temperature of the wood surface and of the bondline and hence the viscosity of the resin (which itself also depends on the already reached degree of hardening) influence the penetration behavior of the resin and hence bond performance [4].

Adhesive penetration and the associated close contact of the adhesive with the internal surface of the substrate plays an important role in wood bonding. Although it is generally known that a certain penetration helps in creating strong bonds [1, 5], it is not clear whether penetration into the lumens or into the cell walls is more critical. Low bond strengths will result from either under- or over-penetration. Under-penetration means that the adhesive is not able to move into the wood substance enough in order to create a large active bonding surface (interface) within the interphase and strong interaction between wood and the adhesive. In contrast, over-penetration occurs if a large portion of the adhesive can penetrate into wood substance causing starved joints; as a consequence, insufficient amount of adhesive remains in the bondline to bridge between the wood surfaces and to establish bond strength. To solve these problems, the viscosity of the adhesive, especially in dependence of the temperature in the bondline during the press cycle, has to be adjusted by proper composition and molecular structure (molar mass distribution, degree of condensation) of the adhesive.

Despite the fact that UF resins have been for many decades the most important type of wood adhesives [6–8] and that resin penetration as such has been reported in the literature quite often [9–11], the influence of the molecular structure (size of the adhesive molecules, degree of condensation) on the depth and rate of penetration of the resin or the adhesive mix has not yet been reported.

The degree of condensation, and hence the molar mass distribution, is one of the most important characteristics of a condensation resin and determines several properties of the resin [12–15], like the viscosity at a given solid content and hence the flowability. The condensation reaction and the increase of the molar masses of a UF resin can be monitored by GPC and from the increase of the viscosity (corrected to the same temperature and solid content) [15–17]. With longer duration of the acidic condensation step, molecules with higher molar masses are formed and the GPC peaks shift to lower elution volumes.

Ferg *et al.* [13] reported that the bond strength increases with the degree of condensation of the UF resin applied. Higher molar masses (higher viscous portions) give a more stable bondline and higher cohesion properties [15]. Already Rice [18] and Nakarai and Watanabe [19] had reported that the resistance of a bondline against swelling in water and redrying increased with the viscosity of the resin.

The reason for this fact most obviously is that resins with advanced degree of condensation preferably stay in the bondline. Rice [18] even found an increase of the thickness of the bondline with an increased viscosity of the resin.

Penetration is highly controlled by the size of the molecules of the adhesive. Low molar mass portions of the adhesive can easily penetrate; this effect is enhanced by the decrease in viscosity due to the increased temperature in the bondline during hot pressing. The higher molar masses remain at the wood surface and form the bondline. In order to achieve good bond strength a certain ratio between low and high molar masses is necessary, as this was mainly determined for PF resins [20–23].

Using UF resins with different degrees of condensation, and hence different viscosities, Scheikl and Dunky [24, 25] showed that higher viscosities of UF resins decreased the penetration into the wood substance. A clear correlation was found between the viscosity of the resins and the initial rate of penetration into the surface, shown as slower decrease of the droplet volume on the surface with time, using the static contact angle measurements, for higher viscosities. Penetration into latewood with narrow lumens was slower than into earlywood with wider lumens.

Brady and Kamke [26] investigated the penetration of PF resins into thin wood flakes using fluorescence microscopy and showed that resin penetration was influenced rather by the natural variability of the wood material than by pressing conditions. Also, it was about three times greater in Douglas Fir earlywood than in latewood. Cell wall fractures enhanced penetration by providing additional paths for hydrodynamic flow.

Sernek and Resnik [27] investigated the penetration behavior of four melamine-urea-formaldehyde (MUF) resins with different contents of melamine and different molar masses, also adding slightly different amounts of extender to the resin mix. The lower the viscosity of the resin mix, the deeper was the penetration into the wood surface.

Stephens and Kutscha [28] investigated the effect of the resin molar mass on the bonding behavior of flakeboards. A commercial PF resin was separated into two molar mass fractions by dialysis, and aspen flakeboards were prepared with the separate fractions as well as with mixtures of the resin fractions. Penetration characteristics of the different fractions were determined microscopically. Results indicate that both low and high molar mass components of the resin are needed to achieve boards with acceptable properties.

The extent of lumen penetration into earlywood, latewood and wood rays is preferably determined by examination of the cross section of the bondline, and many techniques have been successfully used for this purpose: light microscopy [29, 30], transmitted and reflected microscopy [31–33], (epi-)fluorescence microscopy [2, 10, 26, 31, 34–40], fluorescence confocal laser scanning microscopy (CLSM) [34, 41–45], scanning electron microscopy (SEM) [32, 46–48], transmission electron microscopy (TEM) [49–52], SEM in combination with an energy-dispersive analyzer for X-rays (SEM/EDAX) [43, 53], X-ray microscopy [54], autoradiography [11] or combination of different microscopy techniques [55]. Kamke and Modzel [56] and Hass *et al.* [57] recently used microtomography to investigate penetration of various adhesives into wood.

Light visible or fluorescent dyes or pigments can be added especially to light-colored adhesive mixes for enhanced visibility of adhesive distribution in the bond-line and the adjacent wood material under incident visible or UV light [31, 34, 36, 37, 58, 59].

Usually, penetration of adhesives can be described as filling up the lumens by the adhesive material. Cell wall penetration [9, 46], in principle, is possible, but rather applies only to substances with low molar masses, like impregnating resins [42–44, 60–62].

Despite existing results and experience, the lack of more exhausting research on the penetration of UF resins and adhesive mixes with different levels of condensation into wood tissue is quite evident. Therefore, the objective of this study was to evaluate the influence of the degree of condensation on the penetration, and hence on the distribution of UF resins, within the wood substance by means of microscopic investigation. For this purpose, a UF laboratory batch was prepared and samples were taken at three different time spans during the condensation reaction. Hence three different degrees of condensation were obtained. In order to secure the influence of the length of the condensation phase, no low molar mass moieties were added at the end of this condensation phase; the molar ratio was kept at the original high value during condensation. The intention was to emphasize the degree of condensation and to minimize all other chemical influences.

2. Materials and Methods

2.1. Urea-Formaldehyde (UF) Resins

Three laboratory UF resins with different degrees of condensation according to recipes described in the literature [63, 64] were provided by DUKOL Ostrava, s.r.o. (Ostrava, Czech Republic). The degree of condensation (DOC) increases as a consequence of the duration of the acidic condensation step from resin UF I (lowest DOC) to resin UF III (highest DOC). This determines the viscosity of the resin if the solid content is the same. The viscosity of the three resins increased due to the larger molecules present in the resin from 218 mPa s for UF I to 281 mPa s for UF II and eventually to 555 mPa s for UF III, measured within two days after the synthesis of the resins (Table 1). The molar ratio of formaldehyde to urea (F/U) of all resins was 2.0; no urea was added after the condensation step.

The adhesive mixes were prepared by addition of 10 mass% of wheat flour as extender and 0.05 mass% of Safranin (Superlab, Belgrade) as marker, both based on solid resin. Preliminary tests in the laboratory showed that Safranin did not segregate from the resin mix at higher temperatures. Also there was no influence of Safranin on the gel times measured. In order to keep the same gel time of the adhesive mixes the addition of ammonium sulphate as hardener was 0.5% for UF I and 0.3% for UF II and UF III, both values expressed as solid ammonium sulphate

Table 1.
Characteristics of the resins and the adhesive mixes UF I, UF II and UF III

Property	Unit	UF I		UF II		UF III	
		Resin	Adhesive mix	Resin	Adhesive mix	Resin	Adhesive mix
Solid content	%	53.7	54.4	53.7	54.3	53.8	54.6
Brookfield viscosity (20°C)	mPa s	218	545	281	745	555	1644
Gel time	s	58	59	59	60	58	59

on resin solids. An equal gel time was considered to be essential for comparison of penetration of the various adhesive mixes into wood, since the hardening process influences penetration behavior due to increase of the molecular size.

A similar increase in viscosity as for the resins themselves is seen for the adhesive mixes (UF I: 545 mPa s; UF II: 745 mPa s; UF III: 1644 mPa s) just after mixing. The viscosities of the different adhesive mixes were taken in this paper as a measure of the DOC in order to show its influence on the average penetration depth (AP).

For the determination of the solid content 2 g of the sample (either resin or adhesive mix) were dried in a laboratory oven at $105 \pm 2°C$, until a constant mass of the sample had been reached. The duration of this drying step was in the range of 24–48 h. The viscosity of the three UF resins and the adhesive mixes was determined by the Brookfield method. Gel time was determined at boiling water temperature in a glass tube with approximately 2 g of the resin (with addition of hardener as mentioned above) or the adhesive mix by gentle stirring.

Table 1 summarizes the characteristics of the UF resins and the adhesive mixes.

2.2. Bonded Samples, Microtome Slices and Photographs for the Evaluation of the Penetration

Tangentially cut 5 mm thick silver fir (*Abies alba*, Mill.) plies, 100 mm long (parallel to grain) and 30 mm wide, were prepared for radial penetration. In order to have statistically representative results, the silver fir plies were randomly selected. Special effort was made to have as low taper as possible; this should guarantee equal penetration conditions for all samples. Before bonding, the plies were conditioned in standard climate ($T = 20 \pm 2°C$ and RH $= 65 \pm 5\%$), yielding moisture content (MC) of approx. 10%. The prepared UF adhesive mixes were applied to the surface of one of the two plies with an adhesive loading level of 200 g/m^2, and assembling was performed parallel to the grain direction. Pressing was performed in a hydraulic press at 120°C and 1.0 MPa for 15 min. For each of the three UF adhesive mixes ten samples were prepared by assembling and hot pressing.

After hot pressing, the bonded samples were conditioned again in standard climate. Microtome test specimens (20 μm thick) were prepared from three different positions. From each of the microtome slides five photographs were taken using epi-fluorescence microscopy (LEICA DM LS) at different positions across the 30 mm bondline, showing individual sections of the bondline of approx. 1.4 mm width on each photograph (so in total randomly distributed approx. 7 mm of the whole bond-line length of 30 mm). The set of optical filters used consisted of a 450 nm excitation filter, a 510 nm dichromatic mirror and a 515 nm emission filter. The image analysis system included a color video camera (LEICA DC 300) and the image processor with analysis software (IM1000 by LEICA Microsystems, Heerbrugg, Switzerland). These photographs were then evaluated for penetration of the UF resin adhesives (see Section 2.3).

2.3. Determination of Penetration

The individual depths of penetration (μm) were determined from each photograph at 45 positions within the 1400 μm width of the bondline in the photograph, so individual values were obtained at distance of 30 μm each; this distance is similar to the thickness of one cell row. The depth of penetration here is defined as the sum of the distances the resin could penetrate into the two plies starting from the bondline. No separate evaluation was done for the two plies with applied adhesive mix and without it, although small differences might occur. The average value of penetration depth was calculated for each photograph. Further, averages were determined for five photographs per microtome slide, and for three microtome slides per bonded joint. Final 'Average penetration depth' (AP), was calculated from 10 replicates of joints, for each adhesive mix (Table 3). Therefore, each AP result in Table 3 (one for each resin type) is based on 7000 individual measurements.

Two additional measures (mm^2) for the interphase region were calculated from each individual epi-fluorescence microscopy photograph as mentioned (Table 2): (i) total interphase region (IR), including the unfilled lumen area using the maximum value out of the 45 values for the depth of penetration in each photograph and (ii) the sum of all filled lumens or filled rays A; the so-called 'Filled interphase region' (FIR) was expressed as percentage A/IR (%). All FIR data were calculated from each photograph individually and then again averages were calculated per microtome slide, per joint sample, and eventually per resin adhesive (Table 3).

Several microtome slides were also examined with fluorescence confocal laser scanning microscopy (CLSM). The images were obtained using the LSM 510 system with an Axioscope FS2mot microscope (Carl Zeiss). The objectives used were Zeiss Plan-Neofluar 20×/0.30 and Zeiss Plan-Neofluar 40×/1.3 Oil. The Multitrack scan mode was adjusted to two excitation wavelengths, namely 488 nm (obtained by argon laser) and 543 nm (obtained by HeNe laser). The evaluated wavelength ranges were 505–545 nm and >545 nm, respectively. The pinhole was set to an optical slice thickness of 10 μm for the 20× objective and 3 μm for the 40× objective.

Scanning electron microscopy (JEOL JSM-6460LV SEM) on the 20 μm thick microtome test specimens was performed in order to investigate if any penetration through pits had occurred. The specimens were mounted on carbon tapes applied

Table 2.
Average size of the interphase region (IR) and the sum of filled lumens and rays (A) in fir as evaluated
for the adhesive mixes based on the three resins UF I, UF II and UF III

Resin	Average size of the interphase region (IR)		Average sum of filled lumens and rays (A)	
	IR (mm^2)	SD (mm^2)*	A (mm^2)	SD (mm^2)*
UF I	0.51	0.26	0.25	0.13
UF II	0.49	0.15	0.24	0.07
UF III	0.42	0.14	0.21	0.08

* SD, standard deviation.

Table 3.
'Average penetration depth (AP)' and 'Filled interphase region (FIR)' in fir as evaluated for the adhesive mixes based on the three UF resins (UF I, UF II, UF III)

Resin	Average penetration depth (AP)		Filled interphase region (FIR)	
	AP (µm)	SD (µm)*	FIR (%)	SD (%)*
UF I	207	87	48	6
UF II	177	48	50	7
UF III	120	32	52	8

* SD, standard deviation.

to stubs, coating with chromium in a sputter coater device SCD005 and coated
with pure gold. After that, the specimens were viewed in the secondary electron
imaging (SEI) mode at an acceleration voltage of 20 kV and a working distance of
10–35 mm.

3. Results

Adhesive mixes mainly fill lumens of the tracheids as well as rays (Figs 1 and 2).
Comparing the light-colored sections on both sides of the geometrical bondline it is
clearly visible that the UF I adhesive mix penetrates to a greater extent and deeper
into the wood material, due to the smaller size of the molecules before the hardening
reaction starts. Photographs like those in Figs 1 and 2 were used for the evaluation
of AP and FIR.

Table 2 summarizes the size of the interphase region (IR) and the sum of filled
lumens and rays, both expressed as (mm^2).

Table 3 and Figs 3 and 4 summarize the overall results for the 'Average penetration depth' (AP) and the 'Filled interphase region' (FIR) for the three resins with
different degrees of condensation.

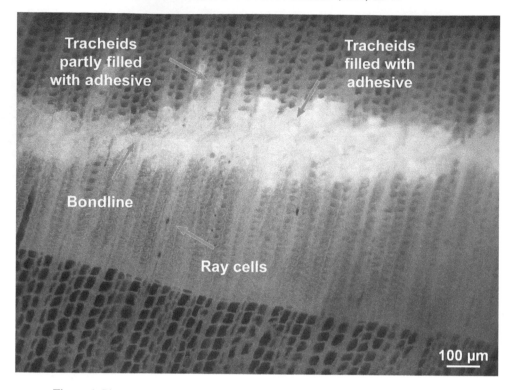

Figure 1. Photomicrograph of UF I (with Safranin) bondline, using epi-fluorescence.

According to the Multiple Comparison Test statistically significant differences are seen for AP between UF I or UF II and UF III; there is no statistically significant difference for AP between UF I and UF II. According to the Kruskal–Wallis test there is no statistically significant difference between the values for FIR between the three resins tested.

The 'Average penetration depth' (AP) and the 'Filled interphase region' (FIR), both as a function of the viscosity of the three resins with different degrees of condensation, are shown in Figs 3 and 4.

The 'Average penetration depth' (AP) shows a clear dependence on the degree of condensation (expressed as viscosity) of the investigated resins. The 'Filled interphase region' (FIR) increases slightly with higher viscosity, but no significant difference between the three values is found.

Figure 5 shows that the combination of Safranin as staining agent of the adhesives and CLSM enabled the adhesive to be sharply differentiated from wood cell walls based on a bright contrast of the colors of the adhesive (appearing light grey) and of the wood cell walls (appearing dark grey).

SEM in combination with secondary electron imaging (SEI) mode adds another important dimension to visualize adhesive penetration into wood tissues. Figure 6 shows tracheids without and with adhesive. In the rays small droplets can be seen;

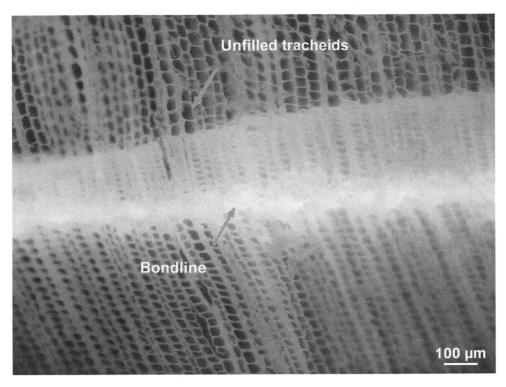

Figure 2. Photomicrograph of UF III (with Safranin) bondline, using epi-fluorescence.

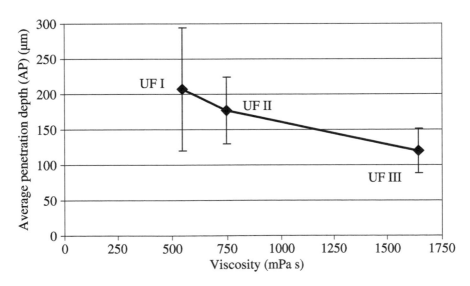

Figure 3. 'Average penetration depth' (AP) as a function of viscosity (different degrees of condensation) of the UF adhesive mixes used.

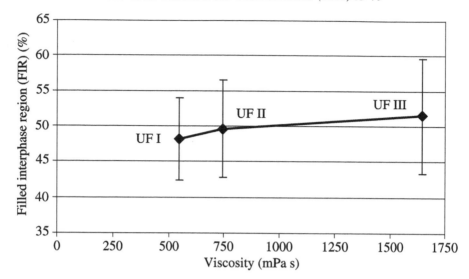

Figure 4. Percentage of 'Filled interphase region' (FIR) as a function of the viscosity (different degrees of condensation) of the UF adhesives used.

additionally the diameter of the pits and the distance between pits in the rays are indicated.

4. Discussion

Penetration preferably occurs when the applied adhesive mix in the bondline is heated up and hence the mobility of the molecules increases (the viscosity of the resin and hence of the adhesive mix drops), as long as this mobility is not hindered by the hardening reaction creating larger molecules and again leading to an increase of the viscosity up to the hardened network.

The higher the degree of condensation the larger are the molecules in size already at the moment when the adhesive mix is applied; even for these higher condensed resins the mobility will increase with higher temperature of the bondline when heating up the joint assembly. Still deeper penetration into wood tissue for the larger molecules will be restricted due to dimension of flow paths.

This difference in flow behavior was examined for the three different UF resins. The average penetration behavior is shown in the results of Tables 2 and 3. Due to less deeper penetration the size of the interphase decreases as expected. It is defined as the sum of the deepest penetration on both sides of the bondline in each photograph examined, multiplied by the width investigated (1.4 mm). The maximum penetration depths are in the range of 335 μm for UF I, 300 μm for UF II and 177 μm for UF III, all numbers expressed as the sum for the penetration into both plies.

The average penetration depth was calculated on the basis of 45 individual measurements from each photograph. This means that a single value as the average penetration depth was calculated per photograph. Also this average penetration

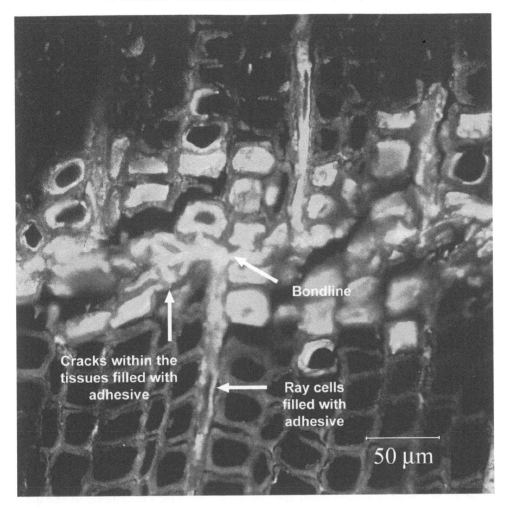

Figure 5. Confocal laser scanning microscopy (CLSM) photograph of UF I (with addition of Safranin) bondline (light grey) showing tracheids with different degrees of filling as well as the deep penetration of adhesive into rays.

depth was defined as the sum of the individual penetrations in each ply. Taking the sum in both directions eliminates possible differences in the penetration into the two plies. In this evaluation it was, however, not distinguished between earlywood and latewood.

All measured penetration depths and the thickness of all parts of the interphase were much smaller than the average width of the year rings of the plies used in the experiments described here (4.2 mm). This means that three cases were observed, based on the large number of samples: (i) penetration rather only into earlywood (the surface of the plies towards the bondline is earlywood); (ii) penetration rather only into latewood (the surface of the plies towards the bondline is latewood); and (iii) the bondline was between the earlywood on the one ply and latewood on the

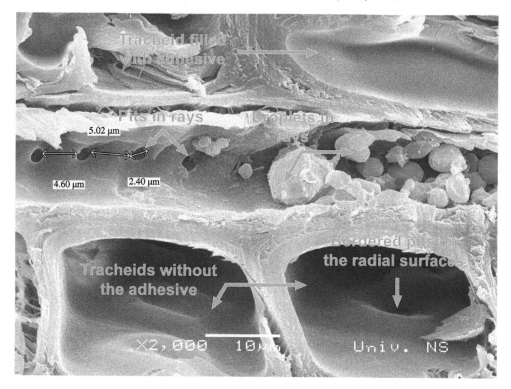

Figure 6. SEM photomicrograph of UF I taken in secondary electron imaging (SEI) mode. Scale bar = 10 μm.

surface of the other ply. Due to the shape of annual rings both earlywood and latewood zones appeared at the bondline. Though a distinct difference in the penetration behavior is known for earlywood and latewood, separate evaluations of the three above-mentioned cases were not performed for the evaluations described here, but will be taken into consideration in the future work.

The photographs in Figs 1 and 2 are good examples to see that penetration of the adhesive mix based on resin UF I (i.e., the resin with the lowest degree of condensation) is significantly deeper into wood tissue, based on the better flow behavior of this resin. As can be seen from Table 3 the penetration is reduced for adhesive mixes based on resins with higher degrees of condensation. This can be easily explained by the higher molar masses present in these resins due to the longer condensation phase during the preparation of the resins.

Since resins with lower DOC contain higher proportions of low molar mass molecules, these resins (and also the adhesive mixes based on these resins) penetrate quicker and deeper into the wood lumens, because the ratio between geometrical sizes of the molecules and the geometry of the flow path is more favorable. The depth of penetration is also influenced and restricted by the increasing size of the resin molecules during hardening. There is a longer time available for smaller molecules to penetrate even when the hardening reaction has already started, compared to already larger molecules at the start of the hardening reaction. Therefore, it was important to guarantee the gelling behavior of the various adhesive mixtures by adding different amounts of hardener. The flow behavior is determined by (i) the ratio of the average anatomical tracheid (lumen) diameter compared to the average size (diameter) of resin molecules and (ii) the proportion of cut cell walls (and, therefore, open lumen cross sections at the wood surface). But also occlusions in the pits or lumens may inhibit flow. Since the viscosity of the resin (and of the adhesive mix) depends on the size of the resin molecules, the measured depth of penetration is correlated with the viscosity of the resin, decreasing nearly linearly with increasing viscosity, i.e., with higher degree of condensation and larger molecular sizes (Table 3 and Fig. 3).

The main flow paths are through the lumens of the tracheids and also through rays. Horizontal flow might happen through the bordered pits (with a diameter of the torus of 5 μm), which are on the radial side of the tracheids; such a flow path, however, depends on the size of the molecules of the liquid phase and would rather be unlikely for adhesive resins. Vertical flow might occur (i) from one tracheid to another one through the bordered pits on the tapered endings of the tracheids and (ii) through the rays and then through the pits on the cross sections between tracheids and ray cells, indicating an open structure of the pit membranes as shown in Fig. 6; again due to the size of the adhesive molecules this still might rather be unlikely but cannot be completely excluded. Radial penetration between longitudinal tracheids is limited, because the pits are concentrated on the radial surfaces. Within the frame of the work reported here, however, it was not possible to clearly confirm if penetration through pits had occurred or not.

With CLSM (Fig. 5) it was possible to visualize the adhesive penetration into the wood tissue more clearly than with epi-fluorescence microscopy, as the unique capabilities of CLSM in optical sectioning through an object and then obtaining a composite image of sequential sections through a considerable depth enabled large tissue area to be brought in the same focal plane as the bondline; this was not possible with epi-fluorescence microscopy [55]. Based on Fig. 5 it can also be assumed that some flow of the adhesive takes place from rays into tracheids through the pits at the overlap between tracheids and rays.

In the preparation for bonding of the wood plies, attention was paid that the axes of most fibers were parallel to the plane of the bondline (thus, avoiding an angle between geometrical bondline and fiber length axis). Nevertheless, it must be assumed that based on the natural irregularities in the wood structure and a length of the joints of 100 mm, a certain portion of these open cross sections will be present, in which resin can penetrate, causing also a certain longitudinal flow in the lumens. Due to the possible angle between fiber length axis and the bondline plane, resin can appear in a cross section of the joint sample at a certain distance away from the geometrical bondline.

To penetrate into lumens of several cell rows away from the bondline, the adhesive would have to pass several times through the pits, then again flowing partly in length direction towards the pits connected with the next cell row. The second flow path is *via* rays and then towards the tracheids, as already mentioned above.

Already Suchsland [49] had mentioned that an increase in the grain angle caused flow of the adhesive mix into the open fiber lumens, especially the earlywood cells.

The photograph in Fig. 6 shows a lumen with resin ('tracheid filled with adhesive') on the right upper part of the photograph and two empty tracheids in the bottom of the photograph ('tracheids without adhesive'). In the middle part of the photograph droplets of different sizes can be discerned in a ray; the droplets look bigger than the size (diameter) of the pits in the photograph. It is not clear if these droplets consist (i) of hardened adhesive, indicating that some penetration might have occurred through the pits, or (ii) of wood inherent chemicals (like natural resins) deposited in the lumens of the rays. In the case of hardened resin, it has to be taken into consideration that at the moment when it passed through the small diameter pits, the adhesive was still liquid. It obviously conglomerated and hardened once reaching the much wider and spatial cavity of the ray, thus, creating adhesive spots larger than the diameter of the pits.

FIR is nearly independent of the degree of condensation of the resin. FIR is the ratio A/IR between the size A of the filled interphase region and the size of the whole interphase region IR. As mentioned above, IR depends on the maximum penetration depth. So for each photograph an individual size of the interphase region was determined. At higher viscosities the resins are concentrated within a smaller (= thinner) interphase, or a bigger portion of the adhesive will stay close to or in the bondline itself. Since now the ratio between these two quantities is calculated and FIR represents a relative number, it can be very similar (constant or even slightly increasing) for all adhesives tested showing no significant correlation with the viscosity of the adhesive mix (degree of condensation of the resin). FIR does not take into consideration how deep the resin penetrates into the wood tissue; it just indicates that a certain portion of the lumens in the interphase is filled.

The question which portion of resin flows through the rays is difficult to answer. It is linked to the question whether AP is more pronounced in the regions where rays exist. Examining Fig. 5 it looks like that flow in the rays is more pronounced compared to flow in tracheids. This appearance, however, might be caused by the different directions of flows in the two cell types: for the tracheids the flow direction is dominantly vertical to the plane of the photograph, whereas flow in rays occurs directly in the plane of the photograph. Due to the limited number of CLSM photographs, showing that the adhesives also pass through the rays in fir, this issue could not be clarified completely.

5. Conclusions

The depth of 'Average radial penetration' (AP) of UF adhesive mixes into fir decreases with increased viscosity of the UF resin used in the mixes, which reflects the different degrees of condensation. The penetration mainly depends on the ratio of the average anatomical tracheid diameter and average size of the adhesive molecules in still liquid state. Also rays were found to be an important flow path into wood tissue.

The effect of adhesive penetration also was expressed by the portion of partly or completely filled cells (lumens or rays) in the interphase region. The filled interphase region (FIR) increases only slightly with an increased adhesive viscosity (higher degree of condensation) but not showing significant differences.

Epi-fluorescence microscopy was shown to be a suitable technique for the determination of adhesive penetration into wood substance.

Acknowledgements

The research work presented in this paper was financed by the Ministry of Science and Technological Development, Project 'Wood biomass as a resource of sustainable development of Serbia' 20070-TP.

We are grateful to Ing. Petr Pastrnak (DUKOL, Ostrava s.r.o.) for the preparation of the UF resins used in the investigation reported in this paper.

The contribution of Dr. Manfred Dunky (Kronospan Lampertswalde, Germany, and the University of Natural Resources and Applied Life Sciences, Vienna, Austria) during discussion of the results and preparation of the paper is gratefully acknowledged.

References

1. A. Marra, *Technology of Wood Bonding Principles in Practice*. Van Nostrand Reinhold, New York, NY (1992).
2. F. Kamke and J. Lee, *Wood Fiber Sci.* **39**, 205–220 (2007).
3. J. Resnik, M. Sernek and F. A. Kamke, *Wood Fiber Sci.* **29**, 264–271 (1997).
4. R. H. Young, E. E. Barnes, R. W. Caster and N. P. Kutscha, *ACS, Div. Polym. Chem., Polymer Prepr.* **24** (2), 199–200 (1983).
5. W. Wang and N. Yan, *Wood Fiber Sci.* **37**, 505–513 (2005).
6. M. Dunky, *Int.J. Adhesion Adhesives* **18**, 95–107 (1998).
7. M. Dunky, in: *Handbook of Adhesive Technology*, 2nd ed., A. Pizzi and K. L. Mittal (Eds), pp. 887–956. Marcel Dekker, New York, NY (2003).
8. M. Dunky and A. Pizzi, in: *Adhesion Science and Engineering, Volume 2: Surfaces, Chemistry and Applications*, D. A. Dillard and A. V. Pocius (Eds), pp. 1039–1103. Elsevier (2003).
9. T. Furuno and T. Goto, *Mokuzai Gakkaishi* **21** (5), 265–334 (1975).
10. W. A. Coté and R. G. Robison, *J. Paint Technol.* **40**, 427–432 (1968).
11. R. M. Nussbaum, E. J. Sutcliffe and A. Hellgren, *J. Coatings Technology* **70** (878), 49–57 (1998).
12. M. Dunky and K. Lederer, *Angew. Makromol. Chem.* **102**, 199–213 (1982).

13. E. E. Ferg, A. Pizzi and D. C. Levendis, *J. Appl. Polym. Sci.* **50**, 907–915 (1993).
14. A. Pizzi (Ed.), *Wood Adhesives, Chemistry and Technology.* Marcel Dekker, New York, NY (1983).
15. J. Billiani, K. Lederer and M. Dunky, *Angew. Makromol. Chem.* **180**, 199–208 (1990).
16. M. Dunky, K. Lederer and E. Zimmer, *Holzforsch. Holzverwert.* **33**, 61–71 (1981).
17. S. Katuscak, M. Thomas and O. Schiessl, *J. Appl. Polym. Sci.* **26**, 381–394 (1981).
18. J. T. Rice, *Forest Prod. J.* **15**, 107–112 (1965).
19. Y. Nakarai and T. Watanabe, *Wood Industry* **17**, 464–468 (1962).
20. S. Ellis, *Forest Prod. J.* **43** (2), 66–68 (1993).
21. S. Ellis and P. R. Steiner, *Forest Prod. J.* **42** (1), 8–14 (1992).
22. W. L.-S. Nieh and T. Sellers Jr., *Forest Prod. J.* **41** (6), 49–53 (1991).
23. B. D. Park, B. Riedl, E. W. Hsu and J. Shields, *Holz Roh. Werkst.* **56**, 155–161 (1998).
24. M. Scheikl and M. Dunky, *Holzforschung* **52**, 89–94 (1998).
25. M. Scheikl and M. Dunky, *Holz Roh. Werkst.* **54**, 113–117 (1996).
26. E. Brady and F. Kamke, *Forest Prod. J.* **38** (11/12), 63–68 (1988).
27. M. Sernek and J. Resnik, in: *Proc. 5th Int. Conf. on Developments in Wood Sci., Wood Technology and Forestry (ICWSF)*, Ljubljana, Slovenia, pp. 145–153 (2001).
28. R. S. Stephens and N. P. Kutscha, *Wood Fiber Sci.* **19**, 353–361 (1987).
29. M. Hameed and E. Roffael, *Holz Roh. Werkst.* **58**, 432–436 (2001).
30. W. Gindl, A. Srenetovic, A. Vincenti and U. Mueller, *Holzforschung* **59**, 307–310 (2005).
31. S. E. Johnson and F. A. Kamke, *J. Adhesion* **40**, 47–61 (1992).
32. D. A. Hare and N. P. Kutscha, *Wood Sci.* **6**, 294–304 (1974).
33. N. P. Kutscha and R. W. Caster, *Forest Prod. J.* **37** (4), 43–48 (1987).
34. M. Sernek, J. Resnik and F. A. Kamke, *Wood Fiber Sci.* **31**, 41–48 (1999).
35. T. Furuno and H. Saiki, *Mokuzai Gakkaishi* **34**, 409–416 (1988).
36. L. Donaldson and J. Bond, *Fluorescence Microscopy of Wood.* New Zealand Forest Research Institute, Rotorua, New Zealand (2005).
37. S. E. Ruzin, *Plant Microtechnique and Microscopy.* Oxford University Press, Oxford (1999).
38. J. Zheng, S. C. Fox and C. E. Frazier, *Forest Prod. J.* **54** (10), 74–81 (2004).
39. V. Rijckaert, M. Stevens, J. Van Acker, M. de Meijer and H. Militz, *Holz Roh. Werkst.* **59**, 278–287 (2001).
40. T. M. Gruver and N. R. Brown, *BioResources* **1** (2), 233–247 (2006).
41. S. Lee, T. F. Shupe, L. H. Groom and C. Y. Hse, *Wood Fiber Sci.* **39**, 482–492 (2007).
42. H. Saiki, *Mokuzai Gakkaishi* **30**, 88–92 (1984).
43. A. J. Bolton, J. M. Dinwoodie and D. A. Davies, *Wood Sci. Technol.* **22**, 345–356 (1988).
44. J. Konnerth and W. Gindl, *Holzforschung* **60**, 429–433 (2006).
45. N. Gierlinger, C. Hansmann, T. Roder, H. Sixta, W. Gindl and R. Wimmer, *Holzforschung* **59**, 210–213 (2005).
46. C. B. Vick and T. A. Kuster, *Wood Fiber Sci.* **24**, 36–46 (1992).
47. B. M. Collett, *Wood Fiber* **2** (2), 113–133 (1970).
48. L. P. Futo, *Holz Roh. Werkst.* **31**, 52–61 (1973).
49. O. Suchsland, *Holz Roh. Werkst.* **16** (3), 101–108 (1958).
50. A. P. Singh, E. A. Dunningham and D. V. Plackett, *Holzforschung* **49**, 255–258 (1995).
51. A. P. Singh, C. R. Anderson, J. M. Warnes and J. Matsumura, *Holz Roh. Werkst.* **60**, 333–341 (2002).
52. A. P. Singh and B. S. Dawson, *IAWA J.* **24** (1), 1–11 (2003).
53. M. S. White, G. Ifju and J. A. Johnson, *Forest Prod. J.* **27** (7), 52–54 (1977).

54. C. J. Buckley, C. Phanopoulos, N. Khaleque, A. Engelen, M. E. J. Holwill and A. G. Michette, *Holzforschung* **56**, 215–222 (2002).

55. A. Singh, B. Dawson, C. Rickard, J. Bond and A. Singh, *Microscopy and Analysis* **22** (3), 5–8 (2008).

56. F. A. Kamke and G. Modzel, in: *Proceedings Int. Conf. Wood Adhesives 2009*, Lake Tahoe, in preparation.

57. P. Hass, M. Stampanoni, A. Kaestner, D. Mannes and P. Niemz, in: *Proceedings Int. Conf. Wood Adhesives 2009*, Lake Tahoe, in preparation.

58. W. Ginzel and G. Stegmann, *Holz Roh. Werkst.* **28**, 289–292 (1970).

59. J. E. Marian and K. Suchsland, *Forest Prod. J.* **7** (2), 74–77 (1957).

60. J. Konnerth, D. Harper, S.-H. Lee, T. G. Rials and W. Gindl, *Holzforschung* **62**, 91–98 (2008).

61. W. Gindl, F. Zargar-Yaghubi and R. Wimmer, *Bioresource Technol.* **87**, 325–330 (2003).

62. W. Gindl, E. Dessipri and R. Wimmer, *Holzforschung* **56**, 103–107 (2002).

63. A. Pizzi, *J. Appl. Polym. Sci.* **71**, 1703–1709 (1999).

64. M. G. Kim, *J. Polym. Sci., Part A: Polym.Chem.* **37**, 995–1007 (1999).

Radial Penetration of Urea-Formaldehyde Adhesive Resins into Beech (*Fagus Moesiaca*)

Ivana Gavrilović-Grmuša [a,*], **Manfred Dunky** [b], **Jovan Miljković** [a] **and Milanka Djiporović-Momčilović** [a]

[a] Faculty of Forestry, University of Belgrade, RS-11030 Belgrade, Serbia
[b] Kronospan GmbH Lampertswalde, D-01561 Lampertswalde, Germany

Abstract
Penetration of adhesives plays an important role in wood adhesion, since wood is a porous material. Objectives of this work were (i) the evaluation of the influence of the degree of condensation of urea-formaldehyde (UF) resins on their penetration into beech, showing the distribution of the adhesive in the cell layers close to the bondline using microscopic detection, and (ii) the comparison of data for beech with results for fir obtained earlier. The degree of penetration mostly depends on the permeability and porosity of the wood surface as well as on the resin type and the size of the resin molecules. The process parameters, which also can affect penetration, were kept constant throughout all tests performed.

The results show a significant correlation between the penetration behavior and the degree of condensation (molecular size, viscosity) of the resins and adhesive mixes based on these resins. The higher the degree of condensation, the lower the possibility for penetration, expressed as "Average penetration depth" (AP). AP into beech is higher than into fir. The portion of filled tracheids and vessels on the whole cross section of the interphase ("Filled interphase region" FIR) increases slightly with the degree of condensation, but this increase is not statistically significant. FIR in fir is higher than in beech.

Keywords
Penetration, beech (*Fagus Moesiaca*), urea-formaldehyde (UF) resin, degree of condensation, Silver Fir (*Abies alba*, Mill.), epi-fluorescence microscopy, fluorescence confocal laser scanning microscopy (CLSM), scanning electron microscopy (SEM)

1. Introduction

The use of condensation resins [1, 2] for bonding wood components like veneers, particles, or fibers plays a dominant role in the wood products industry. Though mechanical interlocking is not the main reason for bonding, still an adequate penetration of the resin into the wood surface enables formation of an enlarged bonding interface; this penetration must take place before curing of the resin has occurred.

---

* To whom correspondence should be addressed. E-mail: ivana.grmusa@sfb.rs

The degree of penetration depends on several parameters related to wood, resin and adhesive type, as well as to processing parameters [3]. Additionally, a good contact of the adhesive with the lumen wall is needed in order to achieve good bonds [4].

Adhesive penetration into wood can be categorized (i) on micrometer level as a result of the hydrodynamic flow and capillary action of the liquid resin from the outer surface into the porous and capillary structure of wood, mostly filling cell lumens, as well as fractures and surface debris caused by processing [5], and (ii) on sub-micrometer level as diffusion penetration into cell walls and micro- fissures. Hydrodynamic flow is initiated by the external compression force as a result of pressure applied to the wood surface to be bonded. The flow then continues into the interconnected network of lumens and pits, with flow moving primarily in the direction of lowest resistance [6]. The extent of utilization of an adhesive may be limited due to excessive penetration into the substrate, since this portion of the applied adhesive is lost within porous substrate structures for the adhesion effect.

The penetration of an adhesive can be investigated using various methods, especially microscopic methods; an overview was given recently [3].

Already Hare and Kutscha [4] had indicated that for a strong bond the adhesive must penetrate deep enough into the wood substrate in order to be able to reinforce weakened cells and to achieve a large contact surface. Penetration easily occurs into fiber cells which are physically ruptured, e.g., during the veneering process. Also Gindl [7] showed using SEM that adhesives penetrate predominantly through cut open tracheids and rays. Bordered pits block the flow of resin from one cell to another cell, whereas simple pits are only a minor obstacle.

Besides the depth of penetration, the evenness of penetration is also important for strong boards. Variations in the depth of penetration will cause weaker zones susceptible to failure [4].

Despite the widespread use of UF resins in the wood working and wood based panels industry, only limited information is available concerning their penetration into wood tissue, especially taking into consideration different degrees of condensation (different molar mass distributions) and different F/U molar ratios.

In an earlier paper [3] for the first time the influence of the degree of condensation of a UF resin on its penetration into a wood surface was investigated using fir as wood material.

Urea-formaldehyde (UF) adhesive resins as the most important type of amino resins were among the first commercially used adhesive systems and still dominate the production of wood based panels [8]. UF resins are based on the reaction of the two monomers urea and formaldehyde; they consist of linear or branched oligomeric and polymeric molecules of various molar masses [9, 10]. The degree of condensation is one of the most important characteristics of condensation resins and determines several of their properties; the viscosity at a certain solid content increases [11], and flowability and hence the equal distribution of the resin on the furnish (particles, fibers) and the penetration into the wood surface are hindered [12, 13].

For phenolic resins various studies concerning the influence of the degree of condensation on the penetration behavior and on bond strength have been performed [3].

Furuno *et al.* [14] showed that preferentially PF resins with low molar mass penetrate into the cell wall of Japanese cedar wood, thereby contributing to the enhancement of dimensional stability and decay resistance. For higher condensated PF resins, only the resin components of low molar masses appear to be present in the cell wall.

Johnson and Kamke [15] used three PF resins with different degrees of condensation in order to investigate the penetration behavior into the cell lumens and also the big vessels of hardwood (yellow poplar). The higher the molar mass, the less was the penetration into the wood flakes. Adhesive penetration into hardwood is likely to be dominated by flow into vessel elements.

In another paper the same authors [16] investigated the effect of the molar mass distribution of a liquid PF resole on the adhesive flow during steam injection pressing and hence the mechanical properties of flakeboards. Deeper penetration of the adhesive likely occurred during exposure to the steam injection environment as the viscosity declined due to an increase of temperature and dilution of the adhesive.

Zheng *et al.* [17] investigated the penetration behavior of PF/polymethylene-diisocyanate (pMDI) hybrid resins into yellow poplar and Southern pine by preparing microtome slides of bondlines, which were then examined using epifluorescence microscopy coupled with digital image analysis. They used the so-called "mean effective penetration" EP (µm) as a measure for the penetration, defined as total adhesive area divided by the bond length in the field of observation.

Wilson *et al.* [18] reported that increased proportions of higher molar mass fractions in PF resins increased the internal bond strength, whereas low and medium molar mass fractions rather tended to overpenetrate into the wood substrate causing starved bondlines.

White [19] varied the time span after addition of the paraformaldehyde hardener to a resorcinol resin in order to achieve different molar masses during the pot-life of the mix. When adhesive mixes which had already become viscous were applied the interphase was very narrow and the amount of adhesive within the interphase was small. Long reaction times and high viscosities clearly reduced penetration and, consequently, bond fracture toughness.

The objective of the work reported here was to expand the investigations of several of the authors [3] concerning the influence of the degree of condensation of UF adhesive resins on the penetration behavior into beech. Since this hardwood species exhibits a completely different wood structure compared to softwood (like fir), it can be expected that penetration of adhesives will occur to a different extent. The structure of beech is characterized mainly by big vessels instead of the rather narrow tracheid cells as it is the case with fir. The resins and adhesive mixtures used in the study reported here were the same as before [3]. The investigation of the penetration behavior was again performed by bonding tests under defined conditions followed by microscopic determination of the penetrated resin. An additional attempt was a comparison of the different penetration behaviors of the two wood species, beech and fir.

2. Materials

2.1. Urea-Formaldehyde (UF) Resins

Three laboratory UF resins with different degrees of condensation according to recipes described in the literature [20, 21] were provided by DUKOL Ostrava, s.r.o. (Ostrava, Czech Republic) as already reported elsewhere [3]. The degree of condensation (DOC) increases with the duration of the acidic condensation step from resin UF I (lowest DOC) to resin UF III (highest DOC). The viscosity of the three resins increased due to the increasing size of molecules present in the resin from 218 mPa s for UF I to 281 mPa s for UF II, and eventually to 555 mPa s for UF III.

A similar increase in viscosity as for the resins themselves is seen for the adhesive mixes (UF I: 545 mPa s; UF II: 745 mPa s; UF III: 1644 mPa s), measured just after mixing. The viscosities of the different adhesive mixes were taken here as a measure of the DOC in order to show its influence on the "Average penetration depth" (AP) in radial direction and on the "Filled interphase region" (FIR).

More details concerning the resins and adhesive mixes used in the investigations reported here have been described elsewhere [3].

2.2. Bonded Samples and Determination of Penetration

The experimental procedure was the same as already reported earlier [3]. Special effort again was taken in order to have the taper as low as possible, this would guarantee equal penetration conditions for all samples.

Individual depths of penetration (μm) were determined from each photomicrograph of the microtome slide at 45 positions within the 1400 μm width of the bondline. The depth of penetration here is defined as the sum of the distances the resin could penetrate into the two plies starting from the bondline. No separate evaluations for the two plies was done during the evaluation of AP and FIR, even though there might be some small difference in the individual penetrations between the ply where the adhesive mix had been applied and the ply without application of adhesive mix. Each of the three AP results (one for each resin type) as shown below is the result of nearly 7000 individual measurements.

In addition to AP (i) the "Total interphase region" (IR), including also the unfilled lumen area, using the maximum value out of the 45 values for the depth of penetration in each photomicrograph, and (ii) the sum of all filled lumens or filled rays A were calculated. The "Filled interphase region" (FIR) was then expressed as percentage A/IR (%).

Several microtome slides were also examined with fluorescence confocal laser scanning microscopy (CLSM). Scanning electron microscopy (JEOL JSM-

6460LV SEM) investigation of the 20 μm thick microtome test specimens was performed in order to investigate if any penetration through pits had occurred. The experimental details for both methods have been described elsewhere [3].

3. Results

The epi-fluorescence photomicrographs in Figs 1–3 show the bondlines formed by the three adhesive mixes based on the three UF resins I to III. The adhesive mixes mainly fill fully or partly the lumens in the earlywood. The depth of penetration apparently decreases in the order UF I, UF II, and UF III. Photomicrographs as shown in Figs 1–3 were used for the evaluation of AP and FIR.

Table 1 summarizes the area of the interphase region (IR) and the sum of filled lumens and rays (A), both evaluated from the photomicrographs shown in Figs 1–3 and expressed as (mm^2).

Table 2 summarizes the overall results for the "Average penetration depth" (AP) and the "Filled interphase region" (FIR) for penetration into beech for the three adhesive mixes based on the resins with different degrees of condensation.

The AP for radial direction in the beech decreases with higher viscosity, as this is a measure for the higher molar masses and the higher degree of condensation (Fig. 4). According to the Bonferroni test for multiple comparisons, statistically significant differences were found for the means of AP for all three investigated degrees of condensation.

Figure 5 shows the "Filled interphase region" (FIR) for beech as a function of molar masses (different degrees of condensation) of the UF adhesive mixes inves-

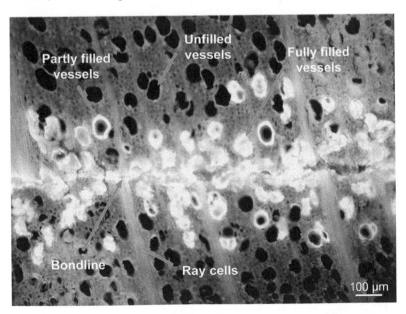

Figure 1. Photomicrograph of UF I bondline (with addition of Safranin), using epi-fluorescence.

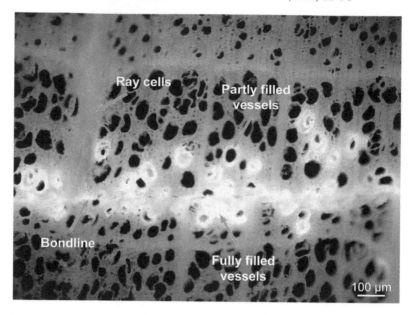

Figure 2. Photomicrograph of UF II bondline (with addition of Safranin), using epi-fluorescence.

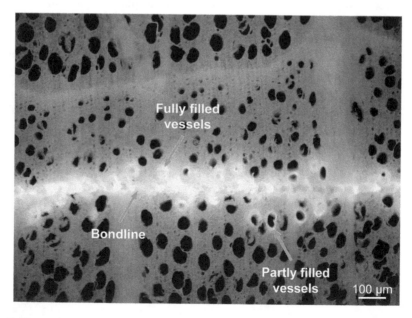

Figure 3. Photomicrograph of UF III bondline (with addition of Safranin), using epi-fluorescence.

tigated here. FIR does not depend on the degree of condensation; no statistically significant difference between the three values for the three investigated adhesive mixes was found based on the results of the Kruskal–Wallis test.

Table 1.
Average area of the interphase region (IR) and the area of filled lumens and rays (A) in beech as evaluated for the adhesive mixes based on the three resins UF I, UF II, and UF III

Resin	Average area of the interphase region (IR)		Average area of filled lumens and rays (A)	
	IR (mm^2)	SD (mm^2)[*]	A (mm^2)	SD (mm^2)[*]
UF I	0.96	0.22	0.22	0.09
UF II	0.79	0.22	0.20	0.11
UF III	0.62	0.18	0.16	0.05

[*] SD: standard deviation.

Table 2.
"Average penetration depth" (AP) and "Filled interphase region" (FIR) in beech as evaluated for the adhesive mixes based on the three UF resins UF I, UF II, and UF III

Resin	Average penetration depth (AP)		Filled interphase region (FIR)	
	AP (μm)	SD (μm)[*]	FIR (%)	SD (%)
UF I	274	98	24	8
UF II	193	81	25	5
UF III	150	53	27	6

[*] SD: standard deviation.

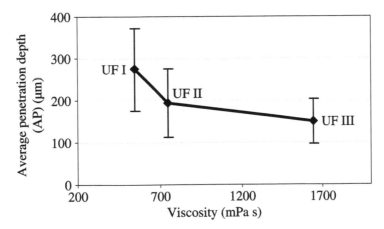

Figure 4. Average penetration depth (AP) into beech as a function of viscosity (different degrees of condensation) of the three UF adhesive mixes used.

The CLSM photomicrograph in Fig. 6 sharply differentiates the adhesives from wood cell walls using the bright contrast of the colors of the adhesive (appearing reddish) and of the wood cell walls (appearing greenish) when adding Safranin as staining agent. It is visible that not only vessels are partially or completely filled

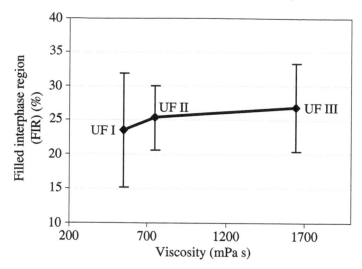

Figure 5. Filled interphase region (FIR) for beech as a function of the viscosity (different degrees of condensation) of the UF adhesives used.

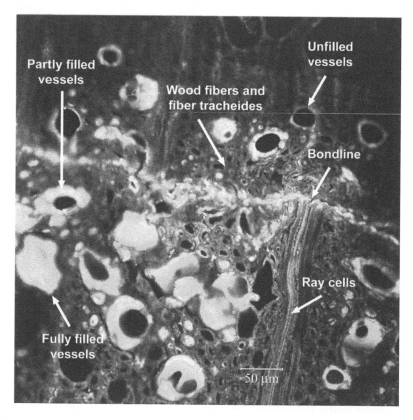

Figure 6. Confocal fluorescence (CLSM) photomicrograph of UF I (with addition of Safranin) bondline (reddish), showing deep penetration of the adhesive into vessels of the beech tissue.

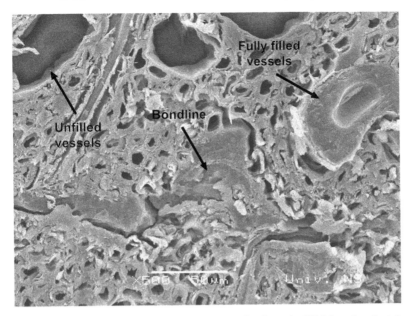

Figure 7. SEM photomicrograph of penetration of the adhesive mix UF I into beech, taken in secondary electron imaging (SEI) mode. Scale bar = 50 μm.

with the adhesive, but also tracheid fibers and wood fibers (characterized by their thick cell walls, small lumens and narrow pits). It is interesting that the adhesive has also penetrated partially or completely into the intercellular areas which are constituted of vasicentric and vascular tracheids, which are similar to small vessels. On the contrary, it appears that the adhesive has not penetrated into ray cells.

Adhesive penetration into wood tissues also can be visualized using SEM in combination with secondary electron imaging (SEI) mode, as shown in Fig. 7 for the penetration of the adhesive mix UF I into beech. Both vessels with and without resin can be discerned.

Figure 8 shows also a further photomicrograph of the penetration of the adhesive mix UF II (with addition of Safranin) into beech wood tissue, using epi-fluorescence microscopy; it especially describes the different penetration into earlywood and latewood. In the upper wood ply on the left side of the photomicrograph mainly earlywood at the beginning of the year ring with big vessels is adjacent to the bondline, causing strong penetration; in the lower wood ply rather small vessels of the latewood zone are close to the bondline, hindering nearly all penetration. The same situation of low penetration also occurs on the right upper half with latewood close to the bondline. Penetration is much deeper into earlywood than into latewood; this effect was to be expected, since earlywood generally has a much lower density than latewood, with larger lumens (larger diameter of the vessels) and thinner cell wall layers. On the contrary, latewood has thicker cell walls and narrower lumens.

Figure 9 again presents difference in the penetration behavior between earlywood (upper ply) and latewood (lower ply). In the left lower corner two vessels (one fully

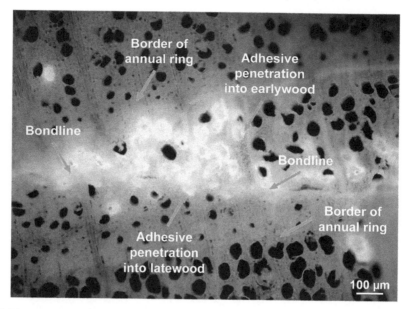

Figure 8. Photomicrograph of the penetration of the adhesive mix based on resin UF II (with addition of Safranin) into beech tissue, using epi-fluorescence microscopy.

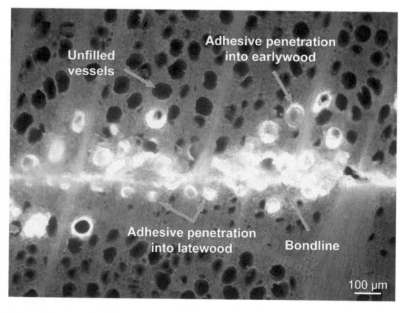

Figure 9. Photomicrograph of the UF II bondline (with addition of Safranin), using epi-fluorescence microscopy and showing difference in the penetration behavior between earlywood (upper ply) and latewood (lower ply) of beech.

and one partly filled with adhesive) are visible with the adhesive having bypassed the latewood and filled the vessels in the earlywood of the lower ply.

Table 3.
Comparison between the average penetration depths (i) for the whole sample (with no distinction between earlywood and latewood), (ii) for earlywood, and (iii) for latewood, as evaluated for the adhesive mixes based on the three resins UF I, UF II, and UF III

Adhesive mix	Wood species	AP (μm) for whole sample[1]	AP (μm) in earlywood	AP (μm) in latewood
UF I	Beech	274	297	96
	Fir	207	237	89
UF II	Beech	193	228	76
	Fir	177	189	64
UF III	Beech	150	166	57
	Fir	120	137	49

[1] No distinction between earlywood and latewood; data for beech taken from Table 2; data for fir were reported elsewhere [3].

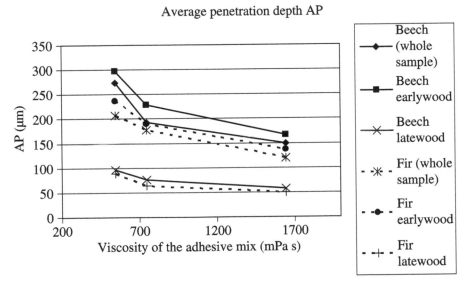

Figure 10. Comparison of the values of the "Average penetration depth" (AP) of beech and fir: (i) comparison between beech and fir for whole samples without distinction between earlywood and latewood; data for fir were already reported elsewhere [3]; (ii) separate evaluations of penetration into earlywood and latewood of beech and fir; (iii) comparison between penetration into earlywood and latewood, respectively, between beech and fir.

Table 3 and Fig. 10 compare the "Average penetration depth" AP for the whole sample without distinction between earlywood and latewood for both wood species for the three adhesive mixes with the penetration evaluated separately for the earlywood and the latewood sections of the samples.

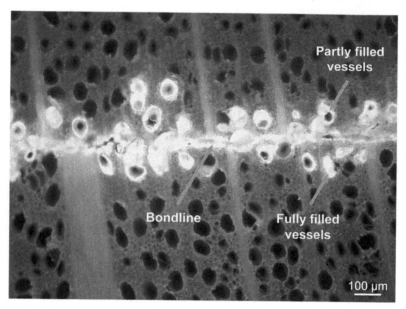

Figure 11. Photomicrograph of a UF III bondline (with addition of Safranin), using epi-fluorescence microscopy and showing only slight penetration into ray cells of beech.

Figure 11 exhibits a part of the beech bondline with the adhesive mix based on resin UF III. The broad part without vessels in the left lower quarter of the figure is radial parenchyma in the ray cells, thus no open lumen of the ray is visible in the photomicrograph. This explains why no adhesive could be determined in the photomicrograph for this ray. It mainly depends on the cutting plane compared to the cell structure in the sample whether penetration into rays is visible in the photomicrograph or not. SEM investigations partly confirm that no or only small penetration occurs into ray cells.

4. Discussion

This paper describes the penetration behavior of UF condensation resins with different molar masses into beech tissue and compares these results with former data for fir [3]. The higher the degree of condensation (DOC), the larger the molecules are present in the resin and hence in the adhesive mix. An increased DOC, hence, creates higher viscosities of the resins and of the adhesive mixes based on these resins, and this reduces the flowability of the adhesive into the wood tissue. Comparing the light colored sections in Figs 1–3 on both sides of the geometrical bondline it is clearly visible that due to the smaller size of the molecules before starting the hardening reaction the UF I adhesive mix penetrates to a greater extent and deeper into the wood material compared to the UF II and especially the UF III based adhesive mixes. The lower molar mass molecules of the resin are able to penetrate deeper into the wood lumens of beech. The average depth of radial penetration (AP) decreases accordingly with increasing degree of condensation in the series UF I to UF III. A similar correlation was found earlier also for penetration into fir [3].

However, from the graph in Fig. 4 it is also evident that even using the same adhesive mix, variations in the penetration behavior occur. This is indicated by the bars showing the standard deviations of AP; nevertheless, a statistically significant difference in AP is shown between the three adhesive types used in the investigations.

FIR for penetration of the investigated adhesive mixes into beech rather behaves independently of the viscosity. No statistically significant difference could be found between the three resins; this finding is similar to that reported for fir [3]. FIR is the ratio A/IR between the area A of the filled interphase region and the area of the whole interphase region IR. IR depends on the maximum penetration depth, which decreases with higher degree of condensation, similar to AP. Also A decreases because the resin is concentrated into a thinner interphase, and a larger portion of the adhesive will stay close to or in the bondline itself. Since FIR represents a relative number, no clear change with the viscosity of the resin, therefore, can be expected. Similar explanations were given also for the penetration of the resins into fir [3].

The comparison of the data for AP for beech as presented here and of the results for fir reported earlier [3], with both using the same adhesive mixes, shows that penetration into beech is higher for all three resins by about 10–30% for the whole sample and for earlywood and 10–20% in the case of latewood (Table 3 and Fig. 10). This different penetration behavior underlines that the anatomical structures of various wood species significantly influence the penetration depth of adhesives. Even the measured values for the average penetration depth AP in the whole sample (without distinction between earlywood and latewood) as well as into earlywood and into latewood are in all cases higher than into fir. The mean values of AP for fir and beech for the whole sample at the various degrees of condensation, however, are not significantly different when applying statistical tests (ANOVA and Bonferroni test). Beech has a more complex anatomical structure with larger variation compared to fir. Bulk flow was observed through the lumens of tracheids in fir wood and through vessels in beech wood. Tracheids of fir (approx. 93% of the wood mass), with small intercellular areas, were found to be principal flow paths. This enables a more uniform distribution of the adhesive across the interphase region and, hence, the concentration of the resin is still closer to the bondline. On contrary, the complex structure of beech consists of diffusely arranged vessels (approx. 55%), with larger intercellular areas compared to fir tracheids. These intercellular areas are built from fiber tracheids and wood fibers, which are much smaller than softwood tracheids and do not support penetration of adhesives due to their thick cell walls, small lumens and narrow pits. Therefore, the adhesive penetrates preferably into the vessels and can flow for a longer distance away from the bondline, increasing the depth of penetration. This is directly influenced by the percentage of vessels present close to the bondline.

The "Filled interphase region" FIR into fir wood is higher than into beech for all three adhesive samples used here. This might be explained by the smaller overall interphase region IR in fir, because the adhesive is distributed more evenly and, hence, stays closer to the bondline compared to beech.

It is commonly known that a clear difference in the penetration behaviors between earlywood and latewood exists with adhesive penetration into latewood being much lower than into earlywood. Figure 8 highlights the different penetration behaviors into earlywood and latewood; in the left upper part of the photomicrograph an earlywood structure is predominant; it can clearly be seen that the resin penetrates for a larger distance measured from the bondline. Several big vessels are completely filled with the adhesive, and some are at least partly filled. But even here it can be observed that penetration shows large variation; some vessels closer to the bondline do not contain adhesive, whereas some vessels at larger distance from the bondline do. Penetration into latewood as seen in the lower ply close to the bondline occurs only very close to the bondline, as it can be seen in Fig. 8. The separate evaluations of the average penetration depth AP for earlywood and for latewood as summarized in Table 3 show that AP as a measure for the penetration is mainly influenced by the earlywood. AP is an average value for all 150 photomicrographs for one combination (one adhesive mix, one wood species) and for 45 positions within the 1400 μm width of the bondline in the photomicrographs. When evaluating AP for the whole sample, no distinction was made between earlywood and latewood. For selected photomicrographs, however, it was attempted to evaluate especially positions with either earlywood or latewood on both sides of the bondline separately; these values are also included in AP of the whole sample. Table 3 and Fig. 10 show that the penetration into earlywood is, therefore, higher than for the whole sample without distinction between earlywood and latewood. The penetration into latewood is much lower, based on the different anatomical structure in latewood for both wood species. AP into latewood is approximately one third of AP for earlywood.

High molar mass molecules as well as occlusions in the pits or lumens may inhibit flow. Since it can be assumed that the penetration depends on the ratio between the average diameter of the vessels in beech (earlywood: 30 μm, latewood: 67 μm) on the one hand and the average size of resin molecules on the other hand, the measured average depth of penetration (AP) can be correlated with the molar mass and with the viscosity of the resin. The same dependence can be expected also for fir, with an average size of the cross section of tracheids of 17 μm × 30 μm in earlywood and 10 μm × 17 μm in latewood, simplifying the cross section of such tracheids as the shape of a rectangle.

The differences in the penetration behaviors have been addressed several times in the literature. Brady and Kamke [22] reported that fractures were abundant along the surface of earlywood flakes, opening additional flow paths. Rays were found to be a more important flow path into fir tissue. Most of the radial penetration of adhesive is probably through the ray cells and between the longitudinal tracheids and rays. Penetration into beech rays is much less than into fir. Therefore, the longitudinal flow through the vessels is important, due to the branching of vessels and the role of inter-vessel pits. Also fracture surfaces within wood and mainly in earlywood open additional penetration paths for adhesives. The authors [22] found similar results as reported here with the average area of penetration for earlywood nearly three times larger than for latewood. Such results also were also found by Smith and Coté [23], White [19] and White *et al.* [24].

The observed differences in penetration are also important in many types of wood-based panels, not only for bonding solid wood. Even two strands in oriented strand board (OSB) or two particles in a particleboard are bonded together *via* a bondline; the only difference is the size of the bonding area. Penetration, however, is independent of the size of the bonding area. Particles from various wood species hence can show different influences on the penetration behavior of an applied adhesive, requiring different resin consumption for the same average bond strength.

It could be presumed that in porous wood tissue an equal quality adhesive bonding could be achieved by higher resin consumption in comparison with les porous wood tissue.

5. Conclusion

The average depth of radial penetration of urea-formaldehyde adhesive resins into beech tissue decreases with higher degree of condensation of these resins. Penetration hence depends on the ratio between the average anatomical diameters of tracheids and vessels and the average size of the adhesive molecules. Epi-fluorescence microscopy, fluorescence confocal laser scanning microscopy and scanning electron microscopy were used for the determination of adhesive penetration into wood tissue.

The measured values for the average penetration depth, AP, in the whole sample (without distinction between earlywood and latewood) as well as into earlywood and into latewood for the various resins into beech are in all cases higher than into fir. Bulk flow was observed through the lumens of tracheids in fir and through vessels in beech. Fir exhibits distribution of the adhesive across the interphase region more evenly and closer to the bondline. For beech with its more complex structure adhesive penetration preferably occurs into the vessels and to a longer distance away from the bondline, hence increasing the depth of penetration.

Acknowledgements

This research work was financed by the Ministry of Science and Technological Development, Project "Wood biomass as a resource of sustainable development of Serbia" 20070-TP.

We are greatly indebted to Ing. Petr Pastrnak of DUKOL, Ostrava s.r.o., for the preparation of the UF resins used in the investigations reported in this paper.

References

1. M. Dunky, in: *Handbook of Adhesive Technology*, 2nd edn, A. Pizzi and K. L. Mittal (Eds), pp. 887–956. Marcel Dekker, New York (2003).
2. M. Dunky and A. Pizzi, in: *Adhesion Science and Engineering, Volume 2: Surfaces, Chemistry and Applications*, D. A. Dillard and A. V. Pocius (Eds), pp. 1039–1103. Elsevier, Amsterdam (2003).
3. I. Gavrilovic-Grmuša, J. Miljković and M. Điporović-Momčilović, *J. Adhesion Sci. Technol.*, doi: 10.1163/016942410X501034 (2010).
4. D. A. Hare and N. P. Kutscha, *Wood Sci.* **6**, 294–304 (1974).
5. A. Marra, *Technology of Wood Bonding Principles in Practice*. Van Nostrand Reinhold, New York (1992).
6. C. R. Frihart, in: *Handbook of Wood Chemistry and Wood Composites*, R. M. Rowell (Ed.), pp. 216–273, Forest Products Laboratory, Madison, WI (2005).
7. W. Gindl, *Holz Roh. Werkstoff* **59**, 211–214 (2001).
8. M. Dunky, *Int. J. Adhesion Adhesives* **18**, 95–107 (1998).
9. M. Dunky and K. Lederer, *Angew. Makromol. Chem.* **102**, 199–213 (1982).
10. J. Billiani, K. Lederer and M. Dunky, *Angew. Makromol. Chem.* **180**, 199–208 (1990).
11. M. Scheikl and M. Dunky, *Holzforsch. Holzverwert.* **48**, 55–57 (1996).
12. M. Scheikl and M. Dunky, *Holz Roh. Werkstoff* **54**, 113–117 (1996).
13. M. Scheikl and M. Dunky, *Holzforschung* **52**, 89–94 (1998).
14. T. Furuno, Y. Imamura and H.Kajita, *Wood Sci.Technol.* **37**, 349–361 (2004).
15. S. E. Johnson and F. A. Kamke, *J. Adhesion* **40**, 47–61 (1992).
16. S. E. Johnson and F. Kamke, *Wood Fiber Sci.* **26** (2), 259–269 (1994).
17. J. Zheng, S. C. Fox and C. E. Frazier, *Forest Prod. J.* **54** (10), 74–81 (2004).
18. J. B. Wilson, G. L. Jay and R. L. Krahmer, *Adhesives Age* **22** (6), 26–30 (1979).
19. M. S. White, *Wood Sci.* **10** (1), 6–14 (1977).
20. M. G. Kim, *J. Polym. Sci., Part A: Polym. Chem.* **37**, 995–1007 (1999).
21. A. Pizzi, *J. Appl. Polym. Sci.* **71**, 1703–1709 (1999).
22. E. Brady and F. Kamke, *Forest Prod. J.* **38** (11/12), 63–68 (1988).
23. L. A. Smith and W. A. Coté, *Wood Fiber* **3** (1), 56–57 (1971).
24. M. S. White, G. Ifju and J. A. Johnson, *Forest Prod. J.* **27** (7), 52–55 (1977).

A Flexible Adhesive Layer to Strengthen Glulam Beams

Maurice Brunner [a,*], **Martin Lehmann** [a], **Sebastian Kraft** [a], **Urs Fankhauser** [b],
Klaus Richter [c] **and Jürg Conzett** [d]

[a] Bern University of Applied Sciences, 2504 Biel, Switzerland
[b] Empa Wood Laboratory, 8600 Dübendorf, Switzerland
[c] Geistlich Ligamenta AG, 8952 Schlieren, Switzerland
[d] Hochschule für Technik und Wirtschaft, 7004 Chur, Switzerland

Abstract

Glulam is manufactured by gluing graded timber boards on top of each other with standard, stiff adhesives to form a beam. The research presented here is concerned with the use of a single flexible adhesive layer at a certain position of the beam cross section instead of the standard, stiff adhesive. In a first step, fundamental theoretical considerations for a favourable redistribution of the stresses over the cross section of the beam are presented and the position of the flexible adhesive layer is optimized in order to obtain the highest bending resistance of the total beam. Further calculations help to determine the requisite mechanical properties of the adhesive layer. They are communicated to the adhesive manufacturer as target values for the development of new adhesives.

The industrial partner was able to produce several adhesive layers with the required mechanical properties. There was a need to select a limited number of the adhesive layers for further tests. The selection was done based on creep and delamination tests.

Finally, several beams were manufactured with three selected adhesive layers. Bending tests were carried out to verify the theoretical predictions. The beams manufactured with two of the selected adhesives did not perform well on the large scale as compared to the small scale tests. One of the adhesive layers, however, gave completely satisfactory results: The beams manufactured with this adhesive layer exhibited much higher strength than the control beams and thus confirmed the theoretical expectations. The authors envision a new strengthening technique for glulam beams with this simple technique of using a single, flexible adhesive layer in the beam cross section.

Keywords

Flexible adhesive layer, stress redistribution, bending resistance

1. Introduction

Structural engineers are keenly aware of the many advantages of flexible systems. Steel and concrete engineers and researchers, for example, have developed sophis-

---

* To whom correspondence should be addressed. E-mail: maurice.brunner@bfh.ch

Wood Adhesives
© Koninklijke Brill NV, Leiden, 2010

ticated structural systems with ductile behaviour to mitigate the effects of earthquakes and other extreme loading situations. The timber industry has been rather
passive in this research field: So far, discussions have been generally limited to steel
connections to bring some flexibility into structural timber elements.

We propose the use of a flexible adhesive layer for glulam beams. Our theoretical
investigations indicate the possibility of a favourable redistribution of the stresses
over the cross section of a glulam beam, which could lead to an increased bending
resistance. Thus this concept of a new strengthening technique for glulam beams
could also help open additional niche markets for the environmentally-friendly
wood material.

Target values for the required mechanical strength and the deformation behaviour of the adhesive layer are derived. We describe how adhesive formulations can
be developed and tested for their potential application based on shear, delamination
and creep tests. With a limited number of suitable adhesive systems, large scale
bending tests were carried out to verify the theoretical predictions. A fully satisfactory behaviour of a single adhesive layer demonstrated the enhanced strength of the
glulam beam.

2. Theoretical Considerations

2.1. The Gamma Method: Bending Resistance of Bipartite Beam with an Elastic Connection

The idea is concerned with a simply supported glulam beam composed of two sections which are connected together with a flexible adhesive layer: the main section
at the top of the beam and a much smaller section at the bottom of the beam.

The load-bearing behaviour of the bipartite beam with an elastic shear connection
was first comprehensively researched by Möhler [1]. It is typically used today in
Germany in the design of timber-concrete composites with an elastic mechanical
connection [2, 3].

Basically, the external bending moment M_{el} applied to the composite beam as a
whole can be simplified into three component moments as follows. The two beam
sections will be subjected to bending moments M_1 and M_2 proportional to their
bending stiffnesses. The flexible shear connection will induce a compressive force
N in the larger member and an equally large tensile force in the smaller member.
These opposing forces, which act at a distance "a" apart, will lead to a third moment, the composite moment $M_V = N \cdot a$, which also contributes to the overall
bending capacity (Fig. 1).

It is well known that the tensile zone is critical in a timber beam test. The compressive zone is less critical because plastification effect can lead to a favourable
redistribution of stresses [4]. The maximum bending resistance will be attained
when the maximum tensile stress in the larger member is exactly equal to that in
the smaller member of the bipartite beam. The detailed calculations are documented
in [5] as well as in the MSc thesis of Donzé [6]. The application of the basic rules

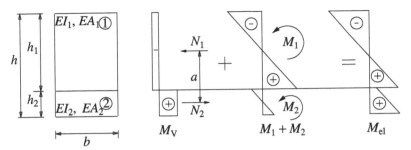

Figure 1. The Gamma-method for a bipartite beam with an elastic shear connection: The external bending moment M_{el} on the composite beam comprises three component moments M_1, M_2 and M_V, which induce the stresses in the two members as shown above.

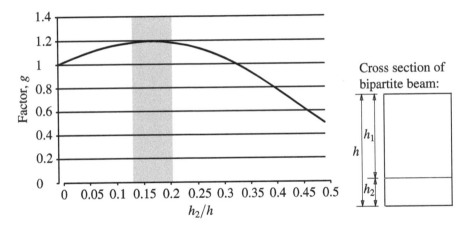

Figure 2. Increment factor g of the bending strength of the bipartite beam as a function of the position of the flexible adhesive layer in the beam cross section. The optimal zone of h_2/h for the greatest strengthening effect is shown as shaded zone.

of the Gamma method leads to the following relationship for the total moment M_{el}, which is the sum of the partial moments M_1, M_2 and M_V as explained above:

$$M_{el} = M_V + M_1 + M_2 = \frac{f_{m,u} \cdot b \cdot h^2}{6} \cdot g\left(\frac{h_2}{h}\right), \qquad (1)$$

$f_{m,u}$ is the bending strength of the glulam beam, h is the total height of the bipartite beam whilst h_2 is the height of the smaller of the two members as shown in Fig. 1. The factor g is a function of (h_2/h) and describes the increase in the bending strength in comparison with conventional glulam:

$$g\left(\frac{h_2}{h}\right) = \frac{1 - 6 \cdot (h_2/h)^2 + 6 \cdot (h_2/h)^3}{1 - 2 \cdot h_2/h + 2 \cdot (h_2/h)^2} \quad \left\{0 \leqslant \frac{h_2}{h} \leqslant 0.5\right\}. \qquad (2)$$

Figure 2 shows that in comparison to a conventional glulam beam, an increase of up to 20% in the bending strength of the bipartite beam is possible, when the adhesive layer is located in the following position in the cross section: $h_2/h \approx 0.2$.

The Gamma method is named after the critical factor Gamma γ, which describes the relative stiffness of the elastic shear connection between the two members which form the bipartite beam. In a normal glulam beam, where the laminates are connected together with very stiff adhesive layers, γ is 1.0; where there is no adhesive at all, and the two laminates are in no way connected to each other, γ is zero. Our calculations indicate that the bipartite beam will attain the above optimal bending resistance only if the γ value of the connection between the two members attains the corresponding ideal value:

$$\gamma_{\text{idéal}} = 1 - 2\frac{h_2}{h}. \tag{3}$$

The theory further explains that the total bending stiffness EI_{ef} of the bipartite beam is the sum of the bending stiffnesses of the two component members and the Steiner component corrected with the γ-factor. The calculations lead to the following result:

$$EI_{\text{ef}} = \frac{Ebh^3}{12} \cdot \left(1 - 6 \cdot \left(\frac{h_2}{h}\right)^2 + 6 \cdot \left(\frac{h_2}{h}\right)^3\right). \tag{4}$$

Figure 3 shows that the bending stiffness of the bipartite beam with the optimized elastic adhesive layer will be reduced to about 80% of that of a classical glulam beam of the same size.

The optimized results depicted in Figs 2 and 3 will only be possible if the adhesive connection exhibits a certain stiffness. The load-bearing behaviour of adhesive layers is normally tested in tensile shear with small beech specimens (Fig. 4). The adhesive layer is typically $b = 20$ mm wide and 10 mm long according to Eurocode 5 [7]. The force (F) is measured in newtons and the shear deformation v in millimetres.

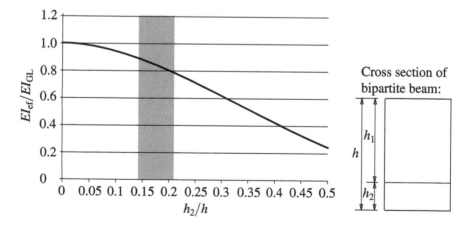

Figure 3. Ratio of the bending stiffness EI_{ef} of the optimized bipartite beam to the stiffness EI_{GL} of a standard glulam beam of the same size. In the shaded zone, where the optimal bending resistance of the bipartite beam will be obtained, the stiffness will be reduced to about 80% that of the standard glulam beam.

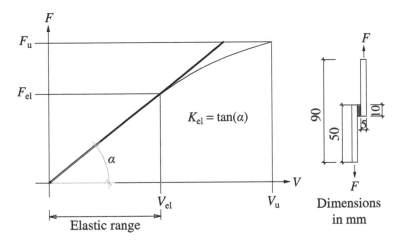

Figure 4. Diagram of force F against the shear deformation v (left) of a classical tensile-shear test which is used to characterize an adhesive layer connecting two beech members (right).

The adhesive manufacturer needed some guidelines as to the results he should aim for when he tested his adhesive mixes with beech specimens in the standard lap shear test. The slope K_{el} in the elastic range of the curve of the measured force F against the shear deformation v was taken as a criterion to classify the tested adhesive layer (Fig. 4):

$$K_{el} = \frac{F_{el}}{v_{el}} \left[\frac{N}{mm} \right]. \qquad (5)$$

The requisite slope for the adhesive layer can be calculated with the rules of the Gamma method. The procedure is based on essentially simple rules, but it involves operations with some rather long equations. The final solution for an elastic adhesive layer is well documented in [5, 6]: $K_{el} = 60$ N/mm.

2.2. Alternative Solution with a Plastic Adhesive Layer

An important finding was that instead of an elastic adhesive layer as required by the Gamma method, a plastic adhesive layer could also lead to the same final strength enhancement of about 20% in comparison to conventional glulam. The calculations for the ideal plastic adhesive layer are documented in Donzé [6]. The adhesive layer would need to exhibit a yield point at a shear stress below 1 N/mm^2, but it is very important that it would fail only after a very large shear deformation (Fig. 5).

The bipartite beam connected by a plastic adhesive would exhibit an initial bending stiffness comparable to classical glulam beams with thin adhesive layers: it would yield only after a certain threshold shear value and allow a redistribution of stresses. The advantage of the plastic adhesive layer over the elastic one is an initial stiffness comparable to that of the classical glulam beam. This factor is very advantageous because in timber engineering practice deflections are often the governing factor for the design. At failure beams with both ideal elastic and ideal plastic adhesive layers would exhibit the same deflections and the same final load (Fig. 6).

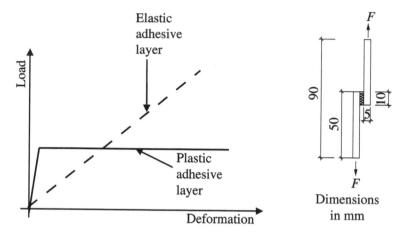

Figure 5. The requisite load-bearing behaviour of the adhesive layers (left), as tested with small tensile-shear beech specimens (right).

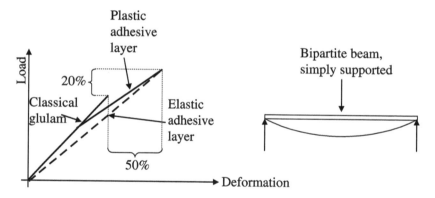

Figure 6. Force-deformation diagram (left) of a simply supported bipartite beam with a plastic or elastic adhesive connection.

2.3. The Requisite Load-Bearing Behaviour of the Adhesive Layers

The mechanical properties of a new adhesive can be characterized in the lap shear test with small beech specimens connected with the new adhesive. The adhesive manufacturer needed some guidelines as to the results he should aim at. The authors presented him the calculated ideal curves of the shear stress τ against the deformation v for both the elastic and the plastic adhesive layers (Fig. 7). The elastic adhesive layer should exhibit a linear relationship for τ and v with a given slope. The plastic (or ductile) adhesive should exhibit an initial linear relationship between τ and v, but it should yield when the shear stress reaches a value between 0.5 and 1.0 N/mm^2.

The adhesive manufacturer prepared several adhesives with different polyurethane mixes (details see Section 3). The adhesives were used in different thicknesses to connect two small beech pieces together. Altogether over two hundred

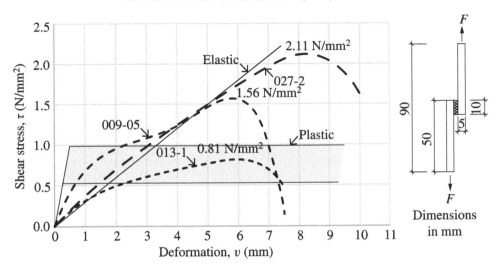

Figure 7. The graph on the left shows the ideal shear stress-deformation curves for small beech specimens (right) with an ideal elastic adhesive connection or with an ideal plastic adhesive connection.
The industrial partner prepared several adhesive layers which exhibited stress-deformation diagrams
close to the theoretical requirements. Three of them are shown above: 027-2, 009-05 and 013-1.

adhesive layers with different thicknesses were tested in tensile shear. In Fig. 7, it
is evident that some test specimens attained shear stress-deformation values close
to those required by the theoretical models. The adhesive layer no. 027-2 for example attains the requisite shear stress-deformation slope and the minimum requisite
deformation as required by the elastic method. The adhesive layer nos 009-05
and 013-1 came close to the theoretical requirements of an adhesive layer which
shows ideal-plastic behaviour; however, the initial required stiffness could not be
attained.

2.4. Numerical Integration of the Load-Bearing Behaviour of the Bipartite Beam

2.4.1. Theoretical Considerations

In order that the bipartite beam should attain the higher bending strength as compared to a glulam beam of the same size and wood quality, the theoretical models
indicate that the adhesive layers must exhibit a very special load-bearing behaviour. The adhesive manufacturer was able to produce some adhesive layers (like
027-2 in Fig. 7) which attained the requisite load-bearing behaviour for the ideal
elastic adhesive layer. However, he was not quite able to fully attain the behaviour
required for the ideal plastic adhesive layer. We decided to perform further tests
with two selected adhesive layers (009-05 and 13-1 in Fig. 7), which came close
to the desired performance needed for the ideal-plastic adhesive layer. There was
a need to estimate the performance of bipartite beams with these adhesive connections. A programme based on the Excel solver function was developed to calculate
the beam behaviour for these and other adhesive layers as follows.

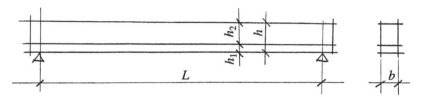

Figure 8. Assumed geometry of the simply supported bipartite beam of total height h, width b and span L, which was modelled in the calculation programme: on the left is the elevation of the beam, on the right the cross section.

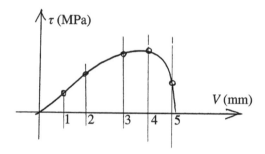

Figure 9. 5 points on the measured τ–v curve of the selected adhesive layer for determining the constants A, B, C, D and E of equation (6).

The programme was developed for a simply supported bipartite beam (Fig. 8). It can calculate a uniform load q (kN/m), which is a common case in engineering practice. It can also calculate two symmetrically placed point loads, which correspond to the beam test set-up which is later described in Section 5. The calculation results could, therefore, be directly compared to the actual test results. The modulus of elasticity of the wood was E. The beam span was L, and the width b. The height of the bipartite beam comprises h_1 for the upper member and h_2 for the bottom member, with the total height being $h = h_1 + h_2$.

The measured shear stress–shear slip (τ–v) curve of the adhesive layer as shown in Fig. 7 above was too cumbersome to be directly integrated into the simple calculation programme. It was first approximated by a 5th order polynomial as shown in equation (6).

$$\tau(v) = Av^5 + Bv^4 + Cv^3 + Dv^2 + Ev. \qquad (6)$$

The constants A, B, C, D and E of equation (6) were determined from 5 points on the measured τ–v curve of the selected adhesive layer, as shown in Fig. 9.

Figure 10 shows that the shear stress τ will induce an axial slip $v = v(x)$ between the top and bottom members of the bipartite beam. The axial slip is assumed to be symmetrical to both sides of the beam axis, measured from the midspan. The x-axis is measured from the beam midspan, where there is no axial slip: $v(x = 0) = 0$. l is the span of the beam. The axial slip v between the top and bottom members of the

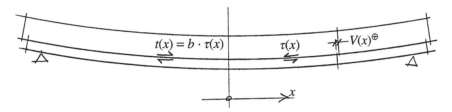

Figure 10. The axial slip $v(x)$ between the top and bottom members of the bipartite beam is caused by the shear stresses $\tau(x)$ acting on the deformable adhesive laxer.

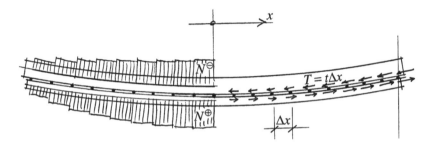

Figure 11. The shear flow t is concentrated at distances of Δx to periodic nodal forces $T(x)$, which induce opposing axial forces N in the top and bottom members of the bipartite beam.

bipartite beam was programmed as a Fourier series with 12 sine functions:

$$v = N_1 \cdot \sin\left(\pi\frac{x}{l}\right) + N_2 \cdot \sin\left(2\pi\frac{x}{l}\right)$$

$$+ N_3 \cdot \sin\left(3\pi\frac{x}{l}\right) + \cdots + N_{12} \cdot \sin\left(12\pi\frac{x}{l}\right). \tag{7}$$

The N-factors of equation (7) were determined by trial and error with a solver function until the assumed moment value was attained.

The axial slip $v(x)$ depicted in Fig. 10 is caused by a corresponding shear stress $\tau = \tau(x)$. For the analysis of the beam behaviour in the longitudinal x-axis, it is practical to consider axial forces rather than two-dimensional stresses. Thus the shear stress τ (N/mm^2) is multiplied by the width of the beam to obtain the so-called "shear flow" $t = \tau \cdot b$, whose units are N/mm. For the numerical handling, the distributed shear flow $t(x)$ is concentrated to periodic nodal forces $T(x) = t(x)$. Δx act at distances of Δx apart and in opposite directions where the adhesive layer is in contact with the top and bottom members of the bipartite beam (Fig. 11). The nodal forces $T(x)$, in turn, induce axial forces N in the beam members. At the beam ends the normal forces are zero: this boundary condition permits the calculation of the axial forces by simple integration.

The axial forces N lead to axial deformations in the top and bottom members of the bipartite beam (Fig. 12):

- Axial deformation of top member: $u_{\text{sup}} = u_{\text{sup}}(x)$.

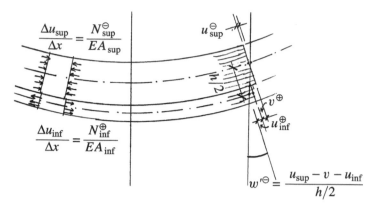

Figure 12. The axial force N in the top and bottom members of the bipartite beam induces axial deformations Δu: these sum up to the slip v between the two members (shown on the left half of the figure). The slip v in turn causes a change w' in the curvature of the beam (shown on the right side of the figure).

- Axial deformation of bottom member: $u_{\text{inf}} = u_{\text{inf}}(x)$.

Since the axial forces N in the two members must be equal and act in opposite directions, the axial deformation (u) of the members will be inversely proportional to their cross-sectional areas (A). And since the width b of the beam is constant, the following relationship between the axial deformations of the two members (u_{inf} and u_{sup}) and the member heights (h_1 and h_2) can be derived:

$$\frac{u_{\text{inf}}}{u_{\text{sup}}} = \frac{A_{\text{sup}}}{A_{\text{inf}}} = \frac{h_2}{h_1}. \tag{8}$$

Apart from the axial slip v, the different axial displacements u of the two beam members will cause the bipartite beam to bend (Fig. 12). The angle change w' is:

$$w' = \frac{u_{\text{sup}} - v - u_{\text{inf}}}{h_1/2 + h_2/2} = \frac{2}{h} \cdot u_{\text{sup}} \cdot \left(1 + \frac{h_2}{h_1}\right) - v. \tag{9}$$

The integration of these changes of curvature will yield the flexural beam curve $w = w(x)$, which is the deflection of the beam at the given position x on the beam axis.

The second integration of the flexural beam curve (w'') multiplied with the corresponding bending stiffness EI, will yield the bending moment induced directly in the two members as follows:

$$M_{\text{sup}} = EI_{\text{sup}} \cdot w''; \qquad M_{\text{inf}} = EI_{\text{inf}} \cdot w''. \tag{10}$$

The axial forces in the beam members can be calculated by differentiating the displacement u (Fig. 13):

$$N_{\text{sup}} = EA_{\text{sup}} \cdot u'_{\text{sup}}; \qquad N_{\text{inf}} = EA_{\text{inf}} \cdot u'_{\text{inf}}. \tag{11}$$

The normal forces N act in opposite directions in the two members of the bipartite beam at a distance $h/2$ apart, where h is the beam height. They thus also induce

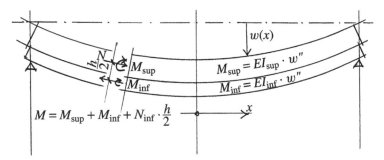

Figure 13. The total moment M acting on the bipartite beam comprises three components: M_{sup} and M_{inf} act directly on the top and bottom members, whilst the normal force N acting on the two members but in opposite directions and at a distance $h/2$ apart induces the third partial bending moment.

a contribution to the total moment acting on the bipartite beam. The total bending moment in the bipartite beam is calculated as:

$$M = M_{sup} + M_{inf} - N_{sup} \cdot \frac{h}{2}. \tag{12}$$

The expected external moment can be calculated for 12 positions along the longitudinal axis of the bipartite beam. The solver function can be activated to calculate the corresponding 12 Fourier coefficients of the axial slip v functions. Thus the deformations w and the internal moments M_{sup} and M_{inf} and normal forces N in the top and bottom members of the bipartite beam can be clearly determined at the 12 positions along the beam axis. The results can be compared to those for a classical glulam beam composed of wood laminates of the same timber grade as those of the bipartite beam. Thus the performance of the special adhesive layer can be evaluated with regard to the possible advantages it might bring.

2.4.2. Numerical Example

The numerical example below will show the performance of a simply supported bipartite beam and compare the possible advantages it might bring in comparison to a conventional glulam beam of the same size and composed of the same wood quality. The following design parameters are representative of timber engineering practice:

Length of girder:	10 m
Cross section $b \times h$:	140 × 750 mm
Cross section of top member of the bipartite beam:	140 × 600 mm
Cross section of bottom member of the bipartite beam:	140 × 150 mm
Modulus of elasticity E:	11 000 MPa
Service load:	7 kN/m
Design load:	20 kN/m
Maximum bending moment by service load:	87.5 kN m
Maximum bending moment by design load:	250.0 kN m

As the program can deal with any variation in the shear modulus as a function of the displacement, the calculations will be done for two different adhesive layers.

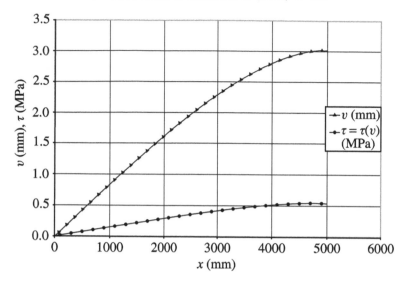

Figure 14. Calculation results for a bipartite beam connected with an elastic adhesive layer: both the shear stress τ in the adhesive layer and the slip v between the top and bottom members of the bipartite beam increase almost linearly from the beam midspan ($x = 0$) towards the supports.

Adhesive Layer with Constant Shear Modulus.   The assumption is that a lap shear test with small beech specimens connected with this elastic adhesive layer will yield the following linear relationship between the shear stress τ and the slip v:

$$\tau \text{ (MPa)} = 0.188 \cdot v \text{ (mm)}. \tag{13}$$

The developed Excel program yielded the curves depicted in Fig. 14 ($x = 0$ at midspan). Both the shear stress τ and the slip v between the upper and bottom members of the bipartite beam are zero at beam midspan and increase almost linearly to maximum values at the beam ends over the supports.

At midspan, the normal stresses in the bipartite beam for the design moment of 250 kN m are:

−21.6 MPa top part of the top member of the bipartite beam,

+15.9 MPa bottom part of the top member of the bipartite beam,

+6.8 MPa top part of the bottom member of the bipartite beam,

+16.1 MPa bottom part of the bottom member of the bipartite beam.

The design stress in a conventional glulam beam of the same size can be readily calculated for the design moment of 250 kN m to be 19.1 MPa. Comparing this value to the critical maximum tensile stress in the bipartite beam (+16.1 MPa), it is evident that the load capacity of the bipartite beam can be increased by 18.6%.

The deformation under service load is 20.9 mm for the bipartite beam, compared to 16.9 mm for the conventional glulam beam. This is an increase of 12.4%.

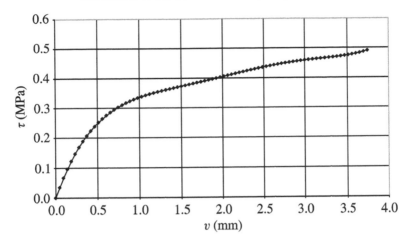

Figure 15. Shear stress–slip graph for the adhesive layer which was used in a bipartite beam.

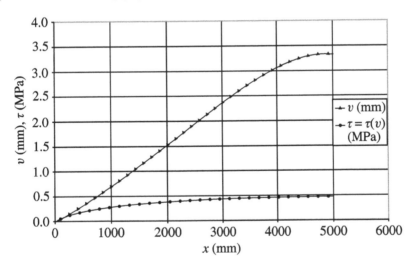

Figure 16. Calculation results for a bipartite beam connected with a plastic adhesive layer: both the shear stress τ in the adhesive layer and the slip v between the top and bottom members of the bipartite beam increase from the beam midspan ($x = 0$) towards the supports. Whilst the slip rises almost linearly, the shear stress soon attains a maximum value.

Adhesive with Variable Shear Modulus   The calculation program was also applied to a bipartite beam connected by an adhesive layer which, in a lap shear test with small beech specimens, yielded the curved relationship between the shear stress τ and the slip v (Fig. 15).

The developed Excel program yielded the curves depicted in Fig. 16 ($x = 0$ at midspan). Both the shear stress τ and the slip v between the upper and bottom members of the bipartite beam are zero at beam midspan. Whilst the slip increases almost linearly to a maximum value at the beam ends over the supports, the shear stress initially increases quickly but soon reaches a maximum value.

At midspan, the normal stresses in the bipartite beam for the design moment of 250 kN m are:

−21.1 MPa top part of the top member of the bipartite beam,

+14.9 MPa bottom part of the top member of the bipartite beam,

+7.7 MPa top part of the bottom member of the bipartite beam,

+16.7 MPa bottom part of the bottom member of the bipartite beam.

The design stress in a conventional glulam beam of the same size can be readily calculated for the design moment of 250 kN m to be 19.1 MPa. Comparing this value to the critical maximum tensile stress in the bipartite beam (+16.7 MPa), it is evident that the load capacity of the bipartite beam can be increased by 14.3%.

The deformation under service load is 18.5 mm for the bipartite beam, compared to 16.9 mm for the conventional glulam beam. This is an increase of 10.9%.

In summary, the calculation results for a bipartite beam with an elastic adhesive layer show that, in comparison to a conventional glulam beam, the increased ultimate load-bearing capacity is combined with an increased deformation under service loads. The calculation results for a bipartite beam with a plastic adhesive layer (characterized by non-linear shear stress–slip behaviour) indicate that it will deform less under service loads, but the increase in ultimate load-bearing capacity will also be slightly smaller than for the bipartite beam with an elastic adhesive layer.

2.5. Possible Applications for the Optimized Bipartite Beam

With the help of a single, flexible adhesive layer, placed at a vantage point in the cross section, the bending resistance of the glulam beam can be increased by 20%, whilst the deformation at failure would be increased by 50% (Fig. 5).

A glulam beam with a modified adhesive layer should find use in situations requiring greater bending resistance combined with greater deformability before failure. In earthquake zones, such properties may be advantageous. Other possible applications may be heavily loaded bridge girders or long-span shed-type buildings. Figure 17 shows a sports arena under construction in Switzerland. Our calculations indicate that the height of the main girders could have been reduced by 20% if, instead of standard glulam, a bipartite beam with an ideal elastic adhesive behaviour had been used.

Steel and reinforced concrete beams have a classical advantage over timber beams because, unlike standard timber beams, they exhibit ductile behaviour which can lead to a favourable redistribution of loads in multi-span beams. This behaviour enhances beam performance in extreme loading cases like earthquakes or an aircraft crash into a building. Our calculations indicate that if the flexible adhesive layer is placed in the top part of a glulam beam over the intermediate support of a multi-span beam (Fig. 18), the reduced stiffness in this section could also lead to

Conventional GL24h

Possibility with flexible adhesive layer

1200

1100
880

220

200

In mm

200

Figure 17. Swiss sports arena (left), where the use of a bipartite beam (right) could have reduced the cross section of the main girders which were actually used (centre).

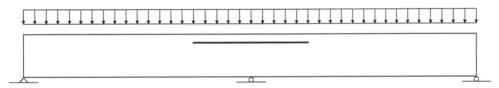

Figure 18. Two-span glulam beam with an ideal, flexible adhesive layer in the top part of the beam over the intermediate support. Calculations indicate that this arrangement could lead to a favourable redistribution of loads and thus enhance beam performance in extreme leading cases.

a redistribution of loads. Thus an inherent weakness of the timber beam could be improved upon with this simple innovation.

3. Development and Chemistry of Requisite Adhesives

3.1. Introduction

The engineers had formulated requirements as to how the desired adhesive layers should deform under a shear loading. These rules were very unusual because normally, engineers make no requirements at all; they just test the adhesives offered by the adhesive company and then report if they are satisfied or not. The adhesive company found it difficult to meet the requirements. The most important characteristics required of the adhesive layer were the flexibility as well as the deformability. Obviously, traditional wood adhesive systems like vinyl acetate or formaldehyde-based adhesives could not be used because they only work in very thin adhesive layers. Furthermore, they are brittle and have virtually no mechanical flexibility. Thus only two technologies were considered as potential candidates: polyurethanes (including silane-terminated types) and epoxies.

Epoxies have some disadvantages when they are used to bond wood: inadequate adhesion properties especially at higher temperature, and application as a two-component system if they should be cured under ambient condition. Finally,

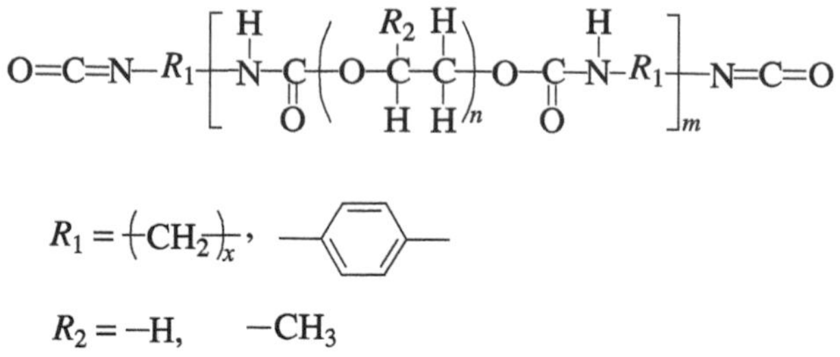

$R_1 = \left(CH_2\right)_x,$ —⟨ ⟩—

$R_2 = -H,$ $-CH_3$

Figure 19. Basic chemical structure of polyurethane prepolymers [8].

those mixtures which are flexible tend to exhibit low mechanical properties (e.g., tear or tensile strength).

Polyurethanes, on the other hand, have good adhesion properties to wood and are applicable as one-component system with fast curing properties at ambient conditions. Furthermore, the mechanical properties can be readily varied with simple manipulation of the chemical formulation. Most important of all, there is considerable experience in using polyurethane to bond wood.

Finally the decision was made to formulate the adhesives based on polyurethane. Four different concepts were studied:

1. Adhesives, based on MDI[1] prepolymers with a high content of isocyanate (15–16%).

2. Adhesives, based on MDI-prepolymers and/or prepolymers comprising aliphatic[2] isocyanates with a low content of isocyanate (2–5%).

3. Adhesives, based on mixtures of 1 and 2.

4. Adhesives, based on different polyurethane prepolymers but the polymer chains were terminated with silane compounds.

On the basis of these polymers, adhesive formulations were developed and produced with the following components: inorganic fillers, organic fibres, plasticizers and adhesion promoters.

3.2. Chemistry and Sample Preparation

3.2.1. Chemistry
The basic chemical structure of polyurethane prepolymers is illustrated in Fig. 19. A good overview on polyurethane adhesive chemistry is given by Habenicht [8].

3.2.2. Samples
For a first evaluation, the adhesives were tested according to EN 204 [9] and EN 205 [10] (lap shear strength) but with thick adhesive layers: 0.5 mm, 1 mm and 3 mm.

---

[1] 4,4′-Methylene diphenyl diisocyanate.
[2] IPDI = Isophorone diisocyante or HDI = Hexamethylene diisocyanate.

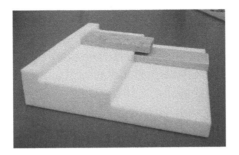

Figure 20. Plastic mould used to produce the lap shear test specimen with a specified, thick adhesive layer.

Contrary to the procedure described in the standard, the samples were prepared as follows:

- Beech wood was cut into pieces measuring $60 \times 10 \times 5$ mm (length × width × thickness).

- Two beech pieces were bonded together with an overlap of 10 mm as follows. The adhesive was applied in sufficient quantity at one end of one beech piece, which was then placed on the lower surface of a PTFE[3] plastic mould shaped like a tiny staircase (Figure 20). The top step of the mould was higher than the lower one by a thickness which corresponded to the total height of the first beech piece and the required thickness of the adhesive. The second beech piece was then pressed on the top step of the mould with an overlap of 10 mm over the lower beech piece. The gap between the two pieces corresponded to exactly the adhesive thickness required.

- For each adhesive mix, three samples were prepared for each of the three thicknesses of 0.5 mm, 1 mm and 3 mm. Thus a total of $3 \times 3 = 9$ samples were produced.

- The samples were cured for 14 days at 23°C and 50% rel. humidity and then tested on a tensile testing machine.

3.3. Formulations

3.3.1. MDI Prepolymers with a High Content of Isocyanate
This type of adhesive was formulated on the basis of MDI-prepolymer with a relatively high concentration of isocyanate in the uncured adhesive (around 15–16%). This also means a low molecular weight (around 500–600 g/mol). This adhesive type is often used to bond wood. In standard practice, the adhesive is applied under pressure so that most of it flows out between the wood members: the thickness of the adhesive layer is virtually zero.

---

[3] Poly(tetrafluoroethylene).

For this application, it was desired to try thicker adhesive layers. Because it contains a large amount of free Isocyanate, this kind of adhesive foams strongly during the curing period. Therefore, it is difficult to specify exactly the properties of the adhesive layer because the cell dimensions, as well as the resulting properties of the foam, are randomly distributed.

The adhesive layers were all subjected to the lap shear test.

Results.   These results represent minimum/maximum values from several dozen formulations:

- Elongation at failure: 2 mm/7 mm.

- Ultimate shear stress in lap shear test: 1 MPa/5 MPa.

- Slope of force–deformation curve: 50/800 N/mm.

The surprisingly high elongation at failure is a result of the foamy structure of the adhesive layer. In general, formulations with a high elongation at failure show a very low slope of the force–deformation curve, which did not fulfil the specifications of the engineers with regard to either the ductile model or the elastic model. Therefore, this research direction was soon abandoned.

3.3.2. MDI Prepolymers with a Low Content of Isocyanate

The adhesives were formulated on the basis of prepolymers with a low content of isocyanate (2–5%). The molecular weights of these prepolymers were between 1500 and 4500 g/mol. With these types, flexible adhesives and sealants can be formulated. Such adhesives are usually applied in thicker adhesive layers (0.5–5 mm). Due to the relatively low isocyanate content, the tendency to foam is very low. Soft, bubble-free adhesive layers were obtained.

Furthermore, prepolymers of these types can be polymerized with aromatic isocyanate or with aliphatic ones. To formulate the adhesive layer required, both types were tested. The best results were obtained with combinations of aromatic and aliphatic isocyanates.

Results.

- Elongation at failure: 2 mm/25 mm.

- Ultimate shear stress in lap shear test: 0.5 MPa/4 MPa.

- Slope of force–deformation curve: 50/300 N/mm.

As expected, these adhesive layers showed the desired, very high elongation at failure. The specifications with regard to the slope of the force–deformation curve according to the elastic model could be readily met. However, the requirements for the ductile model could not be quite met because the initial stiff slope required by this model could not be attained.

3.3.3. Mixture of Prepolymers with High and Low Contents of Isocyanate

As shown above, the adhesives with low isocyanate content exhibited good elongation properties, whilst the adhesives formulated with high isocyanate content prepolymers exhibited an initial high slope of the load-deformation curve for the small beech specimens. Both these properties were required for a good performance of the adhesive layer with regard to the plastic model of the engineers. In order to attain the two desired properties, the two polymer types were mixed.

For concentrations above 30% of high isocyanate prepolymers, the resulting properties are not much different from the formulations with purely high isocyanate content. Below this concentration, the properties of the low isocyanate types were determined. A combination of effects (high elongation at failure, high slope of force–deformation curve) could not be observed.

Results.

- Elongation at failure: 2 mm/20 mm.

- Ultimate shear stress in lap shear test: 0.5 MPa/3 MPa.

- Slope of force–deformation curve: 50/500 N/mm.

3.3.4. Polyurethane-Based Prepolymers Terminated with Silane–Compounds

These prepolymers were based on the same chemistry and the same preparation procedure as all prepolymers mentioned above. But the isocyanate end group was reacted with a silane compound. These prepolymers are isocyanate free and have a different cross-linking mechanism. Adhesives based on these prepolymers are commonly used as sealants. They have good elongation properties, good adhesion to different kinds of materials and are usually more resistant against UV rays. They are not mixable with PU prepolymers of any kind.

Results.

- Elongation at failure: 2 mm/14 mm.

- Ultimate shear stress in lap shear test: 0.5 MPa/4 MPa.

- Slope of force–deformation curve: 50/800 N/mm.

Although they partly showed quite good results from a mechanical point of view, the adhesion properties to wood were completely unsatisfactory: usually adhesion failure occurred.

3.3.5. Overview of Adhesive Formulations

As an overview, Table 1 summarizes the properties of a selected number of the adhesive mixes which were formulated and tested in tensile shear.

Table 1.

Overview of a selected number of adhesive formulations

Formulation number	Chemical basis of the prepolymer	Main formulation components	Best adhesive layer thickness (mm)	Elongation at failure (mm)	Tensile shear strength (MPa)	Slope of force–deformation curve (N/mm)
009	MDI/IPDI	calcium carbonate	0.5	1.55	7.38	140
013	MDI/IPDI	calcium carbonate, plastic fibres	1	0.69	9.94	80
014	MDI/IPDI	calcium carbonate, plastic fibres	1	0.77	9.18	120
027	MDI	calcium carbonate	2	2.44	13.59	80
062	MDI/IPDI	calcium carbonate, plastic fibres	3	1.25	8.85	70
071	MDI	plastic fibres	3	1.40	21.15	80

Table 2.

Adhesive formulations chosen for bending tests. One adhesive (027) satisfied the specifications for an elastic adhesive layer. The other two came close to the requirements for a plastic adhesive layer

Formulation	Mechanical behaviour	Thickness of adhesive layer (mm)
009	Ductile	0.5
013	Ductile	1
027	Elastic	2

3.4. Adhesive Layers Selected for the Manufacture of Test Beams

The most interesting concept was found to be prepolymers with low isocyanate content because they are easy to modify, do not foam during curing, have good adhesion properties and meet the specifications as close as possible.

The specifications according to the elastic model could be readily met by many different formulations. The specifications according to the ductile model were more difficult to meet mainly because of the required initial stiffness. Figure 7 shows some of the adhesive layers which came close to meeting the requirements.

After further tests, which are described in the next section, three formulations were selected for further investigations (Table 2).

4. Adhesive Tests

4.1. Reasons for the Adhesive Tests

As shown in Section 3, the adhesive manufacturer was able to produce several adhesive layers which fulfilled the theoretical requirements with regard to the load-bearing behaviour. The adhesive tests were introduced as screening tools to help eliminate some of the proposed adhesive layers, so that it would be possible to carry out large scale bending tests with a small number of the adhesive layers.

Two tests were selected because of their simplicity and also because they are standard tests which must be passed before any adhesive will be accepted.

- Moisture-related durability of wood-adhesive bonds is characterized in a standard delamination test according to SN EN 302-2:2004 [11].

- The creep behaviour under static shear load and standard climate conditions is assessed over a period of 335 h with small beech specimens.

4.2. Delamination Tests

25 adhesive layers were selected for the delamination test. 15 of them had fulfilled closely the mechanical requirements for the elastic model with regard to the tensile shear tests with small beech specimens. The other 10 adhesive layers had performed close to the requirements for ductile behaviour.

The delamination tests were carried out according to the SN EN 302-2:2004 [11] standard. The procedure for resin type 1 (outdoor use) was used. In order to reduce costs, instead of four small glulam beam samples, only one was produced. Two slices were taken from the sample produced and tested. The adhesives were still in the laboratory stage and produced in laboratory scale, therefore the curing time was neither known nor optimised. Due to the fact that the pot-life and the open-times of the adhesives were not known, each sample was pressed straight after the adhesive was spread on the boards. To avoid failure due to unfinished curing the samples were pressed for 48 h and post-cured for at least 14 days. In order to ensure that the adhesive layer had the required thickness, hardwood spacers were placed at the edges of the boards (Fig. 21). After the curing period, the parts with the spacers were cut out of the beam and 75 mm long slices were cut out from the middle of the sample.

The test procedure involved the following steps. The specimens were placed in a container with water: air was sucked out to create a partial vacuum and thus speed up the soaking process, which lasted 1 h. The soaking process was done twice. The specimens were then dried at 65°C for 24 h. The cycle was repeated three times.

Immediately after the last drying, the extent of delamination was determined by visual inspection as follows. The photo on the right of Fig. 22 shows that the delamination of the adhesive bondlines typically occurred in short lengths on different positions of the beam. The length of adhesive bondlines which had delaminated was measured and summed up for the whole beam as L_1. The total length of adhesive

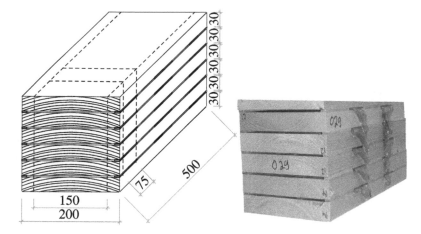

Figure 21. Glulam beam sample produced for the delamination tests. Left: drawing showing that hardwood pieces were placed at the edges to ensure the thickness of the adhesive layer. The delamination tests were performed on the specimens sliced across the wood grain. Right: photo of the glulam beam sample before it was cut.

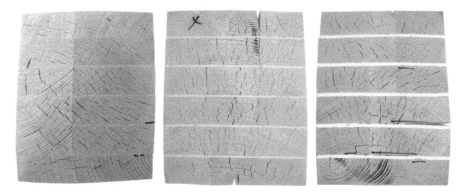

Figure 22. Visual inspection of selected test specimens after the delamination test. The specimens with the adhesive 009 (left) and 013 (middle) showed only a few cracks, whilst the specimen with adhesive 027 (right) showed extensive cracking.

bondlines in the specimen was determined by multiplying the perimeter of the specimen (length and width) with the number of bonding surfaces in the cross section: In the case of the specimen shown in Fig. 21, there are 5 bonding surfaces, the length of the specimen is 500 mm and the width 150 mm. Hence the total length of bonding line is: $L_{\text{total}} = 5 \times (500 + 150) \times 2 = 6500$ mm. The extent of delamination was calculated as a percentage (L_1 / L_{total}).

Only 5 of the 22 adhesive layers fulfilled the standard requirements: they suffered less than 5% delamination. Three of them had exhibited plastic behaviour in the tensile shear test, whilst the other two had exhibited elastic behaviour:

- Adhesive layer no. 009-05: exhibited plastic adhesive behaviour.

- Adhesive layer no. 013-1: plastic adhesive behaviour.

- Adhesive layer no. 014-1: plastic adhesive behaviour.

- Adhesive layer no. 062-3: elastic adhesive behaviour.

- Adhesive layer no. 071-3: elastic adhesive behaviour.

As a control, a standard glulam beam of grade GL24h was also tested. It passed the delamination test. The adhesive layer 027-3 exhibited rather excessive delamination (Fig. 22).

The drying temperature of 65°C is quite high and some of the adhesives which failed showed signs of possible thermal degradation (Fig. 23).

In order to gain some knowledge about the degradation of the adhesive layer during the delamination test, the specimens which successfully passed the test were further tested in compression shear (Fig. 24). For these tests, blocks with a cross section of 50 by 50 mm were cut out of the tested specimens. As a control, specimens of a similar size were cut from the part of the original sample which had not been subjected to the delamination tests. The compression shear tests confirmed that the five adhesive layers which passed the delamination tests did not suffer any degradation during the harsh treatment. Their shear resistance was hardly reduced after the harsh delamination tests.

Figure 23. Specimens with possible thermal degradation of the adhesive layers after the delamination tests: adhesive no. 037 (left) and 052 (right).

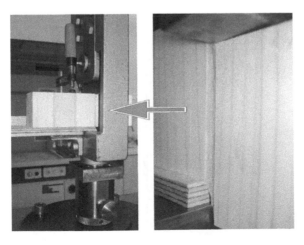

Figure 24. Compression shear tests with a 3 mm thick elastic adhesive layer (no. 027). The left picture shows the test set-up, whilst the right picture shows the extraordinary large shear deformation of the adhesive layer.

4.3. Creep Tests with Small Beech Specimens

4.3.1. Materials and Method

A total of 45 single lap shear specimens made of beech wood were produced by the adhesive manufacturer as described in Section 3.2. The samples were glued with the five adhesive types used in the delamination test: 3 plastic adhesives no. 009, 013 and 014, and 2 elastic adhesives no. 062 and 071. The adhesive layers of the lap shear specimens had the following nominal thicknesses: 0.5, 1.0 and 3.0 mm. For each adhesive type and thickness batch, 3 specimens per batch were tested.

The tensile-shear loading was realised in a set-up involving the three specimens of the same batch as follows. The bottom leg of the top specimen was connected to the top leg of the middle specimen with a bolt with a washer of the same thickness as that of the given adhesive layer in order to assure symmetrical loading. The middle specimen was connected in a similar way to the lowest specimen (Fig. 25).

Each jig was loaded with a lead weight of 6.087 kg, which corresponds to a shear stress level of 0.3 MPa in the bonded area of the individual test specimen. This loading corresponds to 60% of the minimal shear strength results of the adhesives reported in Section 3.3. A millimetre scale was fixed parallel to the adhesive layer of the aligned specimens (Fig. 25). Before loading, a horizontal mark was carved across each adhesive layer to facilitate the measurement of the deformation. Digital photographs with sufficient magnification were taken of each bonded assembly after 0.1, 4, 8, 24, 48, 72, 100, 170, 220 and 335 h. The photographs were analysed with the software Image Access® to determine the longitudinal shear deformation of the adhesive layer to the nearest 0.1 mm. Data from each batch were averaged and

Figure 25. Test set-up for creep deformation test.

results are shown as total creep deformation per adhesive layer. In addition, creep factors ϕ were calculated according to equation (14):

$$\Phi = \frac{f_t - f_0}{f_0}, \tag{14}$$

where:

f_0: Elongation in mm of the specimens immediately after loading (time zero).

f_t: Elongation in mm of the specimens after a time interval of t (seconds).

4.3.2. Results

The average creep deformation against time curves for the different adhesive layers are shown in Figs 26–28. Despite the low shear loading of 0.3 N/mm^2, 10 out of 45 specimens failed within a few minutes after the load was applied. Most of these early failures occurred when the adhesive layer was thick (1.0 and 3.0 mm) and formulated for elastic behaviour (071 and 062). Thus they could be eliminated very early from the list of desired adhesive layers.

The creep development of the various adhesive layers with time is listed in Table 3. The majority of the specimens reached almost 70% of the total creep deformation after 8 h of loading. Among the different adhesives the actual creep deformation varied considerably. For the adhesive layer of 0.5 mm thickness, the adhesive no. 071 showed the lowest creep factor of 0.68 after 335 h (2 weeks), whereas all the other four tested adhesives exhibited creep factors around 2. In the group of 1.0 mm thick adhesive layers, this trend changed: the adhesive no. 062 showed the lowest creep, whilst the other formulations exhibited greater creep, with adhesive no. 009 showing highest creep. The samples of no. 013 with thickness 1.0 mm all failed early. In the group of adhesive layers with thickness 3.00 mm,

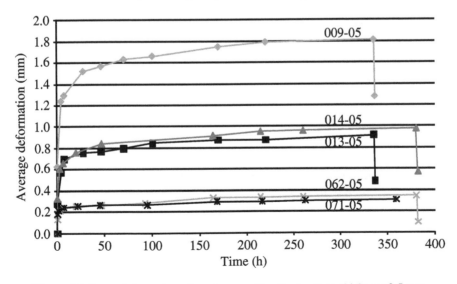

Figure 26. Creep deformation of specimens with adhesive layer thickness 0.5 mm.

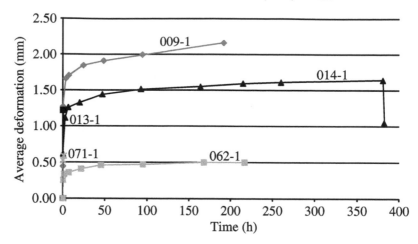

Figure 27. Creep deformation of specimens with adhesive layer thickness 1.0 mm.

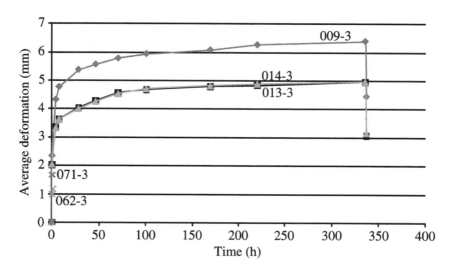

Figure 28. Creep deformation of specimens with adhesive layer thickness 3.0 mm.

adhesives no. 013 and 014 behaved very similarly, whilst no. 009 showed greater creep.

It should be pointed out that the creep behaviour was determined under static loading and constant climate conditions. A variation of load and climate with time could affect the creep deformation and load bearing capacity considerably.

5. Bending Tests

5.1. Overview of the Bending Tests

There were two main reasons for the bending tests:

Table 3.
Increase of creep factors with time, determined for different adhesive layers in the three thickness classes (nd = not determined because of failure)

Time (h)	0.5 mm					1.0 mm				3.0 mm		
	009	013	014	062	071	009	013	014	062	009	013	014
4	0.97	1.09	0.87	0.64	0.17	0.96	nd	0.82	0.34	0.86	0.67	0.64
8	1.06	1.54	1.03	0.91	0.32	1.56	nd	1.06	0.42	1.05	0.80	0.79
24	1.43	1.74	1.35	1.02	0.39	1.62	nd	1.16	0.63	1.31	1.00	0.97
48	1.50	1.78	1.60	1.06	0.43	1.84	nd	1.36	0.84	1.39	1.13	1.10
72	1.60	1.88	1.75	1.23	0.43	1.94	nd	1.48	0.89	1.48	1.27	1.23
100	1.65	2.07	1.84	1.59	0.60	2.06	nd	1.53	1.00	1.55	1.32	1.32
170	1.79	2.16	1.96	1.63	0.60	2.30	nd	1.61	1.00	1.61	1.38	1.37
220	1.86	2.16	1.98	1.67	0.66	nd	nd	1.64	1.10	1.69	1.40	1.41
335	1.89	2.32	2.05	1.72	0.68	nd	nd	1.69	1.18	1.74	1.46	1.45

- First, there was a need to clarify that the adhesive properties were the same for the small laboratory specimens as for the large beams. Experience has shown that when new adhesives are produced in small amounts, the properties could change when they are produced under factory conditions.

- Verify the calculation models which had predicted improved bending performances for the bipartite beams with the optimized adhesive layers.

One adhesive layer with elastic behaviour (027-3) and two adhesive layers with ductile behaviour (009-05 and 013.1) were selected for the large scale four-point bending tests. In addition to these bipartite beams with a single flexible adhesive layer, glulam beams of standard quality (GL24h) were tested as control. All the bipartite and control beams were manufactured with wood lamellas of the same quality S10 according to the German standard [12].

The four-point bending tests were carried out according to the standard SN EN 408:2003 [13]. The cross section was kept constant for all test specimens. On two specimens of each sample set, 4 strain gauges were attached at midspan and 4 directly under the load (Fig. 29). They were placed at the top and bottom parts of the two members of the bipartite beam and they were used to verify the model of the non-linear strain distribution over the beam height. A similar set of gauges was placed at similar positions on two control beams of standard glulam GL24h. A measuring device was placed at one beam end to measure the total longitudinal shear slip between the upper and lower members of the bipartite beam during the bending tests.

Table 4 lists the geometry of the test specimens. Three different spans were tested for each adhesive layer because the calculation model "by integration" had predicted an optimal span for the best results. In the case of the ductile adhesive types

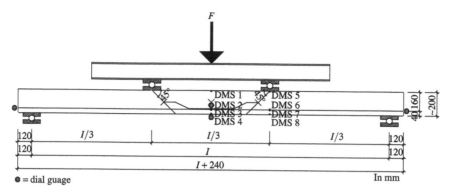

Figure 29. Four-point bending setup used.

Table 4.

Specimens tested in four-point bending

Sample	Number of specimens	Span (m)	Height (mm)	Adhesive type	Thickness (mm)	Mechanical behaviour of adhesive layer
1	3	3	160 + 40	009	0.5	plastic
2	6	3.3	160 + 40	009	0.5	plastic
3	3	3.6	160 + 40	009	0.5	plastic
4	6	4	200	standard glulam beam	n.a.	no special adhesive
5	3	3.7	160 + 40	027	2	elastic
6	6	4.0	160 + 40	027	2	elastic
7	3	4.3	160 + 40	027	2	elastic
8	3	4.9	160 + 40	013	1	plastic
9	6	5.2	160 + 40	013	1	plastic
10	3	5.5	160 + 40	013	1	plastic

009 and 013, the optimal span did not comply with the standard requirement that the span should be about 20 times the height.

5.2. Production of Test Specimens

As raw material for the production of all test specimens, spruce lamellas of quality S10 (for GL24h according to DIN 4074-1 [12]) were used. The reference beams had a cross section of 140/200 mm (width/height) and a standard glulam adhesive (Casco MUF) was used to bind the lamellas together. This was the standard procedure used by the wood company to manufacture glulam beams of standard quality GL24h. The cross section of the bipartite beams is shown in Fig. 30: the upper member had a cross section (width/height) of 140/160 mm, whilst the lower member was composed of a single wood lamella of width 140 mm and height 40 mm.

Figure 31 shows how the two members of the bipartite beam were connected together. The flexible adhesive was applied in a line on the top surface of the lower, thinner member. Before the upper member was placed on top of the lower mem-

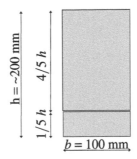

Figure 30. Cross section of the tested bipartite beams.

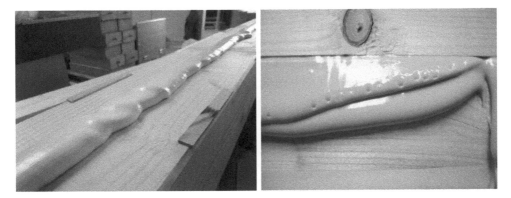

Figure 31. Manufacture of the bipartite beams. Left: the special adhesive layer was spread in a line on the bottom member. Veneer pieces were placed at the edges to ensure the right adhesive thickness. Right: excess adhesive oozed out as the upper member was pressed onto the bottom member.

ber, pieces of veneer were placed at the sides to ensure that the desired adhesive thickness was maintained during the pressing process.

The pressing pressure was supplied by the hydraulic cylinders of the test machine which was adjusted according to the viscosity of the adhesive. The bipartite beams were removed from the machine after two days because it was needed to manufacture more specimens. In order to ensure a full curing and hardening of the adhesive, the pressure was maintained by fixing the bottom member to the top member with screws which were removed after three weeks (Fig. 32). The vertical edges of the beams were sliced off to remove the veneer which was used to ensure the adhesive thickness. The bipartite beams were planed to a final width of 100 mm before the bending tests.

5.3. Results of Bending Tests

The bipartite beams bonded with the adhesive layers 009 and 013 did not perform as expected. The adhesive layers yielded at very low loads when the first beams were tested (Fig. 33). A visual inspection indicated curing problems: the adhesive layer had not fully hardened. Later tests in the laboratory of the adhesive manufacturer confirmed problems with scaling up: whereas the samples produced in the labora-

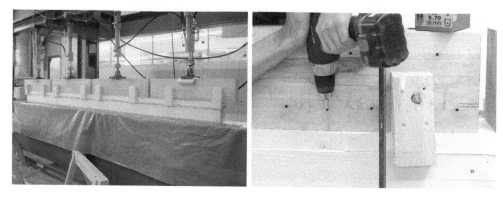

Figure 32. Manufacture of the bipartite beams. Left: pressing process with hydraulic cylinders for two days. Right: The top and bottom members were fixed together with screws before the presses were removed in order to ensure additional hardening under pressure for another three weeks.

Figure 33. Bending test: the longitudinal crack in the beam indicates premature failure of the adhesive layer 009-05 (left). The right picture shows the slip between top and bottom members of the bipartite beam and the adhesive layer is torn.

tory for small specimens had functioned very well, the mass production in a factory had failed to yield adhesives with the same properties. In the case of the adhesive 013, even after 4 months, the adhesive layer in the bipartite beams was still rather wet and had not hardened properly. With the small specimens for the tensile shear tests, they had completely hardened after only a few days and delivered the desired test results.

The elastic adhesive layer 027, on the other hand, had no hardening problems. The simple explanation is that this adhesive was already a commonly used product; hence the manufacturer had considerable experience in its handling. In fact, the small beech specimens for the shear lap tests had already been made with adhesives manufactured under normal factory conditions.

The bipartite beams manufactured with the adhesive 027 could be tested in bending as planned. The adhesive layer was never the cause of failure. The beams all failed on the tensile face, usually by wood fracture at knots (Fig. 34).

Figure 34. Bending tests: the picture on the left shows wood fracture in the bipartite beams bonded by the adhesive 027. The picture on the right shows the strain gauges.

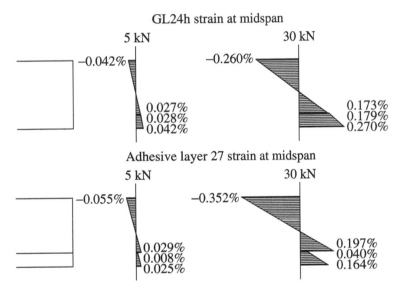

Figure 35. Strain measurements (in %) at beam midspan, taken at two loads of 5 kN and 30 kN. The two drawings at the top are for the reference beam (conventional GL24h): the strains are linearly distributed over the beam height. At the bottom, the results for a selected bipartite beam with adhesive layer 027 are shown: the interrupted linear strain distribution confirms the theoretical prediction of stress redistribution.

The strain measurements confirmed the theoretical model of stress redistribution in the bipartite beam as compared to conventional glulam beam (Fig. 35).

The failure loads of the reference beam GL24h on the one hand, and those of the bipartite beam bonded by adhesive 027 on the other, are listed in Table 5.

The calculation model "by integration" (Section 2.4) had predicted the existence of an optimal span for the bipartite beam with adhesive layer 027. The mean values listed in Table 5 confirm that the maximum bending resistance is attained at a span of exactly 4 m: increasing the span to 4.3 m or reducing it to 3.7 m leads to a reduced bending resistance.

Table 5.

Bending test results of 6 reference beams of 4 m span are listed in the top line as GL24h. The bipartite beam, listed as 027-2 because it was connected with this adhesive layer, was tested at three different spans: three specimens of 3.7 m span, 6 specimens of 4.0 m span and 3 specimens of 4.3 m span

	L (m)	Failure moment (kN m)						Mean (kN m)	St. Dev. (kN m)		Median (kN m)
GL24h	4.00	27.19	17.64	21.21	15.61	20.88	16.97	19.92	4.19	21%	19.26
027-2	3.70	22.02	19.55	24.03				21.86	2.24	10%	22.02
	4.00	20.85	22.47	26.12	23.39	24.12	22.73	23.28	1.77	8%	23.06
	4.30	15.66	20.45	21.19				19.10	3.00	16%	20.45

Table 5 also confirms an increase in bending resistance for the bipartite beam as compared to conventional glulam beam of the same span. The mean values of the failure moments can be compared to yield an increment factor of:

$$\eta = \frac{M_{027-2}}{M_{GL24h}} = \frac{23.28}{19.92} = 1.17. \tag{15}$$

The above value is exactly equal to the value predicted by the calculation model "by integration". The approximate Gamma method had predicted a value of 1.19.

Table 5 further makes a very interesting indication that the bipartite beam yields bending moments with relatively small standard deviations as compared to conventional glulam beam. In engineering practice, materials are designed with the 5% fractile strength values, not the mean values. The 5% fractile strength value is defined as follows: In a bending test with a very large number of specimens, 95% of them must achieve a strength value which is greater than the 5% fractile value. The 5% fractile values of the bending moment attained in the bending test ($M_{5\%}$) can be calculated from the mean value (M_{m}) and the standard deviation (s) with the following formula [14]. The factor γ takes into account the number of specimens tested (six):

$$M_{5\%GL24h} = M_m - \gamma \cdot s = 19.92 - 2 \cdot 4.19 = 11.54 \text{ (kN m)},$$
$$M_{5\%027-2} = M_m - \gamma \cdot s = 23.28 - 2 \cdot 1.77 = 19.74 \text{ (kN m)}. \tag{16}$$

Thus for design purposes, the actual strength of the bipartite beam could be increased to a value much higher than that of conventional glulam beam:

$$\eta = \frac{M_{5\%027-2}}{M_{5\%027-2}} = \frac{19.74}{11.54} = 1.70. \tag{17}$$

5.4. Discussion: the Volume Effect

The engineering community is well aware of the so-called volume effect. Essentially, this theory explains that the tensile strength of a brittle material is dependent on the relative volume of the material which is subjected to large stresses. Weibull's original Volume Theory states the following relationship for two specimens of volumes V_1 and V_2 submitted to constant stresses σ_1 and σ_2 at failure [16]:

$$\frac{\sigma_1}{\sigma_2} = \left(\frac{V_1}{V_2}\right)^{-1/m}. \tag{18}$$

The factor m in the equation above is a material constant.

The Swiss standard SIA 265 [15], for example, lists for a glulam beam the tensile strength at only 60% of the bending strength because in the case of axial stressing, the whole volume of the timber specimen is subjected to the large stress: every point of the beam suffers the same high stress, a weak spot in any position of the beam can induce a chain failure. In the case of the beam subjected to a bending moment, the top half of the beam is under compression stresses, and only the bottom part of the beam suffers the dangerous tensile stresses. Since the stresses are linearly distributed over the cross section (zero stress at the middle of the cross section, maximum stress value at the bottom edge of the cross section), only a small portion of the tensile face is subjected to peak stress values. Assuming that the defects are uniformly distributed over the entire volume of the beam, the chances of a coincidence of a defect with the peak stresses is much lower in a beam subjected to a bending moment than one subjected to an axial tension force: thus it is less likely to fail. Countless tests in various research centres have confirmed that the bending strength of a glulam beam is indeed much higher than the tensile strength.

Figure 36 illustrates how the Swiss standard SIA [15] sees the volume effect with regard to combined loading. The design failure stresses are highest when the beam is subjected to pure bending: they fall to lower design stress values when it is subjected to pure compression or tension. In the case of a compressive force combined with bending, the standard prescribes a parabolic curve to estimate the design stress. In the case of tension combined with bending, however, the standard

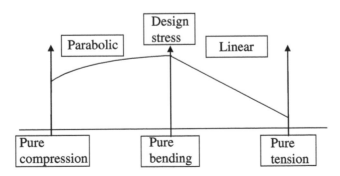

Figure 36. Design stress levels of glulam beams for combined loading according to Swiss standard [15].

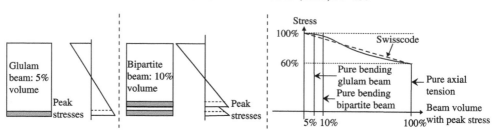

Figure 37. Discussion of the volume effect. In a conventional glulam beam, only about 5% of the beam volume is subjected to peak tensile stresses (see cross section and stress distribution on the left). In the bipartite beam (sketches in the middle) there are two zones with peak tensile stresses and the critical volume is doubled. The test results may indicate a curved relationship between failure stress and beam volume with peak tensile stresses.

prescribes a linear relationship. The latter does not correspond to Weibull's theory, which prescribes a concave curve.

The calculation models presented in this paper do not take the volume effect into account. According to Fig. 36 above, the bipartite beam should have exhibited lower failure stress levels than the glulam beam because of the greater volume under large tensile stresses. Nevertheless, the test results indicate clearly that the bending strength of glulam beam remains unchanged in the bipartite beam. This matter needs to be looked at more closely.

Figure 37 shows the bending stress distribution in a glulam beam and in a bipartite beam. The former has only one tensile peak zone, whilst the latter shows two peaks in the critical tensile zone. Thus the volume of the bipartite beam which suffers high tensile stresses is practically doubled as compared to the glulam beam. Assuming that the beam volume with 90–100% peak stresses to be critical, the timber volume subjected to peak stresses is increased from about 5% to 10%. Since the test results of this study indicate the same failure stresses for both cases, this may be an indication of a curved (and not linear) relationship between the failure stress level and the volume involved in the case of combined loading.

The findings of this paper thus appear to contradict the Swiss design standard SIA 265 for combined loading tension with bending, which is shown in Fig. 36. There may be an alternative model for the design stress levels of timber beams subjected to combined loading, bending with tensile force: the curve on the left side (combined bending with compression) may be carried over into the right hand side of the zone of combined bending with tensile force, before it is turned to join the Weibull concave curve in the zone of bending combined with tensile loading (Fig. 38). This model would satisfy both the Weibull theory and also explain the observation in this paper that the bending strength of conventional glulam beam and that of the bipartite beam composed of the same material grade are both the same, although the volume under peak tensile stresses is much greater in the latter case.

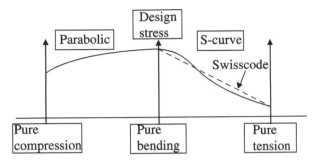

Figure 38. Alternative model for design stress levels of glulam beams for combined loading: in comparison with the Swiss standard shown in Fig. 36, there is no kink in the curve between the compression and tension zones.

6. Conclusion

- The large scale bending tests confirmed the accuracy of the theoretical models which predict an increased bending resistance for a bipartite beam with a flexible adhesive bond.

- The adhesive manufacturer successfully produced adhesives which fulfilled the load-bearing behaviour required for optimal stress redistribution in the bipartite beam.

- An unexpected but very positive discovery was the greatly reduced scatter in the failure loads as opposed to conventional glulam beam. For practical design purposes, the effective design strength of the optimized bipartite beam could be increased by more than 50%.

There is a need for further research to confirm these findings and consolidate the new and relatively simple method to enhance the bending resistance of glulam beams. Large scale creep tests need to be carried out for bipartite beams connected with suitable adhesives: one such creep test is currently being carried out for the adhesive type 027 described above. Potential applications such as arches or long span structures, as well as the possible economic advantages should be studied in more detail.

The main problem encountered in this study was the development of the new adhesives. Whilst the adhesive manufacturer was able to produce the load-bearing behaviour required by the elastic model, he was not quite able to fully attain the conditions required for the plastic model. The creep tests with small specimens as well as the delamination tests revealed many weaknesses. The final tests revealed that the new adhesive mixtures produced in small quantities in the laboratory could not yet be successfully reproduced in large quantities in factory conditions. Much work needs to be done before the new family of adhesives will attain the stability and reliability required for practical applications.

Acknowledgements

The project was generously financed by the CTI, the Swiss Commission for Technological Innovation.

The authors are grateful to Prof. Dr. Antonio Pizzi (ESTIB, France), for his technical advice on the adhesive development. The authors also wish to thank all the collaborators at different research institutions and companies who participated in the work or gave their advice.

References

1. K. Möhler, *Über das Tragverhalten von Biegeträgern und Druckstäben mit zusammengesetztem Querschnitt und nachgiebigen Verbindungsmitteln* (On the load-bearing behaviour of beams and columns with composite sections and flexible connections), Habilitation, University of Karlsruhe, Germany (1956).

2. J. Schänzlin, *Zum Langzeitverhalten von Brettstapel-Beton-Verbunddecken* (On the long-term behaviour of timber-concrete composite decks), Dissertation, Institut für Konstruktion und Entwurf, Universtät Stuttgart, Stuttgart, Germany (2003).

3. J. Schmidt, M. Kaliske and W. Schneider, *Bautechnik* **81**, 172 (2004).

4. M. Brunner, in: *Proceedings of the World Conference on Timber Engineering 2000*, Whistler, Canada (2000).

5. M. Brunner, M. Lehmann, S. Kraft, J. Conzett, K. Richter and U. Fankhauser, Ductile adhesive layer for glulam, research report, Bern University of Applied Sciences, Biel, Switzerland (2010).

6. M. Donzé, *Poutre composée en bois à connexion collée élasto-plastique* (Composite timber beam with an elasto–plastic adhesive connection), *MSc Thesis*, Université Henri Poincaré, Nancy, France (2007).

7. CEN/TC 250 "Structural Eurocodes", SN EN 1995-1-1:2004 Eurocode 5: Design of timber structures — Part 1-1: General — Common rules and rules for buildings, SIA Schweizerischer Ingenieur- und Architektenverein Zürich, Switzerland (2005).

8. G. Habenicht, *Kleben: Grundlagen, Technologien, Anwendungen* (Adhesive bonding: Basics, techniques, applications). Springer Verlag, Berlin, Germany (2008).

9. CEN/TC 193 "Adhesives", *DIN EN 204:2001 Classification of thermoplastic wood adhesives for non-structural applications*, Beuth Verlag GmbH, Berlin, Germany (2001).

10. CEN/TC 193 "Adhesives", *DIN EN 205:2003 Adhesives — wood adhesives for non-structural applications* — Determination of tensile shear strength of lap joints, Beuth Verlag GmbH, Berlin, Germany (2003).

11. CEN/TC 193 "Adhesives", *SN EN 302-2:2004 Adhesives for load-bearing timber structures — Test methods — Part 2: Determination of resistance to delamination*, SNV Schweizerische Normen-Vereinigung, Winterthur, Switzerland (2004).

12. Timber and Furniture Standards Committee, DIN 4074-1 Strength grading of wood — Part 1: Coniferous sawn timber, Beuth Verlag GmbH, Berlin, Germany (2008).

13. CEN/TC 124 "Timber Structures", SN EN 408:2003 Timber structures — construction timber and glued laminated timber — Determination of some physical and mechanical properties, SIA Schweizerischer Ingenieur- und Architektenverein Zürich, Switzerland (2003).

14. L. Sachs, *Angewandte Statistik: Anwendung statistischer Methoden* (Practical Statistics: Application of Statistical Methods). Springer Verlag Berlin, Heidelberg, Germany (1999).

15. Kommission SIA 164 *"Holzbau"*, *SIA 265:2003 Timber Structures*, SIA Schweizerischer Ingenieur- und Architektenverein, Zürich, Switzerland (2003).
16. T. Vallée, T. Tannert, M. Lehmann and M. Brunner, in: *Proceedings of the International Conference on Fracture 12*, Ottawa, Canada (2009).

Properties Enhancement of Oil Palm Plywood through Veneer Pretreatment with Low Molecular Weight Phenol-Formaldehyde Resin

Y. F. Loh [a,b], **M. T. Paridah** [a,*], **Y. B. Hoong** [c], **E. S. Bakar** [c], **H. Hamdan** [d] **and M. Anis** [e]

[a] Institute of Tropical Forestry and Forest Products, Universiti Putra Malaysia, 43400 UPM, Serdang, Selangor, Malaysia

[b] Fibre and Biocomposite Development Center, Malaysian Timber Industry Board, Malaysia, Lot 152, Jalan 4, Kompleks Perabot Olak Lempit, 42700 Banting, Selangor, Malaysia

[c] Faculty of Forestry, Universiti Putra Malaysia, 43400 UPM, Serdang, Selangor, Malaysia

[d] Forest Research Institute Malaysia, 52109 Kepong, Selangor, Malaysia

[e] Malaysian Palm Oil Board, No. 6 Persiaran Institusi, Bandar Baru Bangi, 4300 Kajang, Selangor, Malaysia

Abstract

One of the problems dealing with oil palm stem (OPS) plywood is the high veneer surface roughness that results in high resin consumption during the plywood manufacturing. In this study, evaluation was done on the effects of pretreatment of OPS veneers with phenol-formaldehyde resin on the bond integrity and bending strength of OPS plywood. OPS veneers were soaked in low molecular weight phenol-formaldehyde resin (LMW PF) for 20 seconds to obtain certain percentage of resin weight gain. OPS plywoods were produced using two types of lay-ups (100% outer veneer type and 100% inner veneer type) and two urea-formaldehyde (UF) adhesive spread amounts (200 g/m^2 and 250 g/m^2). The results show that pretreating the veneer with LMW PF could reduce the penetration of the adhesive into the fibres during gluing step. UF adhesive spread amount of 200 g/m^2 is sufficient to produce good quality OPS plywood. The technique used in this study was able to enhance the mechanical properties of OPS plywood as well as reduce the amount of resin consumption.

Keywords

Oil palm stem plywood, low molecular weight phenol-formaldehyde resin, mechanical properties, bond integrity

---

[*] To whom correspondence should be addressed. Tel.: +60389468422; Fax: +60389472180; e-mail: parida_introb@yahoo.com

Wood Adhesives

1. Introduction

Plywood industry has been one of the major contributors to the total timber export in Malaysia. In 2008, export of plywood constituted 28% of the total timber export, only 2% lower than that of wooden furniture. This amount is expected to increase with Malaysia replacing some of Indonesia's share in the plywood market in Japan [1]. With the increasing price of logs, plywood industry has to find alternative raw materials to remain competitive with other panel products such as medium density fibreboard (MDF) and oriented strand board (OSB). One of the most potential raw materials for this industry which is abundant and readily available in Malaysia is oil palm stem (OPS). Oil palm (*Elaeis guineensis*) trees are one type of palms that are grown for oil production. The trees after 20–30 years become old and the oil production declines, then they are considered uneconomic and have to be replanted. The replanting activity generates huge amount of oil palm stems.

From the annual report [2], Malaysia produced about 21.63 million cubic meters of oil palm biomass, including trunks, fronds, and empty fruit bunches. This figure is expected to increase substantially when the total planted hectarage of oil palm in Malaysia will reach 5.10 million hectares in 2020. The total oil palm planted area in Malaysia has expanded from merely 1.7 million hectares in 1990 to 3.37 million hectares in 2002 and to 4.3 million hectares in 2006 [2]. The annual availability of OPS is estimated to be around 13.6 million logs based on 100 000 hectares of replanting each year [3]. Under controlled processing conditions, this amount of OPS could be converted into approximately 4.5 million m^3 of plywood each year.

To date, the usage of OPS plywood has been limited to non-structural applications due to its lower strength and being less stable. This problem may be associated with density variations within the stem itself, as well as in the cell structure found in OPS fibers. Such differences can result in greater variations in density which may affect the performance of plywood. Even though several companies have initiated commercial production of OPS plywood, the problems of strength properties and dimensional stability during post-production and high resin consumption have slowed down a full commercialization of this product.

The OPS itself has some limitations which are responsible for its poor physical and strength properties. Hence, the only way to use OPS efficiently is by treating the stem with either preservatives or chemicals to enhance the performance. Phenolic resin has been used to treat OPS wood to increase its mechanical strength, whilst LMW PF resin was found to be effective in reducing the swelling of wood products. Many studies [4–7] suggested that the low molecular weight resins have smaller molecules and thus can easily be deposited extensively into the wood cell walls and once cured they bind to the cell walls and, consequently, reduce the swelling of the wood more effectively. A low molecular weight resin penetrates easily into the cell walls, resulting in improved strength and dimensional stability, whereas a resin of a higher molecular weight only fills the cell lumens and thus plays a very small role in strength properties.

Because of smaller molecules, a low molecular weight resin penetrates easily into the cell walls and the cell lumens to form an anchor to the cell and thus improves the total strength and dimensional stability of the whole material. A resin with higher molecular weight only partially enter the cell wall and imparts strength only on the surface.

The impregnation of wood by PF resin has been studied since the early half of the twentieth century. Commercial treatment methods using impregnation, also known as Impreg, and compression with impregnation (Compreg) can be used to improve both the mechanical strength and dimensional stability of wood [8]. Phenol-formaldehyde (PF) resin is relatively inexpensive and provides many advantages such as superior mechanical, and electrical properties, heat and flame resistance, as well as less smoke during burning compared to other resins. It is a suitable material to replace wood products that require long term durability in construction [9]. The LMW version of PF (LMWPF) resin will perform comparably to normal PF provided it is properly cured. Since this type of PF has relatively shorter chains and smaller molecules, it can penetrate easily into wood cells and is retained in the voids; once cured, the PF resin in these voids will provide strength to the fibres.

Various techniques like pressurizing, vacuum treatment, acoustic activation, etc. are used to facilitate the impregnation process. Even though these methods can effectively control the absorption rate and penetration depth, there are constraints in using these methods especially in terms of cost. The simplest, inexpensive, precise and rapid method to impregnate resin into the OPS veneers is by merely soaking the veneers in the resin solution. The main reason for introducing the resin into the veneer is to bind the parenchyma tissues so that they become harder and stronger upon curing the resin. Since an OPS veneer behaves like a sponge, it can absorb adhesive easily and rapidly. Once adequate mechanical interlocking and chemical adhesion is produced with the substrate, it will result in much enhanced strength and longevity in service.

The improvement in the quality of OPS plywood was based on an efficient arrangement of veneers through veneer density distribution (outer and inner) and pretreatment of veneers to minimize resin consumption and enhance the mechanical and physical properties of board. It is anticipated that the loosely bound parenchyma tissues will significantly absorb the resin, hence increasing its density, which, in turn, will reduce the density gradient amongst the cells. From this study, PF treated OPS plywood with its good mechanical properties can ensure a good performance in structural use and, in turn, will increase its market value.

2. Experimental

2.1. Experimental Design

A total of 24 boards of size 450 mm × 450 mm were produced in this study with two different adhesive spread amounts (200 g/m^2 and 250 g/m^2) for each lay-up pattern.

Two lay-up patterns were adopted in this study: 100% outer veneer and 100% inner veneer type. Out of 24 boards, half were controls or untreated and bonded at different adhesive spread amounts, while the other half underwent pretreatment of veneer with LMW PF.

2.2. Veneer Production

Two types of peeling machines were used to produce outer and inner veneers. The outer veneers are the first 50% veneer ribbons obtained from the rotary lathe, and subsequently the OPS was fed into a spindleless rotary lathe to peel the softer inner veneer type of OPS. For instance, during the first stage some remaining barks will be removed prior to peeling the peripheral layer (outer layer). The OPS normally has 39 cm diameter and upon first peeling stage, the diameter was reduced to 25 cm. In the second peeling stage, the peeling continued until the diameter of OPS was reduced to 11 cm. One crucial weakness of spindleless rotary lathe is that the peeling process would stop not only when the diameter of stem becomes smaller but also when it is too soft.

2.3. Veneer Pretreatment

Outer and inner OPS veneers were soaked in LMW PF for 20 seconds to obtain a certain level of solid resin uptake. Immediately, the soaked veneer was squeezed using a roller pressing machine to squeeze out the excess LMWPF resin in the veneer to avoid waste and to achieve desired level of solid resin uptake. The veneers were then stacked for further drying at 60°C for two hours to achieve target final moisture content (MC) of $12 \pm 2\%$. After two hours, the dried LMWPF resin treated veneers were weighed again and their MC determined. The MC values of these veneers were maintained between 7–10% to ensure that good adhesive joint will be achieved.

2.4. Plywood Manufacturing

The OPS veneers were then segregated by outer and inner layers. The adhesive mixture was composed of a commercially available urea-formaldehyde (UF) resin (42.5% solid) mixed with industrial wheat flour and hardener (ammonium chloride). The adhesive mixture was spread onto the veneers at different spread amounts: 200 g/m^2 and 250 g/m^2 (double glueline). The assembled veneers were then cold pressed for 10 min and hot pressed at 130°C for 15 minutes. Two types (Type 1: comprised of 100% outer layer veneers and Type 2: comprised of 100% inner layer veneers) of 5-ply, 450 mm × 450 mm × 13 mm plywoods were produced. Then, the plywoods produced were conditioned at RH of 65% and temperature of 22°C for a week prior to cutting into test specimens.

2.5. Properties Assessment

2.5.1. Static Bending Test
The static bending test was carried out according to European Norm EN 310:1993 to determine the properties of OPS plywood. The data were statistically analysed

for the effects of treatment, veneer type and adhesive spread amount on the strength (i.e., modulus of rupture) and stiffness (i.e., modulus of elasticity).

2.5.2. Bond Integrity Test

All the boards were tested under dry (normal) condition and wet (immersion) condition according to European Norm EN 314: 1993 Plywood-Bonding Quality (Part 1: Test Methods). As for immersion test, the specimens were soaked in water for 24 h at 20±°C.

The percentage of shear strength retention was determined using the expression below; whilst wood failure of OPS plywood in this study was determined through visual evaluation to the nearest 5%.

$$\text{Strength retention } (\%) = \frac{\text{Plywood shear strength in wet test}}{\text{Plywood shear strength in dry test}} \times 100.$$

3. Result and Discussion

By soaking the OPS veneers in LMWPF and upon curing, it is anticipated that the partially cured phenolic on the surface of veneer would be able to control penetration of adhesive into the veneers during the gluing process. Normally an untreated OPS veneer can rapidly absorb LMWPF resin through diffusion and bind the resin within the ground tissues and vascular bundles, resulting in an increase in weight. Table 1 shows the resin weight percent gain and moisture content of phenolic treated OPS veneers.

Compared to inner layer, the outer layer veneer type had higher weight gain in grammage 106.9 g *versus* 102.4 g. Since the LMWPF resin has 44.5% solids, the total amounts of solid resin absorbed were 47.6 g in the outer and 45.6 g in the inner veneers. Upon drying, some volatile materials and water evaporated, leaving only solid resin residues in OPS veneers. Resin weight percent gain (WPG) was calculated based on oven-dried method for woody materials. It was noticed that despite higher absorption amount in outer layer veneer, WPG was only 15.9% compared to inner layer veneer which had higher WPG of 19%. The final MC of PF treated veneers was roughly 7–8%.

According to Field and Holbrook [13], liquid movement in a tree trunk may involve pectin hydrogels, which bind cell walls together. When pectin material swells, the pits in the membrane cells will be compressed and, subsequently, will slow

Table 1.
Average resin uptake and moisture content (MC) of LMW PF treated OPS veneers

Veneer type	Initial MC (%)	MC after squeeze-out (%)	Weight gain (g)	Weight of solid resin (g)	Solid resin uptake (%)	MC at bonding (%)
Outer	6.9	22.9	106.9	47.6	15.9	7.1
Inner	8.0	26.4	102.4	45.6	19.0	8.5

down the flow of the water. On the contrary, when the pectin material shrinks, water flow increases and will easily pass through the xylem membranes. The outer veneer of oil palm trunk has relatively lesser pectin material, thus more PF resin can be absorbed due to less degree of obstacle to restrict the resin from diffusing into the woody cells. On the contrary, the veneers obtained from the inner part of oil palm trunk are rich in pectin and thus less resin can be absorbed. The swollen pectin eventually slows down the diffusion movement within the veneers, resulting in lesser weight gain in grammage.

Despite the high absorption of LMW PF by outer layer veneer, the resin WPG was lower compared to inner layer veneer. This showed that inner layer veneer had higher absorption property probably induced by the sponge-like parenchyma tissues which can bind the molecules with the surface of woody cell. Solid resin was bound to the woody cell surface probably *via* physical and chemical bonding. Outer layer veneer might have experienced poor absorption as seen in the lower solid resin uptake which may be attributed to the presence of a high number of silica bodies in the schlerenchymatic sheath of OPS around the vascular bundles and fiber bundles [14, 15]. These cells have the ability to absorb the resin solution but upon drying and compressing the cells retain less solid resin.

Table 2 shows the densities of outer and inner boards produced. The density of OPS veneer was divided into initial density and density after treatment. Untreated veneer showed density between 250–400 kg/m^3, whilst the density for treated veneer ranged between 350–600 kg/m^3. The treated outer veneers were found to give the highest density (586 kg/m^3). After the boards were manufactured, the average density of treated outer layer plywood was 850 kg/m^3, whilst the average density of treated inner layer plywood was 700 kg/m^3. It was observed that the treatment

Table 2.
Interaction effect of LMW PF treated OPS plywood

Condition	Adhesive spread amount (g/m^2)	Veneer layer type	Initial veneer density (kg/m^3)	Treated veneer density (kg/m^3)	Plywood density[1] (kg/m^3)
Untreated	200	Outer	395 (42)	–	749[b] (47)
		Inner	342 (52)	–	607[d] (57)
	250	Outer	339 (86)	–	684[c] (83)
		Inner	287 (46)	–	642[c] (88)
Treated	200	Outer	–	586 (91)	834[a] (42)
		Inner	–	385 (52)	668[c] (10)
	250	Outer	–	544 (64)	879[a] (26)
		Inner	–	400 (29)	751[b] (75)

Note: Means followed by the same letters [a, b, c] and [d] are not significantly different at $p \leqslant 0.05$.
Value in parentheses (·) represents standard deviation.
[1] Density determined after conditioning at 65% RH, 25°C.

of OPS veneers with LMWPF increased the density of the veneers by about 54% in outer veneer type and 25% in inner veneer type and of the resulting plywoods by 19% and 13%, respectively.

In an earlier study [16], the density gradient between the central core and the peripheral zone of OPS was relatively large (minimum 200 kg/m^3 and maximum 600 kg/m^3). After treating the veneers with LMW PF resin, the density was between 385 kg/m^3 and 586 kg/m^3. All the treated inner veneers have density quite close to that of outer veneers. Evidently, this treatment has successfully reduced the density gradient between outer and inner OPS veneers.

3.1. Strength of OPS Plywood

Table 3 summarizes the Analysis of Variance (ANOVA) results for the effects of treatment (untreated and treated), veneer layer and adhesive spread amount. The results show that there were highly significant interactions between these three variables in terms of Modulus of Rupture (MOR) and Modulus of Elasticity (MOE) of OPS plywood. However, it was also observed that the effect of adhesive spread amount was only moderately significant on the bond integrity of OPS plywood in terms both dry and wet shear strengths.

From Table 4, the highest increment in MOR (115%) was obtained for plywood made from outer veneers whilst the lowest increment was for those made from inner veneers (60%) and similarly with MOE increments of 70% and 43%, respectively. This result evidently shows that the parenchyma tissues have some absorption property, particularly for phenolic resin. This property is extremely important in chemical modification of veneers since it influences the final strength and stiffness of the resulting products. The results showed that outer layer veneer produced higher values of MOR and MOE due to wide distribution of mature vascular

Table 3.
Summary of ANOVA results for the effects of treatment, veneer type and adhesive spread amount on mechanical properties of OPS plywood

Source	Significant level			
	MOR	MOE	Shear strength in dry test	Shear strength in wet test
Treatment (T)	***	***	***	***
Adhesive spread amount (ASA)	***	***	*	**
Veneer Type (VT)	***	***	***	***
T × ASA	NS	NS	NS	**
T × VT	***	***	***	***
G × VT	NS	***	NS	*
T × ASA × VT	NS	NS	*	NS

Note: ***$p \leqslant 0.01$, **$p \leqslant 0.05$, *$p \leqslant 0.1$ and NS = not significant.

Table 4.

Effect of treatment and veneer type on MOR and MOE of OPS plywood

Treatment	Veneer layer	MOR (MPa)	MOE (MPa)
Untreated (control)	100% outer	31.0[b]	4298[b]
	100% inner	23.3[d]	2643[d]
LMW PF treated	100% outer	66.6[a]	7325[a]
	100% inner	37.3[c]	3789[c]

Note: Means followed by the same letter (a, b, c and d) in the same column are not significantly different at $p \leqslant 0.05$.

Table 5.

Effect of treatment and veneer layer on bond integrity of OPS plywood

Treatment	Veneer layer	Bond integrity in dry test		Bond integrity in wet test	
		Shear strength (MPa)	Wood failure (%)	Shear strength (MPa)	Wood failure (%)
Untreated	100% outer	1.04[c]	100	0.55[c]	100
(control)	100% inner	0.82[d]	90	0.42[d]	80
LMW PF	100% outer	2.43[a]	100	1.63[a]	100
Treated	100% inner	1.55[b]	90	0.94[b]	100

Note: Means followed by the same letter (a, b, c and d) in the same column are not significantly different at $p \leqslant 0.05$.

bundles. In addition, mature vascular bundles also are related to the density and cause the outer layers of OPS veneers to become denser and thus stronger.

From Table 5, it is observed that PF treated OPS plywood gave better bond performance compared to untreated OPS plywood. PF treated OPS plywood made from outer layer veneer consistently gave superior strength in both dry and wet tests with values 2.433 MPa and 1.626 MPa, respectively. The increase in board density is identified as the main contributor to this large improvement. In the bond integrity test, PF treated OPS plywood gave considerably superior bond strength performance compared to untreated OPS plywood both in dry and wet tests.

Untreated OPS plywood showed decrease in bond strength due to the reaction of UF adhesive with water. Water hydrolyzes the aminomethylenic bond of the UF adhesive, and the action of water on UF-cellulose bonds is first to displace the UF adhesive from adsorption sites on the cellulose. Consequently, leaching of UF in the water results in poor shear strength in the wet test [17].

In terms of wood failure, generally both treated and untreated OPS plywoods with outer layer veneer consistently achieved 100% wood failure in both dry and wet tests, indicating a very good bond property. OPS plywood made from inner layer veneer, both untreated and treated, generally gave acceptable wood failure

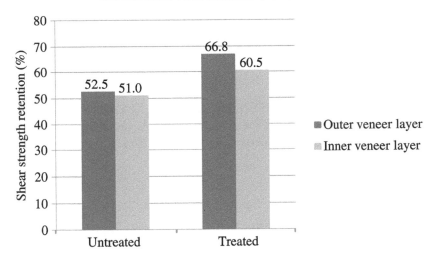

Figure 1. Shear strength retention of LMW PF treated OPS plywood.

percentage (>80%) in both dry and wet tests. Phenolic treatment of OPS veneers apparently improved the bonding of OPS plywood, with outer veneer type giving the highest improvements of 134% in dry test and 197% in wet test. The inner veneers, on the other hand, experienced only a marginal effect when tested in wet condition. This improvement can be attributed to further curing of LMWPF resin during the test, thus imparting some waterproof properties to the plywood test specimens.

According to Fig. 1, untreated inner veneer type OPS plywood generally retained lowest shear strength compared to treated OPS plywood. Highest strength retention (66.8%) was obtained in phenolic treated OPS plywood made from 100% outer layer veneer, whilst inner layer type veneer came second with 60.5% shear strength retained.

4. Conclusion

Pretreatment of OPS veneers with LMWPF resin significantly improved the density, strength, stiffness and bond integrity of oil palm plywood. The effects were more significant on the outer layer veneers of OPS and the treatment was able to improve the MOR by 115%, MOE by 70%, shear strength (dry test) by 134% and shear strength (wet test) by 197%. Whilst the improvement in the inner veneer were 60%, 43%, 90% and 125% in the MOR, MOE, shear strength (dry test) and shear strength (wet test), respectively. Adhesive spread amount did not have any significant effect on the strength and bond integrity of OPS plywood except on MOE. The effect, however, was only significant for outer veneers where an adhesive amount of 200 g/m^2 was sufficient to produce good performance plywood. Higher amount of adhesive would reduce the stiffness of the plywood. Pretreatment of the OPS veneers with LMWPF was able to improve the bond integrity of the plywood, meeting the requirements of European Norm EN-314.

Acknowledgements

The authors would like to thank the Malaysian Timber Industry Board (MTIB) for funding the project, Malayan Adhesive and Chemicals (M) Sdn. Bhd. for providing the PF and UF resins and Business Esprit Sdn. Bhd. for the use of plywood processing facilities.

References

1. Statistics on Exportation of Major Timber Products from Malaysia. Report of the Malaysian Timber Industries Board (2008).
2. Anon, *The Annual Journal of the Malaysian Palm Oil Board*. Ministry of Plantation Industries and Commodities, Malaysia, pp. 31 (2006).
3. M. Anis, in: *Proceeding of The Eighth Pacific Rim Bio-Based Composites Symposium*, Kuala Lumpur, Malaysia, pp. 121–127 (2006).
4. T. Furuno, Y. Imamura and H. Kajita, *Wood Sci. Technol.* **37**, 349–361 (2004).
5. Y. Imamura, M. K. Yalinkilic, H. Kajita and T. Furuno, in: *Proceedings of The Fourth Pacific Rim Bio-Based Composites Symposium*, Bogor, Indonesia, pp. 119–127 (1998).
6. H. Kajita and Y. Imamura, *Wood Sci. Technol.* **26**, 63–70 (1991).
7. M. T. Paridah, L. L. Ong, A. Zaidon, S. Rahim and U. M. K. Anwar, *J. Tropical. Forest Sci.* **18** (3), 166–172 (2006).
8. A. J. Stamm and R. M. Seborg, Report No.1360, USDA Forest Service, Forest Products Laboratory, Madison, WI (1962).
9. B. Goichi and S. Akiko, *Report of the Research Institute of Industrial Technology*. Nihon University, Tokyo, Japan, Number 69 (2003).
10. European Norm EN 310. Wood based panel (1993).
11. European Norm EN 314–1. Plywood-bond quality part 1 (1993).
12. European Norm EN 314–2. Plywood-bond quality part 2 (1993).
13. T. S. Field and N. M. Holbrook, *Cell and Environment* **23**, 1067–1077 (2000).
14. W. Killman and S. C. Lim, in: *Proceedings of the National Symposium on Oil Palm By-products for Agro-Based Industries*, Kuala Lumpur: PORIM Bulletin No. 11, pp. 18–42 (1985).
15. K.-N. Law, W. R. W. Daud and A. Ghazali, *Bio-Resources* **2**, 351–362 (2007).
16. S. C. Lim and K. C. Khoo, *The Malaysian Forester* **49** (1), 3–22 (1986).
17. A. Pizzi, in: *Advanced Wood Adhesives Technology*, pp. 19–66. Marcel Dekker, New York (1994).

Reaction Mechanism of Hydroxymethylated Resorcinol Adhesion Promoter in Polyurethane Adhesives for Wood Bonding

A. Szczurek [a], A. Pizzi [a,*], L. Delmotte [b] and A. Celzard [a]

[a] ENSTIB, Nancy University, 27 rue du Merle Blanc, B.P. 1041, 88051 Epinal, France
[b] IS2M, University of Haute Alsace, 15 rue Jean Starcky, B.P. 2488, 68057 Mulhouse, France

Abstract

Hydroxymethylated resorcinol (HMR) coupling agent is a low condensation resorcinol-formaldehyde resin mix of low molecular weight oligomers. The mechanism that appears to link the hydroxymethylated resorcinol coupling agent to polyurethane adhesives for the reported enhancing of their wood bonding performance was shown to depend on (i) the initial formation of urethane linkages between the methylol groups of HMR and the still active isocyanate groups of the polyurethane adhesive, followed by (ii) slower formation of urethane linkages between the resorcinol phenolic hydroxyl groups and the isocyanate groups of polyurethane adhesive. This mechanism is in addition to the adhesion promoting effect that HMR has on the wood substrate as already reported by other authors.

Keywords

Hydroxymethylated resorcinol, coupling agents, polyurethanes, adhesives, adhesion, resins

1. Introduction

The use of hydroxymethylated resorcinol (HMR) as an adhesion promoter to enhance solid wood bonding is a recent development that holds considerable promise in improving the durability of the bonds between solid wood and adhesives. HMR consists of a mix of methylolated resorcinol monomers and low molecular weight oligomers [1]. Its usefulness as an adhesion promoter has been shown for a number of different wood species and different adhesives: formaldehyde-based resins, epoxies and polyurethanes [2–7].

The precise mechanism on which HMR effectiveness depends is not yet clear, although some notable contributions have been made on this subject. Thus, cell penetration appears to be of some importance in stabilizing the wood substrate

---

* To whom correspondence should be addressed. Tel.: (+33) 329296117; Fax: (+33) 329296138; e-mail: antonio.pizzi@enstib.uhp-nancy.fr

Wood Adhesives
© Koninklijke Brill NV, Leiden, 2010

against its swelling in water, hence decreasing the tendency for interfacial failure [8], as well as stiffening of wood against stress relaxation by cross-linking some of the structural wood constituents [9]. As HMR is a low molecular weight resorcinol-formaldehyde (RF) resin, its capacity for adhesion by secondary forces to the wood substrate is well established, as all RF resins have high adhesion capacity to wood [10].

If a coupling agent has to work its interactions with both the substrate and the adhesive have to be considered. However, the mechanism of interaction between HMR and the adhesive itself has not yet been considered. While its interfacial coreaction with other formaldehyde-based resins can be postulated as its action mechanism, this is not the case for resins such as epoxies and polyurethanes. While epoxies are only little used for wood gluing, due to their high cost, polyurethanes are now extensively used in structural wood bonding [11]. Particularly extensive is the use of one-component polyurethane adhesive systems for exterior grade wood structural laminated beams [12–17]. They have been formally approved and widely promoted for use in structural glulam in several European countries. The main characteristic of these reactive one-component polyurethanes is that they still present reactive isocyanate groups to yield cross-linking during their application.

This paper aims at explaining the reaction between the HMR coupling agent and one-component polyurethane adhesives used for wood lamination.

2. Experimental

The HMR coupling agent was prepared similarly to that reported in the literature [18–20]. An aqueous alkaline mixture of resorcinol (R) and formaldehyde (F) (F/R molar ratio = 1.5) was reacted at room temperature for 4–5 h after adding K_2CO_3 as a catalyst. At the end of the reaction time, 30 min after its preparation, HMR was reacted with a –NCO carrying resin pMDI (polymeric 4,4′ diphenylmethane diisocyanate). The R/pMDI weight ratio was 0.5. HMR solution and pMDI were stirred together and a yellow paste was obtained. A very thin layer of this paste was spread on a glass plate and left at 25°C for 24 h to dry and harden. It was then scraped from the glass and ground for NMR analysis.

A second reaction was also carried out as a control. Resorcinol (5 g) was diluted in water (25 ml), and K_2CO_3 was added as a catalyst (R/Catalyst ratio = 100) to obtain a homogeneous solution. pMDI was then added, the R/pMDI weight ratio was 0.5. The mixture was stirred at ambient temperature and a yellow paste was obtained. A very thin layer of this paste was spread on a glass plate and left at 25°C for 24 h to dry and harden. It was then scraped from the glass and ground for NMR analysis.

The two reaction products were analysed by solid-state CP-MAS ^{13}C-NMR. Spectra were obtained on a Bruker MSL 300 FT-NMR spectrometer at a frequency of 75.47 MHz and a sample spin of 4.0 kHz. The 90° pulse duration was 4.2 s,

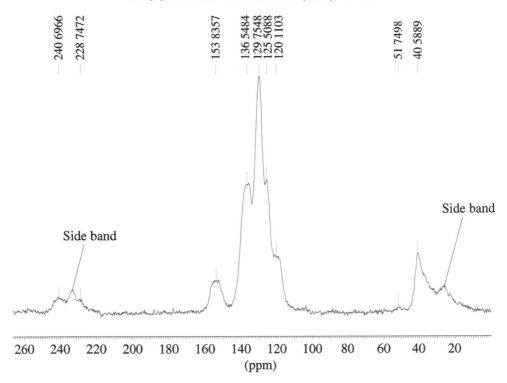

Figure 1. Solid-state CP-MAS ^{13}C-NMR spectrum of hardened reaction product between HMR and pMDI. The spinning side bands are indicated.

contact time was 1 ms, number of transients was about 1000, and the decoupling field was 59.5 kHz. The spinning side bands, which are resonances of major peaks and do not correspond to any additional or different groups, are indicated in Figs 1 and 2.

3. Results and Discussion

The CP-MAS ^{13}C-NMR spectrum in Fig. 1 shows the reactions which appear to occur between the still active isocyanate (–N=C=O) groups of the polyurethane and the groups of the low condensation HMR. Unreacted –N=C=O groups are noticeable in both CP-MAS ^{13}C-NMR spectra between 123 and 126 ppm [21–23]. Both in the HMR/isocyanate reaction in Fig. 1 and in the resorcinol/isocyanate reaction in Fig. 2 it appears that residual unreacted –N=C=O groups are present. In Fig. 1 this is shown by the shoulder at 125.5 ppm while in Fig. 2 it is shown by the peak at 123.5 ppm. Urethane bridges between the two materials have formed in both cases. Thus, in Fig. 1 the urethane group is present at 153.8 ppm (theoretically calculated 153.3 ppm [24]) and in Fig. 2 at 151.4 ppm (theoretically calculated 151.6 ppm [24]). These are different types of urethane linkages formed between different groups. Thus, in the HMR/isocyanate case the urethane bridge is

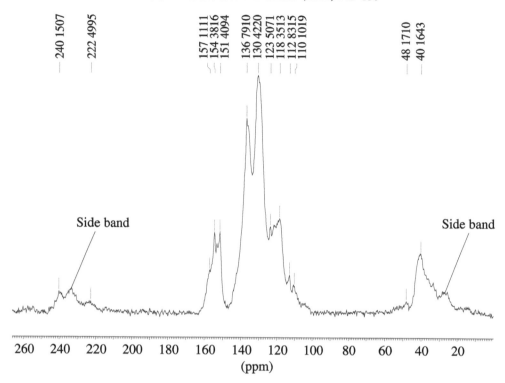

240 1507
222 4995
157 1111
154 3816
151 4094
136 7910
130 4220
123 5071
118 3513
112 8315
110 1019
48 1710
40 1643

Side band   Side band

260   240   220   200   180   160   140   120   100   80   60   40   20
(ppm)

Figure 2. Solid-state CP-MAS ^{13}C-NMR spectrum of hardened reaction product between resorcinol and pMDI. The spinning side bands are indicated.

formed between the HMR methylol groups (–CH$_2$OH) and the isocyanate groups (–N=C=O) to form linkages of the type

That this is the case can be deduced from the rates of reactions that have been reported [23] in water for methylol groups (–CH$_2$OH) of phenolic resins with –N=C=O groups.

In the resorcinol/isocyanate case the urethane bridge formed is the one between the resorcinol hydroxyl groups and and the isocyanate groups (–N=C=O) to form linkages of the type

This is the case because, due to the absence of reactive methylol groups, the only possible reaction for –N=C=O to form a urethane bridge is with some of the resorcinol hydroxyl groups. Such a reaction is rather slow and can occur only at low levels of water or in its absence. At ambient temperature this means that it will occur when the mixture applied on the glass plate is almost dry. The same will occur on a wood substrate where the level of water is low because absorption of water from the adhesion promoter to the substrate is fast.

The other peaks noticed in the spectra are at 40 ppm due to the –CH$_2$– group linking the two atomatic rings of pMDI and the consequent urethane formed; the peaks in Fig. 2 between 110 to 118 ppm are those corresponding to unreacted ortho and para sites in the HMR resin. The peaks between 151–158 ppm are those of aromatic carbons attached to a phenolic hydroxyl group, while between 220 to 240 ppm some bands corresponding to aromatic quinone groups derived by the air oxidation of a few of the resorcinol hydroxyl groups are noticeable. The spinning side bands, that are resonances of major peaks and that do not correspond to any additional or different groups, are also indicated in Figs 1 and 2.

4. Conclusion

To conclude, the mechanism that appears to link the HMR coupling agent to polyurethane adhesives to enhance their wood bonding performance is due to two main reactions. The first is the formation of urethane linkages between the methylol groups of HMR and the still active isocyanate groups of polyurethane adhesive. As the HMR solution dries by losing water to the substrate a second reaction follows. This is the formation of urethane linkages between the resorcinol phenolic hydroxyl groups and the isocyanate groups of polyurethane adhesive. This mechanism is in addition to the adhesion effect of HMR on the wood substrate as reported by other authors [8, 9].

Acknowledgements

The authors gratefully acknowledge the financial support of the CPER 2007–2013 'Structuration du Pôle de Compétitivité Fibres Grand'Est' (Competitiveness Fibre Cluster), through local (Conseil Général des Vosges), regional (Région Lorraine), national (DRRT and FNADT) and European (FEDER) funds.

References

1. C. B. Vick, A. W. Christiansen and E. A. Okkonen, *Wood Fiber Sci.* **30**, 312–322 (1998).
2. C. B. Vick, *Forest Prod. J.* **45** (3), 78–84 (1995).
3. C. B. Vick and E. A. Okkonen, *Forest Prod. J.* **47** (3), 71–77 (1997).
4. C. B. Vick, *Forest Prod. J.* **47** (7/8), 83–87 (1997).
5. C. B. Vick, *Adhesives Age* **40** (8), 24–29 (1997).
6. C. B. Vick and E. A. Okkonen, *Forest Prod. J.* **50** (10), 69–75 (2000).
7. E. A. Okkonen and C. B. Vick, *Forest Prod. J.* **48** (11/12), 81–85 (1998).
8. J. Son and D. J. Gardner, *Wood Fiber Sci.* **36**, 98–106 (2004).
9. N. Sun and C. E. Frazier, *Wood Fiber Sci.* **37**, 673–681 (2005).
10. A. Pizzi, in: *Advanced Wood Adhesives Technology*, A. Pizzi (Ed.), chapter 7, pp. 243–272. Marcel Dekker, New York, NY (1994).
11. B. Radovic and C. R. Rothkopf, *Bauen mit Holz* **6**, 36–38 (2003).
12. F. A. Cameron and E. Scheepers, *Holz Roh Werkst.* **43**, 286 (1985).
13. C. B. Vick and E. A. Okkonen, *Forest Prod. J.* **48** (11/12), 71–76 (1998).
14. B. Radovic and C. R. Rothkopf, *Panels & Furniture Asia*, 34–36 (November 2003).
15. B. Na, A. Pizzi, L. Delmotte and X. Lu, *J. Appl. Polym. Sci.* **96**, 1231–1243 (2005).
16. K. Richter, A. Pizzi and A. Despres, *J. Appl. Polym. Sci.* **102**, 5698–5707 (2006).
17. F. Beaud, P. Nimz and A. Pizzi, *J. Appl. Polym. Sci.* **100**, 4181–4192 (2006).
18. C. B. Vick, K. Richter, B. H. River and A. R. Fried, *Wood Fiber Sci.* **27** (1), 2–12 (1995).
19. A. W. Christiansen, C. B. Vick and E. A. Okkonen, in: *Proceedings Wood Adhesives 2000*, South Lake Tahoe, NV, pp. 245–250 (2000).
20. A. W. Christiansen, C. B. Vick and E. A. Okkonen, *Forest Prod. J.* **53** (2), 32–38 (2003).
21. A. Despres, A. Pizzi and L. Delmotte, *J. Appl. Polym. Sci.* **99**, 589–596 (2006).
22. S. Wieland, A. Pizzi, S. Hill, W. Grigsby and F. Pichelin, *J. Appl. Polym. Sci.* **100**, 1624–1632 (2006).
23. A. Pizzi and T. Walton, *Holzforschung* **46**, 541–547 (1992).
24. ChemWindow 6.5, Biorad Laboratories, Sadtler Division, Philadelphia, PA (1998).

Part 2

Synthetic Adhesives

Optimization of the Synthesis of Urea-Formaldehyde Resins using Response Surface Methodology

João M. Ferra [a], Pedro C. Mena [b], Jorge Martins [a,c], Adélio M. Mendes [a],
Mário Rui N. Costa [d], Fernão D. Magalhães [a] and Luisa H. Carvalho [a,c,*]

[a] LEPAE — Departamento de Engenharia Química, Faculdade de Engenharia, Universidade do
Porto, Rua Dr. Roberto Frias, 4200-465 Porto, Portugal
[b] SONAE Indústria — SIR, Lugar do Espido, Via Norte, 4470-909 Maia, Portugal
[c] DEMad — Departamento de Engenharia de Madeiras, Instituto Politécnico de Viseu,
Campus Politécnico de Repeses, 3504-510 Viseu, Portugal
[d] LSRE — Departamento de Engenharia Química, Faculdade de Engenharia, Universidade do Porto,
Rua Dr. Roberto Frias, 4200-465 Porto, Portugal

Abstract

In the near future, companies will face the need to produce low formaldehyde emission resins, i.e., not above the emission level of natural wood. However, for producing this new generation of urea-formaldehyde resins (UF), it is necessary to optimize the synthesis process.

This work describes an optimization procedure for UF resin synthesis, following an alkaline–acid process, focusing on the conditions of the condensation step. A design of experiments methodology was employed to optimize the 3 selected factors (number of urea additions, time span between urea additions, and condensation pH), in order to produce particleboards with maximum internal bond strength and minimum formaldehyde release.

The condensation pH played a significant role in increasing the Internal Bond (IB) strength and reducing the Formaldehyde Emission (FE). The sequential addition of urea also has a noticeable influence on resin performance. Optimum conditions for production of UF resins have been proposed and tested by the response surface methodology using the desirability function.

Keywords

Urea-formaldehyde resins, internal bond strength, formaldehyde emission, response surface methodology, optimization

1. Introduction

Nowadays, the main goal of urea-formaldehyde (UF) resin industry is to comply with the standards requiring even lower formaldehyde release after curing. This

---

* To whom correspondence should be addressed. Tel.: +351 232 480565; e-mail:
lhcarvalho@demad.estv.ipv.pt

Wood Adhesives
© Koninklijke Brill NV, Leiden, 2010

Table 1.
Current classification of formaldehyde by some organizations

Organization	Current classification
European Chemicals Bureau	Category 3-R40 'limited evidence of a carcinogenic effect'
US Environmental Protection Agency	Probable human carcinogen
International Agency for Research on Cancer	Group 1 — 'there is sufficient evidence in humans & animals for the carcinogeneticity of formaldehyde'

chemical has now been reclassified as carcinogen by different organizations [1] (see Table 1). In California, the State Health and Safety Code mandates the California Air Resources Board (CARB) to develop Airborne Toxic Control Measures (ATCM) to protect public health from airborne carcinogens such as formaldehyde. The ATCM proposed the reduction of formaldehyde emission from particleboard specifically, in a first phase, until September 2009, from 0.30 ppm (specified in ASTM E 1333-96) to 0.18 ppm, and in a second phase, until 2011–2012, to 0.09 ppm [2].

Free formaldehyde is present in UF resins and hydrolytic degradation of UF resins (reversibility of aminomethylene links) under moist and acidic conditions is known to be responsible for formaldehyde emission from wood-based panels. So, formaldehyde emission not only depends on synthesis conditions, but also on the type of bonds in the cured resin [3].

Several strategies have been explored for reducing formaldehyde release. These include the addition of formaldehyde scavengers directly to the resin or wood particles, the treatment of final wood panels with scavengers or impermeable coatings, and the development of improved resin formulations [4].

The main change in the synthesis processes has been the decrease of F/U molar ratio. However, this worsens the mechanical properties of the particleboard formed and, moreover, increases the hardening time using the traditional hardeners.

Different synthesis procedures for UF resins have been found in the literature [5–11] which can be divided into two distinct processes:

(i) The alkaline–acid process [3, 5, 6, 8] is the most common. It involves three steps, usually an initial alkaline methylolation followed by an acid condensation and finally neutralization and addition of the last urea fraction.

(ii) The strongly-acid process was first described in the patents by Williams [4, 11]. It consists of: (1) carrying out the condensation of urea and formaldehyde under a highly acid environment and large excess of formaldehyde; (2) continuing the reaction in alkaline medium after additional urea is added to attain a predetermined F/U molar ratio; (3) carrying out the reaction under a low pH of about 5 to allow further condensation until a desired viscosity is reached; and (4) neutralization of the product and addition of the final urea amount.

Rammon [12] studied the effect of several variables on the synthesis of UF resins by the alkaline–acid process. The results obtained indicated that increasing the pH of the alkaline phase or using a longer acid condensation step leads to a decrease in the amount of ether linkages formed. These linkages can undergo rearrangement and lead to formaldehyde release.

The Williams process [4, 11] entails minimal energy consumption and involves relatively short (3–4 h) reaction times. The reduced formaldehyde emission and increased hydrolytic stability have been attributed to the predominance of the more stable methylene linkages in the cured resin, unlike the alkaline–acid process, leading to a large amount of methylene ether linkages in the cured resin. The main problem with this process lies in the control of the highly acidic condensation step, due to its exothermic character. According to Hatjiissaak and Papadopoulou [7], this process needs careful control, which is difficult to achieve at an industrial scale, to prevent resin gelling in the reactor.

Williams [11] demonstrated that the modification of the resins with the addition of cross-linking agents (e.g., trimethoxymethylmelamine) improves the values of internal bond (IB) strength by 28% for the uncatalyzed resin and 47% for the catalyzed formulation.

Pizzi and coworkers [13–15] reported that multiple additions of urea during resin preparation increases the bond quality, especially for resins with low F/U molar ratios. This helps the build-up of the fraction of polymeric material in the final resin, at the expense of low oligomeric species, thus increasing the condensation degree of the UF resin [14].

Kim and coworkers [16–20] reported on the migration of methylol groups to the last urea added in alkaline medium, leading to the reaction with the free amide groups which were formed in the final stage, producing monomeric methylolureas. This decreases the final viscosity and increases the formaldehyde emission from particleboards. Kim and coworkers [16–20] reported also that this migration continued over a period of 1 month or longer at room temperature and became relatively fast above 50°C.

Recently, Kumar *et al.* [21] have optimized the second step of the strongly-acid process (alkaline methylolation) for producing UF resins. The effect of the 'number of additions' and 'time span between additions' of the second urea on IB strength and formaldehyde emission (FE) was analyzed. The results showed that the sequential addition of urea in the methylolation step plays a significant role in reducing the FE and increasing the IB strength. It was also reported that these observations were due to the conversion of urea and the migration of the hydroxymethyl groups.

Pizzi and coworkers [14, 15] and Kumar *et al.* [21] state that the best strategy for obtaining low formaldehyde emission UF resins, with suitable bond strength to wood, is to produce resins with excess monomeric non-methylolated and methylolated urea species (which lowers formaldehyde emission and, up to a certain extent, improves resin adhesion to the wood substrate). On the other hand, the presence of a certain fraction of non-methylolated and particularly methylolated polymeric species ensures that cross-link density and cohesion, and hence an acceptable mechanical strength, will be obtained.

The present work describes the optimization of an alkaline–acid process, mainly focusing on the condensation step, considering 3 factors: the number of urea additions, the time span between additions, and the pH. A response surface methodology (RSM) with a central composite design (CCD) was used in order to evaluate the effect of independent variables on the process performance and to optimize the operating conditions (maximize the IB strength and minimize the FE). Second-order models were fitted using the CCD results, which describe the effect of the operating conditions on the process responses. These models were used for interpolating predicted values for the experimental conditions and, therefore, for comparing them with the experimental ones. On the other hand, the models were used to evaluate the effect of the critical operating conditions on the responses and to obtain the operating conditions that maximized the objective function.

2. Materials and Methods

2.1. Experimental Design

Response surface methodology (RSM) is a combination of mathematical and statistical tools which is effective for studying and modelling processes where responses are dependent on several operating variables [22]. The model parameters are estimated using the least squares method. In this work a CCD was selected, which is the most used method for fitting second-order models.

Pizzi [15] reports that the most important synthesis parameters with an influence on the properties of UF resins are: temperature, intermediate F/U molar ratios, and pH of the condensation step.

In the present work, focussing on an alkaline–acid synthesis process, three factors were analyzed, while maintaining the global F/U molar ratio constant: (a) total number of urea additions during the condensation step, (b) time span between consecutive additions of urea during the condensation step, and (c) pH of the condensation step. The pH of the methylolation step was also initially considered, but preliminary experiments, performed with a pH range between 8.6 and 9.4, indicated only small modifications in the final product characteristics.

The levels of the three selected factors are given in Table 2. The pH values selected were based on the range used in industrial production.

The properties selected for evaluating the resins produced were: internal bond strength, formaldehyde content measured by the perforator method, and the fraction of insoluble aggregates (FIA) measured by Gel Permeation Chromatography (GPC)/Size Exclusion Chromatography (SEC). However, for all the resins pro-

Table 2.
Experimental levels of the three factors

Factors	Levels		
	−1	0	1
No. of U additions in the condensation step (A)	2	3	4
Time span between additions of U in the condensation step (B)	10	15	20
pH of the condensation step (C)	5.6	5.9	6.2

Table 3.
Central composite design matrix of experiments generated by the DOE tool

Run	Experimental values		
	A — No. add. U	B — Time span (min)	C — pH
1	4	10	5.6
2	2	10	6.2
3	3	15	5.9
4	4	20	5.6
5	3	10	5.9
6	4	15	5.9
7	3	15	6.2
8	3	15	5.9
9	2	20	6.2
10	2	15	5.9
11	4	20	6.2
12	3	15	5.6
13	2	10	5.6
14	4	10	6.2
15	3	20	5.9
16	2	20	5.6

duced, several other properties were measured, namely, gel time, final pH, viscosity, solids content, and stability at 25°C.

For generating the matrix of experiments a CCD was employed [22] with 2 central points, which resulted in 16 experiments (Table 3). Even though the number of urea (U) additions is a discrete variable, it was assumed as a continuous variable so that the RSM approach could be used directly. JMP 5 software (SAS, Cary, NC, USA) was used for generating the matrix of experiments and for analyzing the results.

2.1.1. Statistical Analysis

The analysis of variance (ANOVA) was used to determine the adequacy of the model to describe the observed data. The R^2 statistic indicates the degree of variability of the optimization parameters that are explained by the model.

Three-dimensional response surface plots were generated for each quality parameter. Calculation of optimal synthesis conditions for optimum IB strength and FE was performed using a multiple response method designated as desirability [22–24]. This optimization method incorporates desired values and priorities for each variable.

The natural variables (A, B and C) are associated with coded variables (X_1, X_2 and X_3) which are dimensionless, according to [22]:

$$X_1 = \frac{A - 3}{1}; \qquad X_2 = \frac{B - 15}{5}; \qquad X_3 = \frac{C - 5.9}{0.3}. \qquad (1)$$

The quadratic equation for the variables is defined as follows:

$$Y = \beta_0 + \Sigma \beta_i X_i + \sum_i \sum_{j \geq i} \beta_{ij} X_i X_j, \qquad (2)$$

Y is the predicted response; β_0 is a constant; β_i are the linear coefficients; β_{ii} are the squared coefficients; and β_{ij} with $i \neq j$ are the cross-product coefficients. The above quadratic equation was used to build response surfaces for the dependent variables.

2.2. Synthesis of UF Resins

In the preparation of UF resins, industrial-grade raw materials were used, provided by EuroResinas — Indústrias Químicas S.A., Portugal, namely, urea, 50% formalin, sodium hydroxide solution, and acetic acid solution. The synthesis was carried out in a laboratory-scale 5 l glass reactor.

The reaction of urea and formaldehyde consists basically of a three-step process (alkaline–acid process):

 (i) Methylolation under alkaline conditions;

 (ii) Condensation under acidic conditions;

(iii) Neutralization and addition of the so-called final urea.

Methylolation step: after the required amount of urea was added to produce an initial F/U molar ratio of 2.15, the reaction mixture was held at the final temperature ($T = 95°C$) for about 30 min under alkaline conditions (pH $= 9.0$). The pH was afterwards adjusted to the pH of the condensation step using acetic acid.

Condensation step: the second urea was added sequentially to the previous reaction mixture, varying the number of intermediate additions and the time interval between additions. The F/U molar ratio after this step was 1.8. After the desired viscosity was reached, the reaction was stopped by alkalinisation with sodium hydroxide solution and cooling. The third (final) urea was then added at 60°C, yielding a final F/U ratio of 1.12.

The UF resins were stored at 25°C for three days and then were analysed and incorporated into wood panels.

2.3. GPC/SEC Analysis

A Gilson HPLC system equipped with a Gilson Differential RI detector and a Viscotek Dual Detector (differential viscosity and Right Angle Laser Light Scattering (RALLS)) was used. A Rheodyne 7125 injector with a 20 μl loop was used for sample introduction. The column used was a Waters Styragel HR1 5 μm. Dimethylformamide (DMF) was used as the mobile phase. The column was conditioned at 60°C using an external oven and the flow rate was 1 ml/min. The Universal Calibration was based on poly(ethylene glycol) standards from Polymer Laboratories, Germany, with molecular weight between 106–12 140. The RALLS detector was not used for the GPC/SEC molecular weight calculations owing to the very weak response at lower molecular weights. However, the RALLS signals were qualitatively taken into account.

The samples for GPC/SEC analysis were prepared by dissolving approximately 100 mg sample of resin in 3 ml of dimethylsulfoxide (DMSO), then stirring vigorously and filtering through a 0.45 μm Nylon syringe filter.

2.4. Preparation of Laboratory-Made Particleboards

Wood particles were provided by a particleboard producer (Sonae Industria, Oliveira do Hospital Plant, Portugal), and a standard mix of wood particles was used for the face layer and core layer. The standard mix included about 30% maritime pine (*Pinus Pinaster*), 15% eucalypt (*Eucalyptus Globulus*), 25% pine sawdust and 30% recycled wood. Wood particles with 4% of moisture content were blended with the resins and paraffin in a laboratory glue blender. The UF resins were applied at 8% resin solids in both the face and core layers, based on the oven-dry weight of the respective particles. The resin formation was more catalyzed in the core layer (3% solids based on oven-dry weight of resin) than in the face layer (1% solids based on oven-dry weight of resin).

A three-layer particleboard was hand-formed in a metallic container with dimensions 220 mm × 220 mm × 80 mm. The total percentages of board mass were: 20% for the upper face layer, 62% for the core layer and 18% for the bottom face layer. Boards were then pressed in a laboratory scale hot-press, controlled by a computer and equipped with a displacement sensor (LVDT), a load cell, thermocouples and pressure transducers. The mat was pressed at 195°C for 2.8 min to produce a board with a target density of 650 kg/m^3 and with 17 mm thickness.

After pressing, boards were conditioned for 3 days at normal conditions (20°C, 65% RH).

2.5. Particleboard Testing

Samples were then tested accordingly to the European standards and the following physico-mechanical properties were evaluated: density (EN 323), moisture content (EN 322) and IB strength (EN 319).

The formaldehyde content of all samples was determined according to EN 120 (perforator method) and the formaldehyde emission was determined for some samples according to EN 717-2 (gas analysis method).

3. Results and Discussion

3.1. Characteristics and Performance of the UF Resins Produced

In the resin synthesis, the condensation reaction step is stopped when the desired viscosity is attained for the reaction mixture. Figure 1 shows the time necessary for the condensation reaction as a function of the pH in this step. There is a strong influence of pH on the reaction rate. Interestingly, it was observed that the other two factors, time span between urea additions and number of urea additions, had no influence on the reaction time.

Table 4 reports the characteristics measured for all the UF resins produced in this work, namely, gel time, viscosity, solids content and final pH.

All resins were analysed by GPC/SEC three days after synthesis, to evaluate the fraction of insoluble molecular aggregates. Table 5 presents the resins properties considered for the experimental design. Figure 2 shows a typical chromatogram for a UF resin synthesised in this work. The peak with larger retention volume

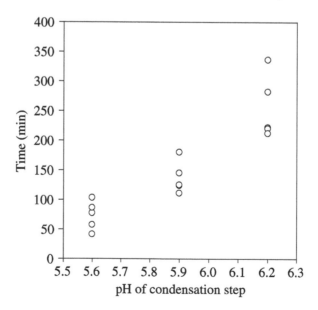

Figure 1. Condensation reaction time *versus* pH of the condensation step for the resins produced.

Table 4.
Characteristics of UF resins produced

Run	Gel time[1] (s)	Viscosity[2] (25°C) (mPa s)	Solids content[3] (%)	pH (25°C)
1	67	190	63.92	9.05
2	77	170	63.86	8.66
3	94	170	64.36	8.73
4	115	200	64.04	8.69
5	66	240	63.59	9.10
6	69	260	64.22	8.18
7	60	130	64.73	9.40
8	93	160	64.99	8.63
9	74	130	64.30	8.02
10	80	150	63.86	9.26
11	80	170	65.35	9.54
12	68	130	63.37	9.18
13	71	130	63.32	8.87
14	58	130	65.02	9.16
15	62	190	64.69	8.90
16	80	210	63.52	9.30

[1] Gel time at 100°C with 3 wt% of NH_4Cl (20 wt% solution).
[2] Brookfield viscometer.
[3] 120°C, 3 h.

corresponds to free urea, methylolureas and oligomers (MW < 400), and the intermediate zone of the chromatogram (5.8–8.2 ml) corresponds to polymer with moderate molecular weight (400 < MW < 12 140). The leftmost portion of the chromatogram would correspond to polymer with unrealistic high molecular weight (MW > 12 140). This fraction may actually represent insoluble molecular aggregates (colloids) smaller than 0.45 µm and not actually individual molecules, as suggested by Hlaing *et al.* [25] and Despres and Pizzi [26]. The presence of these colloidal particles will influence the resin performance during the bonding process [27]. The fraction of insoluble aggregates is computed here as the ratio of the area of the chromatogram corresponding to elution volumes below 5.8 ml to the total area. This segmentation of UF resin chromatogram into three zones has been previously suggested by different authors [25, 27, 28].

In Table 5, it is possible to observe quite distinct values of IB strength and FE, indicating that the produced resins have different properties, caused by the changes in the operating conditions.

Note that for the particleboards produced with all resins, other physical properties were also measured, namely, density, moisture content, and thickness swelling. All the results obtained comply with standard specifications for particleboards (EN 312).

Table 5.

Experimental results for the three measured responses

Run	IB strength[1] (N/mm^2)	FE[2]		FIA[3]
		Perforator method (mg/100 g dry board)	Gas analysis method (mg/m^2 h)	
1	0.56	6.5	–	0.079
2	0.59	4.1	3.61	0.093
3	0.72	6.7	–	0.051
4	0.37	5.7	–	0.054
5	0.39	5.7	–	0.057
6	0.80	4.5	4.24	0.098
7	0.62	5.8	–	0.062
8	0.63	6.2	–	0.065
9	0.72	7.4	–	0.054
10	0.59	6.2	–	0.052
11	0.58	5.3	–	0.069
12	0.38	6.6	–	0.036
13	0.50	6.5	–	0.049
14	0.41	4.3	3.62	0.076
15	0.57	5.8	–	0.074
16	0.44	6.1	–	0.061

[1] IB strength — internal bond strength.
[2] FE — formaldehyde emission.
[3] FIA — fraction of insoluble aggregates.

According to the standards specifications, the minimum acceptable value for IB strength is 0.35 N/mm^2 (type P2 board, EN 312). Although for formaldehyde content, the limit for class E1 is 8 mg/100 g oven-dry board, perforator value (EN 312; EN 13986), particleboard producers are compelled by customers to lower the limits to 4 mg/100 g oven-dry board (EN 120). Some of the resins (runs 3, 7, 8 and 9) show excellent values of IB strength, while others have very low FE (runs 2 and 14), while maintaining acceptable values of IB strength. In particular, one of the resins (run 6) produced panels with very good properties, i.e., a high value of IB strength (0.80 N/mm^2) and a low value of FE (4.5 mg/100 g oven-dry board).

It is known that the release of formaldehyde from wood panels is caused by three factors: (a) residual formaldehyde trapped as gas in the board structure, (b) formaldehyde dissolved in the retained water, and (c) hydrolysis of weakly bound formaldehyde, in the form of methylols, acetals and hemiacetals and, in more severe cases, hydrolysis of methylene ether bridges (at high relative humidity) [3].

Figures 3 and 4 show how the fraction of insoluble aggregates may be related to IB strength and FE. Despite the scatter in the data, it seems acceptable to estab-

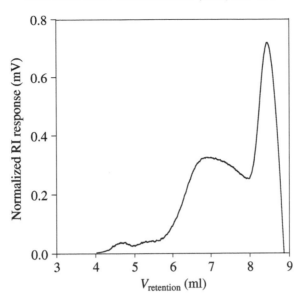

Figure 2. Normalized response of RI sensor for UF resin run 5 diluted 3% in DMSO.

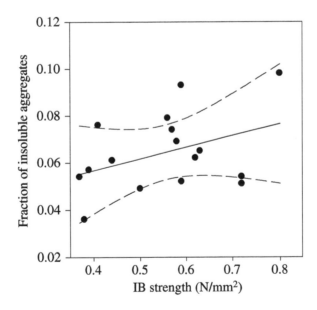

Figure 3. Relation between insoluble aggregates and internal bond strength. The dashed lines represent the 90% confidence intervals.

lish a trend such that as FIA increases, IB strength increases and FE decreases. The contribution of the dispersed insoluble phase to the strengthening of the adhesive bond, by acting as a reactive reinforcing filler, has been previously suggested [27, 29]. On the other hand, the association of this disperse phase to a lower formalde-

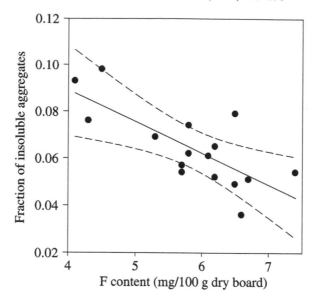

Figure 4. Relation between insoluble molecular aggregates and formaldehyde (F) content. The dashed lines represent the 90% confidence intervals.

hyde release, as suggested by Fig. 4, can be interpreted in terms of the presence of polymer with higher content in methylene linkages.

These results point a way to produce UF resins with improved performance, by obtaining high fractions of insoluble aggregates combined with low molecular weight polymer. Dunky [3] reports that higher molecular weight species improve the cohesive strength, while the low molecular weight species contribute to the wetting and penetration of the resin in the substrate.

3.2. Model Fitting

The application of RSM gave the regression equations shown below (equations (3)–(5)), which are empirical relations between the three responses and the three test variables. ANOVA indicated that the second-order polynomial model (equation (2)) was adequate to represent the FE (p-value $= 0.05$ and $R^2 = 0.86$). For IB strength (p-value $= 0.25$ and $R^2 = 0.73$) the analytical method employed for quantifying IB strength did not have enough resolution, within the range considered for each factor, resulting in a smaller confidence level of the fitted model. The parameters of the ANOVA indicated that the model was not adequate to represent the FIA.

It must be remarked that in the cases where the error in equations (3)–(5) was equal or higher than the corresponding coefficient, the associated variable was not included in the models, as usual [22].

As can be seen in Fig. 5, the values predicted by the second-order models for response IB strength and FE agree reasonably well with the experimental data.

After removing the negligible parameters from the original fitting polynomial

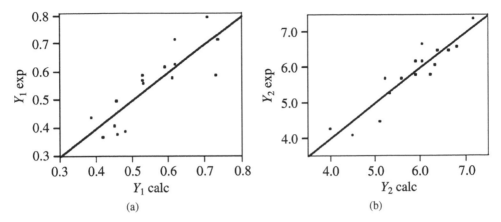

Figure 5. Experimental and calculated results for the responses considered. (a) Y_1 — internal bond strength and (b) Y_2 — formaldehyde emission.

equations, one obtains:

$$IB\ strength = 0.6208621(\pm0.051411) + 0.067(\pm0.03434)X_3$$
$$+ 0.06875(\pm0.038393)X_2X_3 + 0.1012069(\pm0.06688)X_1^2$$
$$- 0.113793(\pm0.06688)X_2^2 - 0.093793(\pm0.06688)X_3^2, \quad (3)$$

$$FE = 6.0810345(\pm0.256739) - 0.4(\pm0.17488)X_1 + 0.32(\pm0.17488)X_2$$
$$- 0.45(\pm0.171488)X_3 - 0.3375(\pm0.19173)X_1X_2$$
$$+ 0.6875(\pm0.19173)X_2X_3 - 0.546552(\pm0.333988)X_1^2, \quad (4)$$

$$FIA = 0.0598276(\pm0.008309) + 0.0067(\pm0.00555)X_1 + 0.0075(\pm0.00555)X_3$$
$$+ 0.0142586(\pm0.01081)X_1^2 - 0.011741(\pm0.01081)X_3^2, \quad (5)$$

where X_i are the coded variables (equation (1)) for each factor: X_1 is the number of urea additions in the condensation step, X_2 is the time span between successive urea additions in the condensation step, and X_3 is the pH of condensation step.

Considering just the first-order effects of each variable in equation (3), it is clear that the main factor that affects IB strength is the pH of the condensation step. For the formaldehyde emission property (equation (4)), the number of additions of urea and the pH of condensation step are the main factors, but the time span between urea additions also plays a noticeable role.

3.3. Effect of All Three Factors on the Measured Responses

Figures 6 and 7 depict the response surfaces, showing the effect of the number of urea additions, the time span between consecutive urea additions, and the pH of the condensation step on the IB strength and FE of the panels produced.

The polynomial expression in equation (4) was used to calculate the response surface illustrated in Fig. 6. It can be seen from Fig. 6(a) that depending on the

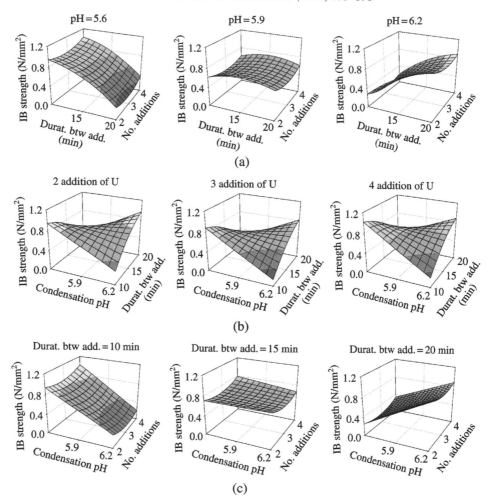

Figure 6. Response surface for internal bond strength as a function of: (a) time span between urea additions and number of urea additions (for different pH values of condensation step), (b) pH of condensation step and time span between urea additions (for different numbers of urea additions), and (c) pH of condensation step and number of urea additions (for different time spans between urea additions).

pH of condensation step, the time span between urea additions may have different effects on the IB strength. For the lower condensation pH (5.6), increasing the time span affects negatively the IB strength, whereas for the higher pH (6.2) the effect is reversed. As the time span between urea additions increases, if the reaction rate is sufficiently high, the urea might be depleted before the next addition takes place. As we can see from Fig. 1, for pH 5.6 the reaction is about 5 times faster than for pH 6.2. Hence, there is a high probability that urea added too late during the condensation step will not be totally consumed and will be carried on to the last step. At pH = 5.9 there is a mixed effect on the IB strength value. By allowing a higher

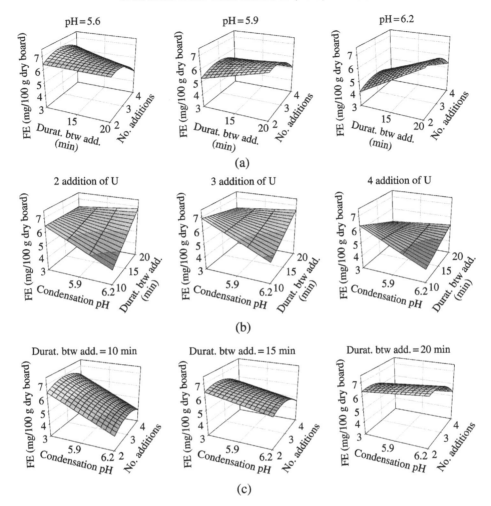

Figure 7. Response surface for formaldehyde emission as a function of: (a) time span between urea additions and number of urea additions (for different pH values of condensation step), (b) pH of condensation step and time span between urea additions (for different numbers of urea additions), and (c) pH of condensation step and number of urea additions (for different time spans between urea additions).

conversion of urea in this reaction step, a higher fraction of polymer is obtained in the final resin, in contrast to oligomeric species. This leads to an improved bond quality, as previously discussed by Pizzi [15].

Figure 6(b) shows that the number of urea additions does not change the effect of the other two factors on IB strength. Once again, it can be seen that each of the two factors has a positive or negative effect on IB strength, depending on the value of the other factor (cross-effects).

The response surfaces illustrated in Fig. 7 show the effects of the three factors on the FE. It is particularly evident from Fig. 7(b) that FE tends to decrease as pH is increased and the time span between additions is decreased.

The lower emission at the higher pH might be related to the formation of the more stable methylene linkages instead of methylene ether linkages during the condensation process. The former can rearrange to methylene bridges by splitting off formaldehyde [3].

The cross-effects discussed above justify the use of design of experiments tools for optimization of the synthesis of UF resin.

3.4. Optimization of Operating Conditions

The optimization consisted in the simultaneous maximization of the IB strength value and minimization of the FE value. This was done based on the desirability function methodology [23], assigning preponderance factors of 0.5 and 1 to the IB strength and FE responses, respectively, on a 0–1 scale. The optimum formulation found is presented in Table 6.

These synthesis conditions were reproduced in order to validate the predictions. Table 7 shows the characteristics obtained for this resin.

The experimental values of the three responses for this optimized resin are presented in Table 8, together with the values predicted by the empirical models (equations (3)–(5)). The particleboard properties (IB strength and FE) are sufficiently close to the predicted values, considering the inevitable variability induced by the use of industrial grade reagents, complex control of synthesis conditions (namely the pH history and monitoring of the viscosity in the condensation step) and the natural heterogeneity of the wood mix used for particleboards.

Table 6.

Operating conditions that produce the minimum formaldehyde emission

Factor	Optimum values
No. of U additions in the condensation step (A)	4
Time span (min) between U additions in the condensation step (B)	13
pH of the condensation step (B)	6.1

Table 7.

Characteristics of the UF resin optimized for minimizing the formaldehyde emission

Run	Gel time[1] (s)	Viscosity[2] (25°C) (mPa s)	Solids content[3] (%)	pH (25°C)
Optimum 1	74	150	64.22	9.27

[1] Gel time at 100°C with 3 wt% of NH_4Cl (20 wt% solution).
[2] Brookfield viscometer.
[3] 120°C, 3 h.

Table 8.
Predicted and experimental values for the three responses

Response	Optimum 1	
	Predicted	Experimental
IB strength[1] (N/mm^2)	0.65 ± 0.11	0.51
FE[2] (mg/100 g dry board)	4.5 ± 0.5	4.2
FIA[3]	0.84 ± 0.018	0.076

[1] IB strength — internal bond strength.
[2] FE — formaldehyde emission.
[3] FIA — fraction of insoluble aggregates.

The measured value of insoluble molecular aggregates is slightly lower than the predicted value (the fitted model has a small *p*-value). Also, this result can be explained by some variation in the final viscosity during condensation step. The value of insoluble molecular aggregates is significantly affected by the duration of the condensation step.

The optimized UF resin has a good quality for a class E1 resin (the value measured by the perforator method should be below 8 mg/100 g oven-dry board). The IB strength is well above the minimum acceptable value of 0.35 N/mm^2 for a P2 type board (EN 312). However, resin reactivity (gel time) has to be enhanced.

4. Conclusions

An alkaline–acid synthesis process was studied using a design of experiments methodology, in order to optimize the adhesion performance and formaldehyde emission of UF resins. It was concluded that the pH and the time span between consecutive urea additions in the condensation step have a strong influence on the analysed properties.

The fraction of insoluble molecular aggregates in the produced resin, detectable by GPC/SEC analysis, is related to the particleboard performance. A larger amount of dispersed phase seems to lead to higher internal bond strength and lower formaldehyde emission.

In order to optimize the resins performance, in terms of internal bond strength and formaldehyde emission, the desirability method was used. An optimized resin was identified and produced. The properties obtained were within or close to the values predicted by the empirical models employed.

Acknowledgements

The authors wish to thank Margarida Nogueira (EuroResinas — Indústrias Quími-cas S.A., Portugal) for helping in the preparation of some of the resins and Brigida Pinto, Filipe Silva and João Pereira for helping in the preparation and evaluation of some particleboards. João Ferra wishes to thank FCT for the PhD grant SFRH/BD/23978/2005.

References

1. E. Athanassiadou, S. Tsiantzi and C. Markessini, *COST Action E49 'Measurement and Control of VOC Emissions from Wood-Based Panels'*. Braunschweig, Germany (2007).
2. CARB, *Airborne Toxic Control Measure to Reduce Formaldehyde Emissions from Composite Wood Products*. California Environmental Protection Agency (2007).
3. M. Dunky, *Intl. J. Adhesion Adhesives* **18**, 95 (1998).
4. J. H. Williams, US Patent No. 4482699 (1984).
5. P. Christjanson, T. Pehk and K. Siimer, *J. Appl. Polym. Sci.* **100**, 1673 (2006).
6. L. Graves and J. Mueller, US Patent No. 5362842 (1994).
7. A. Hatjiissaak and E. Papadopoulou, WO Patent No. 138364 (2007).
8. H. Kong, US Patent No. 4603191 (1986).
9. H. Spurlock, US Patent No. 4381368 (1983).
10. S. Vargiu, S. Giovanni, G. Mazzolen and U. Nistri, US Patent No. 3842039 (1974).
11. J. H. Williams, US Patent No. 4410685 (1983).
12. R. M. Rammon, The influence of synthesis parameters on the structure of urea-formaldehyde resins, *PhD Thesis*, Washington State University, Pullman, WA, USA (1984).
13. A. Pizzi, in: *Wood Adhesives: Chemistry and Technology*, A. Pizzi (Ed.), p. 59. Marcel Dekker, New York, NY (1983).
14. A. Pizzi, L. Lipschitz and J. Valenzuela, *Holzforschung* **48**, 254 (1994).
15. A. Pizzi, in: *Handbook of Adhesive Technology*, A. Pizzi and K. L. Mittal (Eds), 2nd edn, p. 635. Marcel Dekker, New York, NY (2003).
16. M. G. Kim, *J. Polym. Sci. Polym. Chem.* **37**, 995 (1999).
17. M. G. Kim, *J. Appl. Polym. Sci.* **75**, 1243 (2000).
18. M. G. Kim, *J. Appl. Polym. Sci.* **80**, 2800 (2001).
19. M. G. Kim, B. Y. No, S. M. Lee and W. L. Nieh, *J. Appl. Polym. Sci.* **89**, 1896 (2003).
20. M. G. Kim, H. Wan, B. Y. No and W. L. Nieh, *J. Appl. Polym. Sci.* **82**, 1155 (2001).
21. R. N. Kumar, T. L. Han, H. D. Rozman, W. R. W. Daud and M. S. Ibrahim, *J. Appl. Polym. Sci.* **103**, 2709 (2007).
22. D. C. Montgomery, G. C. Runger and N. F. Hubele, *Engineering Statistics*. Wiley, Hoboken, NJ (2001).
23. G. C. Derringer, *Quality Progress* **27**, 51 (1994).
24. G. E. P. Box and K. B. Wilson, *J. Royal Statistical Soc. Series B* **13**, 1 (1951).
25. T. Hlaing, A. Gilbert and C. Booth, *British Polym. J.* **18**, 345 (1986).
26. A. Despres and A. Pizzi, *J. Appl. Polym. Sci.* **100**, 1406 (2006).
27. J. Ferra, A. Mendes, M. R. Costa, L. Carvalho and F. D. Magalhães, *J. Appl. Polym. Sci.*, in press.
28. G. Zeppenfeld and D. Grunwald, *Klebstoffe in der Holz- und Möbelindustrie*. Drw Verlag, Weinbrenner, Germany (2005).
29. J. Ferra, J. Martins, A. Mendes, M. R. Costa, L. Carvalho and F. D. Magalhães, in: *Proc. 3rd International Conference on Environmentally-Compatible Forest Products*, J. Caldeira (Ed.), Porto, Portugal, p. 17 (2008).

Characterization of Urea-Formaldehyde Resins by GPC/SEC and HPLC Techniques: Effect of Ageing

João M. Ferra [a], Adélio M. Mendes [a], Mário Rui N. Costa [b],
Fernão D. Magalhães [a] and Luisa H. Carvalho [a,c,*]

[a] LEPAE — Departamento de Engenharia Química, Faculdade de Engenharia,
Universidade do Porto, Rua Dr. Roberto Frias, 4200-465 Porto, Portugal
[b] LSRE — Departamento de Engenharia Química, Faculdade de Engenharia,
Universidade do Porto, Rua Dr. Roberto Frias, 4200-465 Porto, Portugal
[c] DEMad — Departamento de Engenharia de Madeiras, Instituto Politécnico de Viseu,
Campus Politécnico de Repeses, 3504-510 Viseu, Portugal

Abstract

During the last 40 years, several analytical techniques have been developed/adapted to characterize urea-formaldehyde (UF) resins. However, a great part of the research about this kind of wood adhesives has been performed by industrial producers and, thus, the main part of the existing knowledge is retained within those companies.

This work describes a methodology for determining the molecular weight distribution (MWD) of UF resins using Gel Permeation Chromatography (GPC)/Size Exclusion Chromatography (SEC) with 2 detectors (differential refractive index (RI) and differential viscosity). This method permitted to characterize/distinguish commercial UF resins produced with different F/U molar ratios and to monitor the molecular weight and MWD with ageing.

An HPLC method was additionally used to evaluate the fraction of unreacted urea, monomethylolurea and dimethylolurea present in commercial UF resins and measure the evolution of these three compounds with ageing.

Keywords

UF resins, characterization, polymerization reaction, ageing, GPC/SEC, HPLC

1. Introduction

The complex physics and chemistry of urea-formaldehyde (UF) resins has been the subject of several studies. These works have yielded further knowledge regarding these systems, but still many issues remain concerning their structure as well as the kinetics and mechanisms of their formation. The variety of randomly linked structural elements such as methylene bridges, ether bridges, methylol and amide groups,

---

* To whom correspondence should be addressed. Tel.: +351 232 480565; e-mail:
lhcarvalho@demad.estv.ipv.pt

and possible cyclic derivatives makes their analysis a tough challenge. Moreover, these highly reactive chemical systems have tendency to change during preparation for the analysis, or during the analysis itself. The difficulty to find suitable solvents for these resins is an additional problem [1]. Fortunately, the availability of modern spectroscopic and chromatographic methods has led to considerable improvements in the characterization of these resins: more specifically, ^{13}C NMR [2] and FT-IR [3] for the structure and GPC/SEC [4–13] and even more recently MALDI-TOF-MS [14] for the determination of the detailed chemical constitution, although the true molecular weight distribution (MWD) remains elusive. More recently, Minopoulou et al. [15] explored the capabilities of FT-NIR spectroscopy [15, 16] for on-line monitoring of the amino resin synthesis. The cured system has been investigated by solid state ^{13}C CP MAS NMR [17], FT-IR [18] and Raman spectroscopy [19]. The reaction kinetics has been studied experimentally [20] and theoretically [19, 21, 22].

In this work, the characterization of UF resins is focused on the determination of the MWD. The mechanical and bonding properties of an adhesive are strongly dependent on its MWD [23, 24]. This can be done by GPC/SEC, but the low solubility of the colloidal fraction of these resins introduces unique features in the chromatograms that must be taken into account [25].

GPC/SEC is a controlled separation technique in which molecules are separated on the basis of their hydrodynamic molecular volume or size [26]. With proper column calibration or using molecular weight-sensitive detectors, such as light scattering, viscosimetry, or mass spectrometry, the MWD and the statistical molecular weight averages can be readily obtained. In their review, Barth et al. [26] mentioned that the GPC/SEC is the premier technique to evaluate these properties for both synthetic polymers and biopolymers.

The main problem using GPC/SEC is the choice of the proper solvent and mobile phase to ensure complete resin solubility. It is necessary to use dimethylformamide (DMF) or even dimethylsulfoxide (DMSO) to dissolve the higher molecular mass fractions. Salts such as lithium chloride (LiCl) or lithium bromide (LiBr) are often used to increase the solvent polarity and consequent polymer solubility, thus, minimizing the formation of aggregates. Another problem is related to the complex nature of the polymer present in UF resins, where linear and branched fractions coexist. The calibration standards cannot represent this accurately, leading to inaccuracies in the measured distribution of molecular masses. In contrast to methods such as the nowadays seldom used analytical ultracentrifugation, no exact molecular mass distribution can be extracted from chromatographic traces without several assumptions. But even for linear fractions of UF polymers, as there are no commercial standards for molecular mass calibration (UF compounds with a single molar mass and molecular structure), oligomers would have to be synthesized in the laboratory. GPC/SEC together with light scattering detection, which should avoid the need for external calibration, is not a viable solution, as only the high molar masses are detected by the light scattering sensor [1] and for this reason, the calculation of averages such as weight and number molar masses is still much affected by the absence of a reliable calibration.

Some partial success has been claimed for the use of GPC/SEC in the analysis of UF polymers in previous research works. For instance, Dankelman *et al.* [5] have reported that GPC/SEC can estimate the ratio of low to high molecular mass components as well as the amounts of some oligomers. Billiani *et al.* [4] used this technique to characterize UF resins synthesized with different degrees of condensation. They found that measured average molecular mass increased with the duration of the condensation steps, from a few thousand up to more than 100 kDa.

In the present work, preliminary studies have pointed out that the Right Angle Laser Light Scattering (RALLS) signal is too weak in the low to moderate molecular weight fractions of the chromatograms obtained with UF resins. Additionally, preliminary tests with polystyrene standards indicated that this RALLS system was not able to detect molecular weights below about 7000 g/mol. This implies the need for using a traditional, universal calibration technique, with two detectors: differential refractive index and differential viscosity. The information provided by the RALLS detector was used only in a qualitative way.

The GPC/SEC technique is useful for the characterization of MWD, but does not give complete information about the composition of the low molecular weight species present. For this purpose, HPLC can be effectively used for identifying low molecular weight components in UF resins [27, 28], and may contribute to a deeper understanding of UF chemistry.

Grunwald [29] mentioned the combination of HPLC with GPC techniques as a relevant area for future R&D on UF resins. In our present work HPLC was used successfully for determining urea, monomethylolurea and dimethylolurea in different resins.

2. Materials and Methods

2.1. Resins Preparation

All resins characterized in this work were produced according to the alkaline–acid process, which consists basically of three steps: methylolation under alkaline conditions, condensation under acidic conditions, and neutralization and addition of the so-called final urea or last urea.

Samples of UF-R5 and UF-R2 were supplied by EuroResinas (Sonae Indústria, Portugal), while sample of UF-Exp17 was prepared in our laboratory according to procedure described elsewhere [30]. Table 1 shows the technical data collected for these three resins. All of them are UF resins in water solution with low amounts of melamine and hexamine. The main differences in the synthesis are the duration of the condensation step — leading to different kinds of polymers formed — and the final amount of urea added — leading to different final F/U ratios. Resin UF-R2 has the longest condensation step and the largest amount of final urea, while resin UF-R5 has the shortest condensation step and the lowest amount of final urea. Resin

Table 1.
Technical data on UF resins used

Resin	Molar ratio F/U	Solids content[1] (%)	Gel time[2] (s)	pH range (25°C)	Viscosity[3] (25°C) (mPa s)
UF-R5	1.30	63 ± 1	35–55	7.5–8.5	150–350
UF-R2	1.00	64 ± 1	40–100	8.0–9.5	150–300
UF-Exp17	1.12	64 ± 1	40–100	8.0–9.5	150–300

[1] 105°C, 3 h.
[2] Gel time at 100°C with 3 wt% of NH_4Cl (20 wt% solution).
[3] Brookfield viscometer.

Table 2.
Technical data on UF-resins used from different producers

Resin	Molar ratio F/U	Solids content[1] (%)	Gel time[2] (s)	pH value (25°C)	Viscosity[3] (25°C) (mPa s)
UF-A	1.03	67.4	56	8.25	220
UF-B	1.12	63.6	64	8.78	210
UF-C	1.11	69.0	54	8.33	310
UF-D	1.15	68.1	43	8.40	400
UF-E	n.a.	64.0	57	8.30	258

[1] 105°C, 3 h.
[2] Gel time at 100°C with 3 wt% of NH_4Cl (20 wt% solution).
[3] Brookfield viscometer.

UF-Exp17 has a sequential addition of urea during the condensation step and the preparation procedure is completely described by Ferra *et al.* [30].

Some commercial resins from several major European producers were also studied. The principal characteristics are presented in Table 2.

2.2. GPC/SEC Analysis

The main instrument used was a Gilson HPLC system equipped with a Gilson Differential RI detector and a Viscotek Dual Detector (differential viscosity and a light scattering detector RALLS). A Rheodyne 7125 injector with a 20 µl loop was used for injection. The column used was a Waters Styragel HR1 5 µm column. DMF was used as the mobile phase. The column was conditioned at 60°C using an external oven and the flow rate was 1 ml/min. The universal calibration was done using poly (ethylene glycol) standards from Polymer Laboratories Ltd., UK, with molecular weight between 106–12 140. The RALLS detector was not used for the determination of molecular weight, because of no response of the detector to lower molecular weights. However, the RALLS signals were analysed qualitatively.

The samples for GPC/SEC analysis were prepared by dissolving the resin in DMSO, then vigorously stirring and filtering through a 0.45 µm Nylon syringe filter.

The addition of LiCl and the use of an ultrasonic bath in the preparation of the samples were also tested, but the differences found in chromatograms were slight or nonexistent.

2.3. HPLC Analysis

A JASCO HPLC system equipped with a JASCO Differential RI detector and a Rheodyne 7725i injector with a 100 μl loop was used. The column used was a Waters Spherisorb silica column. A mixture of acetonitrile and water (90/10) was used as the mobile phase. The column was conditioned at 30°C using an external oven and the flow rate was 1.5 ml/min.

The samples for HPLC analysis were prepared by dissolving the resin in 1 ml of DMF. After vigorous agitation for 1 min, it was diluted in 2 ml of mobile phase. When the mobile phase was added, flocculation occurred and the sample was allowed to rest for 10 min. The supernatant was finally withdrawn with a micropipette.

3. Results and Discussion

3.1. Characterization of UF Resins

3.1.1. Determination of Molecular Weight Distribution
In Fig. 1 one can see the GPC/SEC chromatograms of resins UF-R5 and UF-R2. In both cases, at least two samples were prepared and analysed in order to verify the reproducibility of the results.

Figure 1(a) shows the normalized weight fraction (Wt Fr) for resins UF-R5 and UF-R2 after 5 days. The general features of the chromatograms are similar to those found in the literature on UF resins. Three zones can be identified in the chro-

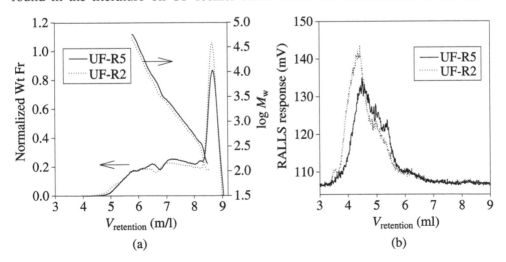

Figure 1. Chromatograms for UF-R5 and UF-R2 diluted 3% in DMSO and stored for 5 days at 25°C. (a) Normalized weight fraction (Wt Fr); (b) RALLS response.

matograms, as discussed in a previous work [25]. Zone I (elution volume between 8 and 9 ml) corresponds to the low molecular weight species. Zone II (elution volume between 5.8 and 8 ml) corresponds to intermediate molecular weight species, with molecular weights ranging from about tens of thousands Da to about 600 Da. Zone III (elution volume below 5.8 ml) would correspond to species with quite high molecular weights, eluting before the exclusion limit of the GPC/SEC column. It has been suggested that these are molecular aggregates and not individual polymer molecules [9, 31]. These aggregates would be insoluble in the original aqueous medium, probably forming larger colloidal structures which become partially disaggregated in the DMSO solvent.

Since the chromatograms may reflect the presence of molecular aggregates, a straightforward computation of the average molecular weights would be misleading. Two different approaches were, therefore, followed to quantitatively represent the chromatographic data. On the one hand, assuming that Zone III corresponds essentially to insoluble material, molecular weights were computed by neglecting this portion of the chromatograms. On the other hand, the two following parameters were introduced in order to complement the description of the particular features of these chromatograms:

$$f_1 = \frac{\text{area of Zone I}}{\text{total area of chromatogram}}, \tag{1}$$

$$f_2 = \frac{\text{area of Zone III}}{\text{areas of Zone II} + \text{Zone III}}, \tag{2}$$

f_1 reflects the amount of low molecular weight species in the sample, while f_2 indicates the fraction of high molecular weight species, probably in the form of molecular aggregates in the polymerized material.

The main difference between the chromatograms for resins UF-R5 and UF-R2 (Fig. 1(a)) is observed in the zone of low molecular weights, which probably originated from the larger amount of the urea added in the last step of the reaction for resin UF-R2. In addition, resin UF-R2 shows a more pronounced tail for low elution volumes, corresponding to Zone III. This probably reflects the higher condensation state of this resin, which induces higher molecular aggregation [32]. Other than this, the two chromatograms are generically very similar. However, as will be discussed below, ageing will introduce more pronounced differences.

The RALLS response (Fig. 1(b)) gives qualitative information on the insoluble particles present in solution [32]. One can see that the RALLS chromatograms are similar for the two resins, but the trace for resin UF-R2 is more intense at lower elution volumes, once again indicating a more significant aggregation.

The quantitative data shown in Table 3 confirm the previous analysis: resin UF-R2 presents a higher fraction of lower molecular weight species (f_1) as well a higher fraction of insoluble aggregates (f_2). Resin UF-R2 is the most condensed of the resins studied in this work, but this feature is not evidenced from the molecular weight results, because of the larger addition of urea in the last step of the

Table 3.
Values of M_n, M_w, polydispersity (M_w/M_n), and parameters f_1 and f_2, obtained by SEC for UF-R5 and UF-R2 stored for 5 days at 25°C

Resin	M_n	M_w	M_w/M_n	f_1	f_2
UF-R5	3.77×10^2	3.59×10^3	9.5	0.356	0.171
UF-R2	2.90×10^2	3.44×10^3	11.9	0.410	0.193

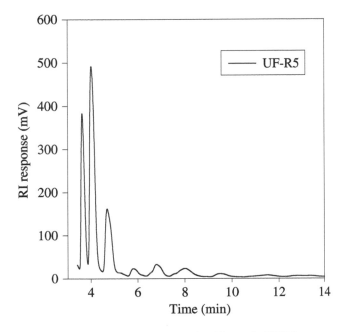

Figure 2. Chromatogram obtained for resin UF-R5.

reaction. Nevertheless, this resin shows the highest value of parameter f_2, which is indeed relatable to the highest degree of condensation.

3.1.2. Determination of the Fractions of Urea and Methylolureas

Figure 2 presents a typical chromatogram obtained for a UF resin in HPLC. The three first peaks correspond to urea, monomethylolurea and dimethylolurea, respectively. This was confirmed by injection of the isolated compounds. The other peaks in the chromatogram correspond to oligomeric species.

Figure 3 shows the distribution of the urea, monomethylolurea and dimethylolurea present in solution for the two resins stored at 25°C for 5 days after synthesis. UF-R2 has a much larger fraction of unreacted urea than resin UF-R5. The last urea was added in order to react with the free formaldehyde present, but as the added amount was large, most of the urea remained unreacted in the final resin. However, this unreacted urea may play another role, since it may form a solvation

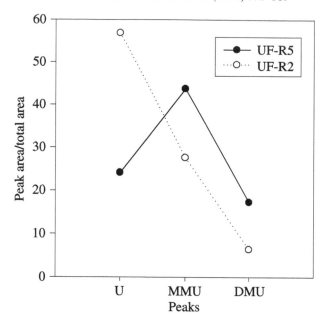

Figure 3. Peak areas normalized by total chromatogram area for UF-R5 and UF-R2 stored at 25°C.

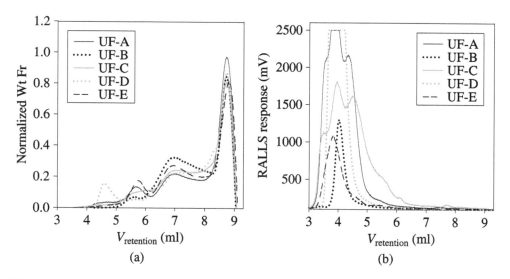

Figure 4. Chromatograms for five UF resins from different manufactures in Europe. (a) Normalized weight fraction; (b) RALLS response.

layer surrounding the colloidal aggregates surface, contributing to its stabilization against agglomeration [25].

3.1.3. Analysis of Commercial UF Resins

Five commercial UF resins from different European producers were analysed. Figure 4(a) shows the normalized Wt Fr obtained by GPC/SEC. One can see that

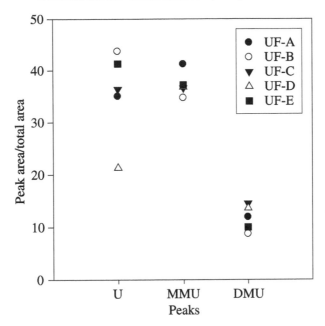

Figure 5. Ratios of peak areas/total area of urea (U), monomethylolurea (MMU) and dimethylolurea (DMU) for five UF resins from different producers.

UF-D presents a distinctly larger fraction of insoluble aggregates (Zone III) and a higher fraction of oligomers in the elution volume range 7.8–8.2 ml. On the other hand, resin UF-B has the lowest fraction of insoluble aggregates and a larger fraction of polymer with moderate molecular weight. Resins UF-A, UF-C and UF-E present similar chromatograms, but some differences in the three zones of the chromatograms were found, namely the large amount of polymer with low molecular weight present in UF-A and a large amount of insoluble aggregates existing in UF-E.

The chromatographic trace from the RALLS detector shown in Fig. 4(b) agrees qualitatively with the previous analysis concerning the presence of insoluble aggregates in the different resins.

The distributions of urea and methylolureas present in the five commercial resins are shown in Fig. 5. The main difference among the resins is the fraction of unreacted urea. In particular, the fraction of urea in resin UF-D is approximately half of the value for the other resins. The final percentages of urea and methylolureas as shown above are related to the amount of last urea added and the free formaldehyde present in the final condensation step.

The results obtained by GPC/SEC and HPLC indicate that each producer had likely used different processes for the production of UF resins.

3.2. *Monitoring the Ageing of UF Resins*

Two UF resins were analysed after ageing, using GPC/SEC and HPLC techniques. The pH and viscosity of resins were also monitored.

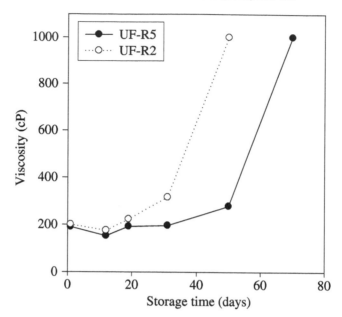

Figure 6. Viscosity of UF-R5 and UF-R2 with storage time at 25°C.

Figure 6 depicts the evolution of viscosity for resins UF-R2 and UF-R5 during storage at 25°C. The initial slight decrease in viscosity is related to the migration of hydroxymethyl groups (methylolureas) from the polymeric UF resin components to the last urea as reported by Kim [33]. Resin UF-R2 gels faster than UF-R5, but both are stable up to 30 days, which is the normal specification for UF resins. The storage time limit for UF-R2 is about 40 days, while for UF-R5 it is about 60 days. The higher degree of condensation of the resin UF-R2 can explain this behaviour.

UF-R5 and UF-R2 resins were monitored at six different ageing periods (5, 12, 19, 32, 50 and 53 days). Figures 7 and 8 show the chromatograms for some selected storage times. Note that no significant changes in the molecular weight distributions were observed for the first two storage times (5 and 12 days). This is consistent with the stable viscosity measurements for the first 20 days. Kim *et al.* [34] reported similar results by monitoring the storage of UF resins using ^{13}C NMR. They showed that the degree of polymerization remained stable for about 15 days but then increased rapidly until 30 days and remained constant afterwards until gelling.

On analysing the chromatograms for 4 selected storage times, we find that urea and methylolureas (peak at 8.5 ml) decrease with ageing while the peaks at 8.2 and 7.8 ml increase. This suggests that urea and methylolureas react during storage to produce a polymer with a narrow range of molecular mass/size (peak at 8.2 ml), which, in turn, reacts to produce polymer eluting in the vicinity of 7.8 ml, visible as a broad peak. Globally, there is a decrease in Zone I of the chromatogram, yielding an increase in moderate to high molecular weight polymer (Zone II) and in molecular aggregates (Zone III). Interestingly, a well-defined separation between

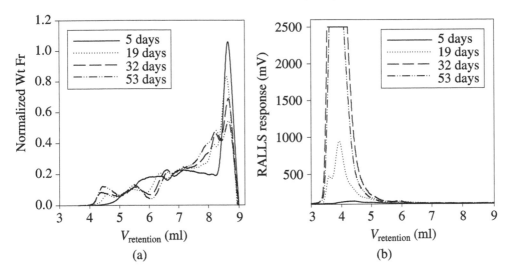

Figure 7. Chromatograms for UF-R2 diluted 3% in DMSO, for different storage periods at 25°C. (a) Normalized weight fraction; (b) RALLS response.

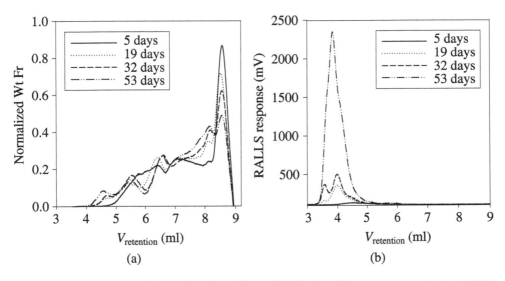

Figure 8. Chromatograms for UF-R5 diluted 3% in DMSO, for different storage periods at 25°C. (a) Normalized weight fraction; (b) RALLS response.

Zones II and III becomes apparent (at an elution volume of about 6 ml), which was also visible in the commercial resins (Fig. 4(a)). This might be associated with the process of agglomeration of smaller aggregates into larger particles, shifting towards the left portion of the chromatogram.

RALLS responses show a sharp increase in insoluble aggregates between 19 and 32 days for resin UF-R2 and between 32 and 53 days for resin UF-R5. This is related to the viscosity evolution previously measured for both resins (Fig. 6): the

Table 4.
Values of M_n, M_w, polydispersity (M_w/M_n), and parameters f_1 and f_2, obtained by GPC/SEC for UF-R2 and UF-R5 stored for different days at 25°C

Storage period	M_n	M_w	M_w/M_n	f_1	f_2
UF-R2					
5	2.90×10^2	3.44×10^3	11.9	0.410	0.193
32	3.79×10^2	2.50×10^3	6.6	0.315	0.201
50	4.14×10^2	2.03×10^3	4.9	0.303	0.217
UF-R5					
5	3.77×10^2	3.59×10^3	9.5	0.356	0.171
32	4.55×10^2	2.30×10^3	6.6	0.268	0.190
50	5.10×10^2	2.75×10^3	5.4	0.225	0.185

pre-gelling increase in viscosity for each resin is associated with a significant formation of insoluble molecular aggregates or, as mentioned above, to agglomeration of existent aggregates into larger particles. Zanetti and Pizzi [35] and Despres and Pizzi [31] reported that the continuing formation of colloidal structures followed by the formation of 'superaggregates' (globular masses) were the normal steps for physical gelation of MUF and UF resins.

From the data in Table 4 for resins UF-R2 and UF-R5 at different ages, one can see that the value of polydispersity decreases with time, due to the condensation of the low molecular weight species to form polymers with moderate and high molecular weights. This solubilized polymer might then form insoluble molecular aggregates, but the portion of the chromatogram corresponding to the insoluble molecular aggregates is not included in the molecular weight calculation, as mentioned above. The decrease of the low molecular weight fraction (f_1) and the increase of insoluble aggregates fraction (f_2) with ageing corroborates the idea that condensation progresses with storage time, consuming urea and oligomers. The decrease in f_2 observed for resin UF-R5 after 50 days is related to the large increase in fraction of the polymer with moderate molecular weight (Zone II).

It is also interesting to look at the RALLS responses obtained for resins UF-R2 and UF-R5 with ageing. Figures 7(b) and 8(b) show a peak located roughly in the region corresponding to Zone III, assigned to molecular aggregates present in the samples. When resins were 'fresh' the peak magnitude was very low, but with ageing it increased sharply for both resins. This seems to indicate that the ageing process produces a significantly higher concentration of aggregated material. These aggregates might actually be agglomerated into larger particles in the resin.

Figures 9 and 10 show the evolution with ageing of the fractions of urea, monomethylolurea, dimethylolurea and three more oligomeric species present in the HPLC chromatograms, for resins UF-R2 and UF-R5, respectively. According to the results described by Ludlam *et al.* [28], who used the same analysis conditions (silica columns with NH_2 groups, mobile phase and sample prepara-

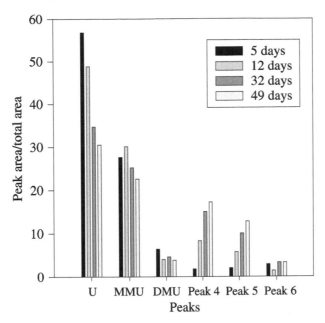

Figure 9. Evolution of the ratios of peak areas/total area of the urea (U), monomethylolurea (MMU), dimethylolurea (DMU) and three other oligomeric species for UF-R2 stored for various periods at 25°C.

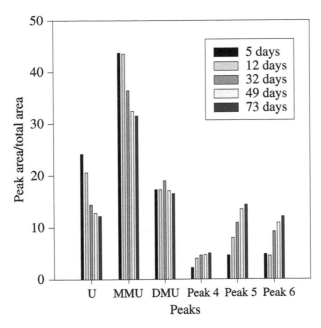

Figure 10. Evolution of the ratios of peak areas/total area of the urea (U), monomethylolurea (MMU), dimethylolurea (DMU) and three other oligomeric species for UF-R5 stored for various periods at 25°C.

tion) for the identification of oligomeric species present in UF resins, the three unknown peaks can be identified as monomethylolmethylenediurea, monomethyloloxymethylenediurea and dimethylolmethylenediurea, respectively.

In both cases, the fraction of urea in the solution decreases significantly up to 30 days. It ends up becoming stable as the formaldehyde present in the solution is consumed. It is interesting to note that the evolution of dimethylolurea is different for the two resins. It remains almost constant for resin UF-R2 but it goes through a maximum at about 30 days for resin UF-R5. This behaviour is related to the existence of free formaldehyde in the solution for resin UF-R5, which reacts with urea and mostly with monomethylolurea (that is in excess in solution) forming dimethylolurea. These results indicate that the polymerization reactions between free formaldehyde, urea and methylolureas continue during the storage of the resin. Similar observations were reported by Kim *et al.* [34], which reported that the amount of monomethylolurea during storage could increase or decrease depending on the amount of free formaldehyde present in the solution.

3.3. Determination of Water Tolerance

UF resins are colloidal suspensions that tend to flocculate as they are diluted in water [25]. Figure 11 shows the GPC/SEC chromatograms obtained for the original resin UF-R5 after 5 days storage at 25°C and for the resin flocculated with a large excess of water. In this case the supernatant was collected after sedimentation of the precipitate and analysed. The corresponding chromatogram shows similar con-

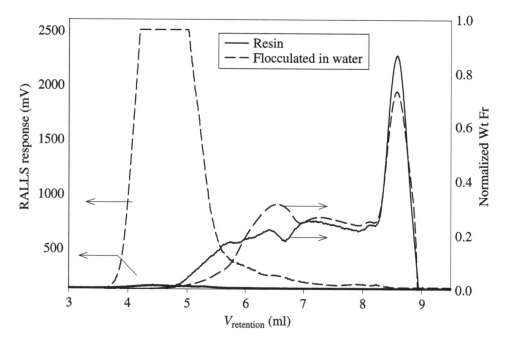

Figure 11. Chromatograms for UF-R5 aged for 5 days, diluted in DMSO and very diluted (flocculated) in water.

centrations of moderate and low weight molecules (elution volumes between 7.0 to 9.0 ml), but a higher concentration of the zone that corresponds to high/moderate molecular weight (elution volumes between 5.8 to 7.0 ml) and a lower concentration of insoluble molecular aggregates (Zone III). Apparently, the transfer of molecules from the high to the intermediate molecular weight zone is related to disaggregation of the larger particles due to the dilution. It can also be seen from Fig. 11 that insoluble aggregates originally corresponding to Zone III form aggregates of larger dimensions after flocculation, thus, eluting earlier. These aggregates only are detected by the RALLS sensor due to their low concentration. However, it is necessary to use some caution in analysing the results from RALLS detector because it saturates in the zone between 3.5 to 4.5 ml of the chromatogram. A similar behaviour was observed for other UF resin, namely UF-R2.

These results demonstrate that the GPC/SEC is an interesting technique to evaluate the water tolerance. The evaluation of water tolerance using the common method, which consists in the addition of small amounts of water until the resin flocculation occurs, is very difficult and inaccurate. A high value of water tolerance confers good washdown properties to the product and allows easy cleaning of the apparatus used for production and storage of UF resins.

4. Conclusions

Different UF resins were characterized by GPC/SEC and HPLC techniques.

The GPC/SEC analysis encompasses information on the MWD of the soluble polymer as well as on insoluble molecular aggregates that constitute the original dispersed phase and have not been completely dissolved in the DMSO solvent. The information obtained by GPC/SEC is useful for characterization of the resins and allows to distinguish resins obtained from different production processes. The information from the RALLS detector complements qualitatively the information on the insoluble material.

GPC/SEC and HPLC methods permitted to verify that UF resins produced by the European companies had distinct characteristics (MWD and relative amounts of U, MMU and DMU). These results suggest that each producer uses a particular process for the production of UF resin.

The GPC/SEC analyses of the resins at different ageing periods indicated that both polymer condensation and aggregation/agglomeration proceed during storage. This technique permits to monitor accurately the ageing of UF resins.

Acknowledgements

The authors wish to thank EuroResinas — Indústrias Químicas S.A. (Portugal) for providing the resin samples. João Ferra wishes to thank FCT for the PhD grant SFRH/BD/23978/2005.

References

1. M. Dunky, in: *Proc. 1th European Panels Products Symposium*, C. Loxton (Ed.), p. 217. Llandudno, North Wales, UK (1997).
2. R. M. Rammon, W. E. Johns, J. Magnuson and A. K. Dunker, *J. Adhesion* **19**, 115 (1986).
3. S. S. Jada, *J. Macromol. Sci. Chem.* **A27**, 361 (1990).
4. J. Billiani, K. Lederer and M. Dunky, *Angew. Makromol. Chem.* **180**, 199 (1990).
5. W. Dankelman, J. M. H. Daemen, A. J. J. D. Breet, J. L. Mulder, W. G. B. Huysmans and J. D. Wit, *Angew. Makromol. Chem.* **54**, 187 (1976).
6. M. Dunky and K. Lederer, *Angew. Makromol. Chem.* **102**, 199 (1982).
7. M. Dunky, K. Lederer and E. Zimmer, *Holzforsch Holzverw.* **33**, 61 (1981).
8. D. Grunwald, *Kombinierte Analytische Untersuchungen von Klebstoffen für Holzwerkstoffe*. Mensch & Buch Verlag, Berlin (2002).
9. T. Hlaing, A. Gilbert and C. Booth, *British Polym. J.* **18**, 345 (1986).
10. C. Y. Hse, Z. H. Xia and B. Tomita, *Holzforschung* **48**, 527 (1994).
11. S. Katuscak, M. Tomas and O. Schiessl, *J. Appl. Polym. Sci.* **26**, 381 (1981).
12. P. R. Ludlam and J. G. King, *J. Appl. Polym. Sci.* **29**, 3863 (1984).
13. G. Zeppenfeld and D. Grunwald, *Klebstoffe in der Holz- und Möbelindustrie*. Drw Verlag, Weinbrenner, Germany (2005).
14. H. Mandal and A. S. Hay, *Polymer* **38**, 6267 (1997).
15. E. Minopoulou, E. Dessipri, G. D. Chryssikos, V. Gionis, A. Paipetis and C. Panayiotou, *Intl. J. Adhesion Adhesives* **23**, 473 (2003).
16. E. Dessipri, E. Minopoulou, G. D. Chryssikos, V. Gionis, A. Paipetis and C. Panayiotou, *European Polymer J.* **39**, 1533 (2003).
17. S. Tohmura, C. Y. Hse and M. Higuchi, *J. Wood Sci.* **46**, 303 (2000).
18. G. E. Myers, *J. Appl. Polym. Sci.* **26**, 747 (1981).
19. L. M. H. Carvalho, M. R. P. F. N. Costa and C. A. V. Costa, *J. Appl. Polym. Sci.* **102**, 5977 (2006).
20. A. F. Price, A. R. Cooper and A. S. Meskin, *J. Appl. Polym. Sci.* **25**, 2597 (1980).
21. M. R. P. F. N. Costa and R. Bachmann, in: *Handbook of Polymer Reaction Engineering*, T. Meyer and J. Keurentjes (Eds), vol. 1. Wiley-VCH (2005).
22. A. Kumar and A. Sood, *J. Appl. Polym. Sci.* **40**, 1473 (1990).
23. M. Dunky, *Intl. J. Adhesion Adhesives* **18**, 95 (1998).
24. A. Pizzi and K. L. Mittal (Eds), *Handbook of Adhesive Technology*, 2nd edn. Marcel Dekker, New York, NY (2003).
25. J. Ferra, A. Mendes, M. R. Costa, L. Carvalho and F. D. Magalhães, *J. Appl. Polym. Sci.*, in press.
26. H. G. Barth, B. E. Boyes and C. Jackson, *Anal. Chem.* **70**, 251 (1998).
27. K. Kumlin and R. Simonson, *Angew. Makromol. Chem.* **68**, 175 (1978).
28. P. R. Ludlam, J. G. King and R. M. Anderson, *Analyst* **111**, 1265 (1986).
29. D. Grunwald, COST-E13 Wood Adhesion and Glued Products — Working Group 1: Wood Adhesives (2001).
30. J. Ferra, J. Pereira, B. Pinto, F. Silva, J. Martins, A. Mendes, M. R. Costa, L. Carvalho and F. D. Magalhães, in: *Proc. International Panel Products Symposium*, M. Spear (Ed.), p. 97. Espoo, Finland (2008).
31. A. Despres and A. Pizzi, *J. Appl. Polym. Sci.* **100**, 1406 (2006).
32. J. Ferra, J. Martins, A. Mendes, M. R. Costa, L. Carvalho and F. D. Magalhães, in: *Proc. 3rd Int. Conf. on Environmentally-Compatible Forest Products*, J. Caldeira (Ed.), p. 17. Porto, Portugal (2008).

33. M. G. Kim, *J. Appl. Polym. Sci.* **80**, 2800 (2001).
34. M. G. Kim, H. Wan, B. Y. No and W. L. Nieh, *J. Appl. Polym. Sci.* **82**, 1155 (2001).
35. M. Zanetti and A. Pizzi, *J. Appl. Polym. Sci.* **91**, 2690 (2004).

Formaldehyde-Free Dimethoxyethanal-Derived Resins for Wood-Based Panels

M. Properzi [a,*], **S. Wieland** [b], **F. Pichelin** [a], **A. Pizzi** [c] **and A. Despres**

[a] Bern University of Applied Sciences, Architecture, Wood and Civil Engineering,
Solothurnstrasse 102, CH-2504 Biel, Switzerland
[b] Salzburg University of Applied Sciences, HTB-Wood Processing Technology & Construction,
5331 Kuchl, Austria
[c] ENSTIB-LERMAB, University of Nancy 1, 27 Rue du Merle Blanc, BP 1041,
F-88051 Epinal, France

Abstract

Here we report on the results of a research project focused on the reduction of formaldehyde emissions from wood-based panels, by using a novel set of adhesive formulations based on dimethoxyethanal (DME)-derived resins. The investigated adhesives were evaluated on laboratory scale in order to study their technical performances, their gluing parameters, their reactivity as well as their formaldehyde emissions. It was found that all formulations met the requirements of the current standards EN 319:1993-08 and that for class P2 particleboards for general uses. From the technical point of view, major advantages of the resin systems tested were found to be: colorless, low toxicity, easy handling, and high stability at room temperature (long shelf-life, pot-life and open-time). The formaldehyde emissions of the boards produced were found to be comparable to those of natural wood (F****JIS A 1460:2001 standard). The laboratory results obtained with these formulations were validated on industrial scale. The technical properties as well as the formaldehyde emissions were measured. The new formulations were shown to be able to satisfy the requirements of standards with very low levels of formaldehyde emissions. However, to fulfill the requirements of the wood industry, the reactivity of the adhesive needs to be enhanced.

Keywords

Dimethoxyethanal, DME, glyoxylic acid, formaldehyde, particleboards, aminoplastic resins, wood adhesives, UF

1. Introduction

The impact of volatile gas emissions from wood-based panels on the indoor environment is attracting more and more attention. In several cases the emissions to the indoor air have become an essential product selection criterion for both the

---

[*] To whom correpondence should be adressed. Tel.: (41-3) 2344-0344; Fax: (41-3) 2344-0391; e-mail: milena.properzi@bfh.ch

end-users and producers. Urea-formaldehyde (UF) resins are one of the most important adhesives in the wood panels industry. In Europe about 4 millions tons of UF resins are annually used in the production of particleboards and fiberboards. This corresponds to 80%–85% of the global European wood adhesive consumption. Compared to other wood adhesives, UF resins have some advantages such as water solubility, fast curing, good performances and low price. However, in spite of these advantages UF resins release formaldehyde that is considered to be dangerous to the human health. Formaldehyde is either generated during board manufacturing or during service life of the products. Principal reasons responsible for the formaldehyde emissions are well known to be: the unreacted free formaldehyde as well as the low stability and susceptibility to hydrolysis of the amino-methylene linkages. Thus, during the last years, reduction of formaldehyde emissions has become a key issue for both chemical and wood industries. Within the possible approaches, formaldehyde emissions have been minimized by: reduction of the formaldehyde/urea (F/U) molar ratio; addition of formaldehyde scavenger to the UF liquid resin or to the wood particles prior to gluing; post-treatment of the boards; and use of sealing agents and coatings [1, 2, 10].

However, in spite of these improvements, the increasing market demand has fostered research in the direction of novel products for use as alternatives to UF resins. One possible way is the substitution of formaldehyde with other aldehydes. However, research done in this direction showed that the use of alternative aldehydes, such as furfural, glyoxal, glutaraldehyde, leads to certain problems due to low solubility, excessive volatility, and toxicity [1, 5, 8, 11].

Recently a new colorless, low toxicity, low volatility and water soluble aldehyde was developed as cross-linking agent [3]. The product named dimethoxyethanal (DME) is obtained by the controlled reaction of methanol with glyoxal in acidic conditions. In comparison with glyoxal, where the two adjacent aldehyde groups impart a very high reactivity and functionality, DME exhibits the aldehyde monofunctionality of formaldehyde. Consequently, DME has potential to replace formaldehyde in the areas of melamine and urea-formaldehyde resins. Like formaldehyde, DME can react with melamine and urea in similar pH ranges, but its reactivity level is much lower. Moreover, while DME reacts with melamine (M) and urea (U) to form M-DME and U-DME aminoresins precursors, the subsequent condensation and cross-linking reaction does not occur unless the reaction is catalyzed during resin preparation by the addition of glyoxylic acid ($C_2H_2O_3$) [4, 5]. Research results displayed that the use of glyoxylic acid during reaction allows the formation of different oligomers derived by condensation of glyoxylic acid with two molecules of melamine to form dimers. These formaldehyde-free aminoresins precursors showed potential for use as adhesives in the wood-based panels manufacturing [3–5].

Within the frame of a project focused on the reduction of formaldehyde emissions from wood-based panels a number of M-DME and U-DME derived resins have been tested [5, 12]. The aim of the project was the development of new environmentally-friendly wood adhesives, behaving in many applications as formaldehyde-based adhesives, while not containing any formaldehyde. To promote the curing reaction of the aminoresins precursors, pMDI and latex were added to the systems. This article reports on the performances of a novel set of DME-based adhesive formulations for particleboards. It also summarizes the results of the optimization of both adhesive formulation and manufacturing parameters to increase the adhesive performances.

2. Material and Methods

2.1. Adhesive Mix Preparation

Three different adhesive formulations based on DME-derived resins were investigated (Table 1). Urea-dimethoxyethanal (U-DME, Fig. 1(a), U : DME = 1 : 2) and melamine-dimethoxyethanal (M-DME, Fig. 1(b), M : DME = 1 : 2) aminoresin precursors were synthesized according to the protocol prescribed by Despres *et al.* [4]. For the formulation number 1, the aminoresin precursor U-DME (68.9% solid content), was mixed with different amounts of pMDI (polymeric 4,4′-diphenylmethane diisocyanate, Desmodur VKS20, Bayer). Ammonium sulfate $((NH_4)_2SO_4$, 35% solid content), was used as hardener. The formulations number 2, were obtained by mixing M-DME (67.5% solid content) with pMDI (Desmodur VKS20, Bayer) and latex (Appretan N92131, Clariant), an acrylic ester copolymer of predominantly ethyl acrylate, with minor amounts of styrene and methacrylic acid, having latex solid content of 45% and pH of 6. Glyoxylic acid $(C_2H_2O_3$, 50% solid content) was used as hardener. The formulation number 3, was obtained by mixing together M-DME (67.5% solid content) with latex (Appretan N92131, Clariant). Glyoxylic acid $(C_2H_2O_3$, 50% solid content) was used as hardener.

2.2. Particleboards Production and Testing

Single-layer boards were produced using core layer particles having moisture content (MC) of 1%–3%. Softwood, Norway spruce (*Picea abies* L.), was used for the

Table 1.
Tested formulations. *Ammonium sulfate $((NH_4)_2SO_4)$, percentages on the resin solid content. **Glyoxylic acid $(C_2H_2O_3)$, percentages on the resin (M-DME + Latex) solid content

Formulation number	Formulation type	U-DME [%]	M-DME [%]	Latex [%]	pMDI [%]	Glyoxylic acid [%]**	Ammonium sulfate [%]*
1	U-DME/pMDI	80	–	–	20	–	3.5
2a	M-DME/Latex/pMDI	–	68.5	26.5	5	5	–
2b		–	63.5	26.5	10	5	–
2c		–	57	26.5	16.5	5	–
3	M-DME/Latex	–	70	30	–	5	–

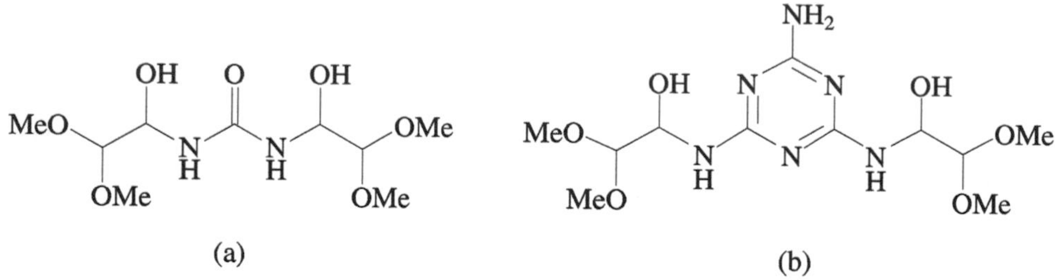

Figure 1. (a) Urea-dimethoxyethanal (U-DME molar ratio U : DME = 1 : 2) and (b) mela-mine-dimethoxyethanal (M-DME molar ratio M : DME = 1 : 2).

Table 2.
Molecular assignments for the ^{13}C NMR spectrum (Fig. 2) of U-DME resin (U : DME molar ratio = 1 : 2)

Molecules	Structures	Assignment (ppm)
Free urea	$NH_2-\underline{C}(=O)-NH_2$	161.050
Free DME	$(\underline{C}H_3-O)_2-CH-CH(OH)_2$	54.321
	$(CH_3-O)_2-\underline{C}H-CH(OH)_2$	104.099
	$(CH_3-O)_2-CH-\underline{C}H(OH)_2$	87.907
U-(DME)$_n$ with $n = 1, 2$	$NH_2-\underline{C}(=O)-NH-CH(-OH)-CH-(O-CH_3)_2$	158.819
	$-(HO-)HC-NH-\underline{C}(=O)-NH-CH(-OH)-$	156.770
	$-NH-\underline{C}H(-OH)-CH-(O-CH_3)_2$	72.359
	$-NH-CH(-OH)-\underline{C}H-(O-CH_3)_2$	103.360
	$-NH-CH(-OH)-CH-(O-\underline{C}H_3)_2$	54.321

wood particles. For the particleboards produced at laboratory scale, a rotary drum blender and an automated hot press were used to produce the panels. Five panels were produced per tested formulation and per gluing/manufacturing parameter. The board dimensions were 40 cm × 40 cm with 14 mm and 16 mm thickness, for a target density of 630–670 kg/m^3 (Tables 2–4). After pressing and prior to testing, the boards were conditioned in an environmental chamber (RH 65%, T 20°C) for 7 days (Tables 4–6).

For the industrial trial (Table 7) a rotary drum blender and an automated hot press were used to produce the boards. 30 m^3 panels were manufactured using the formulation number 3. The board dimensions were 418 cm × 192 cm with 16 mm thickness, for a target density of 670 kg/m^3. The internal bond (IB) strength of the boards was tested immediately after panel cooling as well as after 24 h (Table 7) using the method described in the EN 319:1993-08 standard [6], while the panel density was measured according to the EN 323:1994 standard [7].

Table 3.
Molecular assignment for the quantitative ^{13}C NMR spectrum (Fig. 3) of M-DME resin (M : DME molar ratio = 1 : 2)

Molecules	Structures	Assignment (ppm)
DME	($\underline{C}H_3-O)_2-CH-CH(OH)_2$	53.802
	$(CH_3-O)_2-\underline{C}H-CH(OH)_2$	103.806
	$(CH_3-O)_2-CH-\underline{C}H(OH)_2$	87.645
M-(DME)$_n$	$-N=\underline{C}-NH_2$	164.995
with $n = 0-3$	$-N=\underline{C}-NH-CH(-OH)-CH-(O-CH_3)_2$	164.068
	$-N=C-NH-\underline{C}H(-OH)-CH-(O-CH_3)_2$	72.395
	$-N=C-NH-CH(-OH)-\underline{C}H-(O-CH_3)_2$	102.995
	$-N=C-NH-CH(-OH)-CH-(O-\underline{C}H_3)_2$	54.090

Table 4.
Manufacturing parameters and IB strength of the panels produced at laboratory scale using the formulations numbers 1 and 2. Panel thickness of 14 mm. *MC = moisture content (%). To increase the blending MC, water was added to the system. **Pressing factor: time of pressing per mm of panel thickness (s/mm)

Sample number	Formulation number	Resin loading (%)	Blending MC* (%)	Open-time (min)	Pressing time (min)	Pressing factor** (s/mm)	Pressing temperature (°C)	Average target density (kg/m^3)	IB strength (N/mm^2)
1	1	8	12	15	4.7	20	193	629	0.38
2	1	8	12	15	6.4	27	193	609	0.40
3	1	8	10	15	4.7	20	193	602	0.37
4	1	8	10	15	6.4	27	193	609	0.40
5	1	8	8	15	4.7	20	193	611	0.47
6	1	8	8	15	6.4	27	193	593	0.45
7	1	8	8	15	4.7	20	200	609	0.44
8	1	8	8	15	7.5	32	200	623	0.52
9	2a	10	7	15	10	43	200	620	0.34
10	2b	10	7	15	10	43	200	633	0.42
11	2c	10	7	15	10	43	200	640	0.55

2.3. Determination of Formaldehyde Emissions

Formaldehyde emission tests were performed on the boards glued with the most promising formulations. Two boards per formulation type were tested. The emissions were measured according to the specifications of the JIS A 1460:2001 standard [9]. The determination of the formaldehyde emitted by the boards was done by the desiccator method. The emitted quantity of formaldehyde was obtained from

Table 5.
Manufacturing parameters and IB strength of the panels produced at laboratory scale using the formulation number 3. Panel thickness of 14 mm. *MC = moisture content (%). **Pressing factor: time of pressing per mm of panel thickness (s/mm)

Sample number	Formulation number	Resin loading (%)	Blending MC* (%)	Open-time (min)	Pressing time (min)	Pressing factor (s/mm)	Pressing temperature (°C)	Average target density (kg/m^3)	IB strength (N/mm^2)
12	3	10	8	15	5	21	200	650	0.33
13	3	10	8	15	7	30	200	642	0.38
14	3	10	8	15	10	43	200	643	0.44
15	3	10	8	30	5	21	200	651	0.36
16	3	10	8	30	7	30	200	644	0.43
17	3	10	8	30	10	43	200	639	0.42
18	3	10	8	15	5	21	220	640	0.25
19	3	10	8	15	7	30	220	603	0.24
20	3	10	8	15	10	43	220	647	0.28
21	3	10	8	30	5	21	220	636	0.26
22	3	10	8	30	7	30	220	–	–
23	3	10	8	30	10	43	220	638	0.32

Table 6.
Manufacturing parameters and IB strength of the panels produced at laboratory scale using the formulation number 3. *Pressing factor: time of pressing per mm of panel thickness (s/mm)

Sample number	24	25	26
Formulation number	3	3	3
Resin loading (%)	10	10	10
Blending moisture content MC (%)	8	8	8
Panel thickness (mm)	16	16	16
Pot-life (min)	15	150	15
Open-time (min)	15	15	150
Pressing time (min)	11.5	11.5	11.5
Pressing factor (s/mm)*	43	43	43
Pressing temperature (°C)	200	200	200
Average target density (kg/m^3)	660	660	660
Temperature of the wood chips (°C)	55	30	30
IB strength (N/mm^2)	0.17	0.37	0.38

the concentration of formaldehyde absorbed in a litre of distilled water, and was calculated as follows:

$$G = F(Ad - Ab)1800/S, \tag{1}$$

where:

Table 7.
Manufacturing parameters, IB strength (N/mm^2) and formaldehyde emissions (mg/l) for panels produced on industrial scale. *Pressing factor: time of pressing per mm of panel thickness (s/mm)

Sample number	27	28	29
Formulation number	3	3	3
Resin loading (%)	10	10	10
Blending moisture content MC (%)	8	8	8
Nominal panel thickness (mm)	16	16	16
Pot-life (min)	150	150	150
Open-time (min)	150	150	150
Pressing time (min)	11.5	9.1	7.5
Pressing factor (s/mm)*	43	34	28
Pressing temperature (°C)	200	200	200
Average target density (kg/m^3)	670	670	670
Temperature of the wood chips (°C)	30	30	30
IB strength (N/mm^2) tested immediately after panel cooling	0.36	0.32	0.26
IB strength (N/mm^2) tested 24 h after cooling	0.40	0.35	0.28
Formaldehyde emissions (mg/l)	0.20	0.20	0.25

G: concentration of formaldehyde in the solution (mg/l),

Ad: colorimetric absorbance of the solution inside desiccators containing test pieces,

Ab: colorimetric absorbance of the background water sample. This water sample was used in an empty desiccator without test pieces to absorb the formaldehyde background of the laboratory/desiccator air,

F: slope of calibration curve obtained using standard solutions of formaldehyde (mg/l),

S: surface area of the test piece (cm^2).

2.4. Nuclear Magnetic Resonance

The liquid ^{13}C NMR spectrum of the M-DME used was obtained on a Bruker Avance 400 FT NMR spectrometer. Chemical shifts were calculated relative to DSS (sodium 2,2-dimethyl-2-silapentane-5-sulfonate) dissolved in D$_2$O. The spectrum was recorded at 100 MHz for approximately 3000 transients. All the spectra were recorded with a relaxation delay of 2 s and the chemical shifts were accurate to 1 ppm.

3. Results and Discussion

Melamine-DME and urea-DME resins precursors (Fig. 1) were prepared by the co-reaction of melamine and urea with two moles of dimethoxyethanal (DME). The

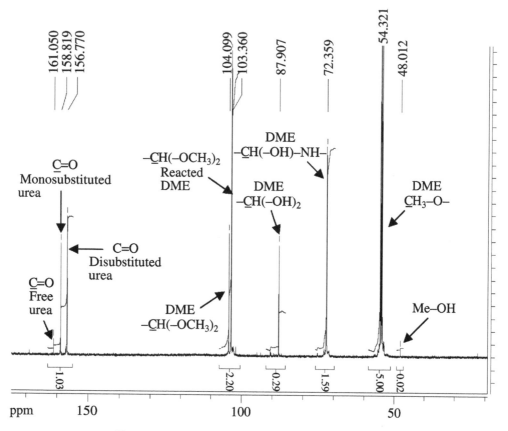

Figure 2. ^{13}C NMR spectrum of U-DME resin (U : DME molar ratio = 1 : 2).

^{13}C NMR spectrum of the U-DME resin with molar ratio U : DME = 1 : 2 is shown in Fig. 2 and assignments are presented in Table 2. The results (Table 2) show the carbonyl groups (C=O) of the free, mono and disubstituted ureas that transmit in the region of 156.77–161.05 ppm. The small intensity of the signal at 161.05 ppm indicates that the majority of the urea had reacted. The peaks at 54.321 ppm, 87.907 ppm and 104.099 ppm correspond to the carbons of the free DME. The ^{13}C NMR also shows the absence of carbon corresponding to DME reacted with two moles of urea (49 ppm). This can be interpreted as the absence of polymers in the U-DME resin and it results in low reactivity of the resin and poor mechanical strength of the cured U-DME.

The ^{13}C NMR spectrum of the M-DME resin with molar ratio M : DME = 1 : 2 shown in Fig. 3 is very similar to that of the U-DME. Major difference is represented by the C=N groups of the triazine ring of the reacted melamine that did not appear in the U-DME spectrum (Table 3). As melamine presents three reactive amine groups and due to the resin molar ratio (M : DME = 1 : 2), a large majority of the DME seemed to have reacted. However, no trace of DME reacted with two

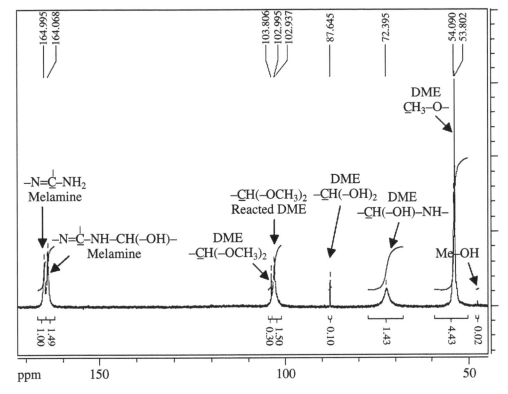

Figure 3. ^{13}C NMR spectrum of M-DME resin (M : DME molar ratio = 1 : 2).

molecules of melamine is evident. It is known that M-DME resins are able to polymerize. The reactive hydroxyalkyl groups (on the carbon in the α-position to the nitrogen atom) allow them to form ether bridges between two adducts. However, the absence of polymers in the M-DME spectrum indicates that condensation reactions occur only under specific conditions. Thus, to increase the adhesive performance of the U-DME and M-DME aminoresins precursors, pMDI and latex were added as cross-linking agents in the glue mix; and to foster condensation reactions, glyoxylic acid was used as catalyst.

Tests carried out on the formulation type 1 (U-DME/pMDI) showed that all produced boards were able to satisfy the specifications of the EN standard for particleboards class P2 (IB strength > 0.35 N/mm^2) (Table 4). The increase of the blending moisture content from 8% to 12% did not display any positive effect on the internal bond (IB) strength of the boards. Results suggest that the moisture coming from the wood particles (1%–3% RH), as a well as the water from the glue —
mix is sufficient to produce boards able to reach the target IB strength. Results also showed that while pressing at the temperature of 193°C, an increase of the pressing time from 4.7 min to 6.4 min had only a slight influence on the final IB strength of the boards. On the contrary, the increase of press temperature to 200°C in combination with a long pressing time led to a substantial increase in the IB strength. The

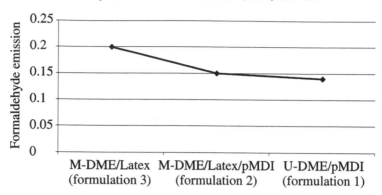

Figure 4. Formaldehyde emission test results (mg/l) using the desiccator method according to JIS A 1460:2001 standard [9]. F**** designation: formaldehyde emission < 0.3 mg/l, corresponding to the emission of natural wood.

best IB strength of 0.52 N/mm^2 was reached with a resin loading of 8%, blending moisture content of 8%, and 7.5 min of pressing at 200°C temperature (Table 4, sample number 8).

The investigation carried out on the type 2 formulations (M-DME/pMDI/Latex) displayed very positive results. The focus of the study was the reduction of the pMDI amount and of the global resin loading, as well as of the pressing time. Table 4 illustrates the IB strength of the boards produced with different combinations of glue mix and manufacturing parameters. The best result (0.55 N/mm^2) was reached with the introduction of 16.5% of pMDI into the adhesive mix. Reduction of the pressing time from 10 min to 5.5 min was also possible; but there was a decrease in the IB strength from 0.55 N/mm^2 to 0.40 N/mm^2. Also the amounts of pMDI and the total resin loading could, up to a certain extents, be reduced. However, at least 10% of pMDI and long pressing time (10 min) were needed to reach the target IB strength. Boards produced with shorter pressing times were not able to fulfill the requirement of the EN standard for particleboards (class P2).

Results also showed that a slight increase of the latex content (from 26.6% to 30%) in the M-DME-based adhesive mix could provide sufficient IB strength without any addition of pMDI. Table 5 displays the results of the parametric study carried out on the type 3 formulation (M-DME/Latex). With a resin loading of 10%, a positive correlation between pressing time and IB strength of the boards was found. Thus the best result (0.44 N/mm^2) was achieved after 10 min of pressing. However, within the investigated parameters, it was found that 7 min of pressing time was sufficient to produce boards able to pass the standard. Moreover, a slight positive influence of the open-time on the final panel's IB strength was found in most cases. With the aim to reduce the production time, the pressing temperature was increased to 220°C. However, boards produced even at such a temperature were not able to fulfill the requirement of the EN standard for particleboards (class P2).

Formaldehyde emissions tests were carried out on the most promising adhesive formulations (Fig. 4). The best result 0.14 mg/l was obtained with the formulation number 1 (U-DME/pMDI), but all other formulations also gave boards able to pass the requirements of the F**** designation of the Japanese standard for formaldehyde-free emissions for wood-based panels. Such standard requires the emission levels measured with the desiccator method to be lower than 0.3 mg/l.

Due to the absence of pMDI in the glue mix, and based on the positive results achieved in terms of IB strength and formaldehyde emissions, the formulation number 3 was selected for the industrial trial. Thus, additional studies were done at laboratory scale to verify the industrial suitability of the chosen formulation. For this purpose, panels of 16 mm thickness were produced under different conditions and tested. Table 6 summarizes the results of the pre-industrial study. It was found that an increase of the board thickness from 14 mm to 16 mm resulted in only a slight decrease of the panel quality. However, the boards produced with the optimized parameters easily satisfied the requirements of the standard (Table 6). Based on the results of a previous study carried out on different release agents [5], the product Würtz PAT 1667/D (Würtz GmbH & Co., Germany) was selected as the most efficient agent for the M-DME/latex formulation. Table 6 also shows that the evaluation of the influence of wood chips temperature revealed that prior to gluing the high temperature of the wood particles (55°C) had a negative influence on the final quality of the boards (IB strength 0.17 N/mm^2). It was concluded that this temperature would need to be lowered to below 30°C prior to gluing. The study carried out on boards pressed 2.5 h after the open-time and 2.5 h after the pot-life did not display any significant change in the IB strength of the boards. This confirmed the high stability at room temperature of the selected formulation (Table 6).

Table 7 summarizes the manufacturing parameters for the industrial trials. The first series of boards showed an average IB strength of 0.36 N/mm^2. However, due to the long pressing time (11.5 min), board manufacturing in a continuous production process was not possible. The production had to be stopped and started again each time to refill the particle bunker on the press line. Due to this, some variations in density were observed. To increase the productivity and to overcome these difficulties, the pressing time was reduced to 9.1 min and 7.5 min. The IB strength of the boards tested immediately after cooling showed that a reduction of the pressing time was possible but resulted in a decrease of the IB strength. These results confirmed the previous finding at the laboratory scale. Table 7 also shows an increase in the board strength 24 h after pressing, suggesting that the adhesive was probably not fully cured immediately after pressing. The low reactivity of the M-DME aminoresin precursor would also explain the decrease of the IB strength with the increase of the panel thickness. It is known [4] that while DME reacts with melamine (M) and urea (U) to form M-DME and U-DME aminoresins precursors, the subsequent condensation and cross-linking reaction does not occur unless the reaction is catalyzed during resin preparation with glyoxylic acid ($C_2H_2O_3$). The acidity of the glyoxylic acid during M-DME resin preparation is known to play a major role in promoting the oligomerization of the aminoresin precursor. However, within the frame of the investigated parameters, the simple addition of glyoxylic acid to the

M-DME resin during boards production seemed not to be sufficient to promote the adhesive curing. Therefore, to foster condensation and cross-linking reactions, either the M-DME is pre-reacted with glyoxylic acid, or a longer pressing time or a higher resin loading is used [4]. Formaldehyde emission tests on the industrial boards confirmed the laboratory scale results. The formaldehyde emissions were found to be always lower than 0.3 mg/l (F**** designation, JIS standard) (Table 7). This corresponds to formaldehyde emission levels of solid untreated wood.

4. Summary and Conclusions

Within the scope of the project several new adhesive formulations were developed and tested. The formulations were based on DME (dimethoxyethanal) resin and the products derived from it to produce formaldehyde-free wood adhesives. The investigated adhesives were evaluated on laboratory scale in order to study their technical performances, their gluing parameters, their reactivity as well as their formaldehyde emissions. It was found that all formulations met the requirements of current standards EN 319:1993-08 and that for class P2 particleboards for general uses. From the technical point of view, major advantages of the tested systems were found to be: colourless, low toxicity, easy handling, and high stability at room temperature (long shelf-life, pot-life and open-time). The formaldehyde emissions of the boards produced were found to be comparable with those of solid untreated wood (F****JIS A 1460:2001 standard).

The addition of pMDI increases the IB strength of the boards (formulation 2). However, when glyoxylic acid is used as hardener, the combination of M-DME and latex without any pMDI provided good performances of wood adhesives (formulation 3). The laboratory results obtained with this formulation were validated on industrial scale and 30 m^3 of particleboards were produced. The technical properties as well as the formaldehyde emissions were measured. The new formulation was shown to be able to satisfy the requirements of standards with low levels of formaldehyde emissions. However, to fulfill the requirements of the wood industry, the reactivity of the adhesive needs to be enhanced. A possible way, already explored by other authors [4], is the co-reaction of M-DME with glyoxylic acid prior to adhesive application.

Acknowledgements

The French author gratefully acknowledges the financial support of the CPER 2007-2013 "Structuration du Pôles de Compétitivité Fibres Grand' Est" (Competitiveness Fibre Cluster), through local (Conseil Général des Vosges), regional (Région Lorraine), national (DRRT and FNDT) and European (FEDER) funds.

References

1. A. Pizzi (Ed.), *Wood Adhesives: Chemistry and Technology*. Marcel Dekker, New York (1983).

2. B.-D. Park, E.-C. Kang and J.-Y. Park, *J. Appl. Polym. Sci.* **110**, 1573–1580 (2008).
3. C. Vu, A. Pizzi and A. Despres, Patent, Pub. no. WO/2007/099156 (2007).
4. A. Despres, A. Pizzi, C. Vu and H. Pasch, *J. Appl. Polym. Sci.* **110**, 3908–3916 (2008).
5. A. Despres, *PhD Thesis*, University HP Nancy 1, Epinal, France (2006).
6. EN 319:1993-08, Particleboards and fibreboards: determination of tensile strength perpendicular to the plane of the board (1993).
7. EN 323:1994, Wood-based panels: determination of density (1993).
8. H. M. Mansouri and A. Pizzi, *J. Appl. Polym. Sci.* **102**, 5131 (2006).
9. JIS A 1460:2001, Building boards, determination of the formaldehyde emissions — desiccator method (2001).
10. S. Kim, *Construction Building Materials* **23**, 2319–2323 (2009).
11. S. Wang and A. Pizzi, *Holz Roh Werkst.* **55**, 9 (1997).
12. S. Wieland, *PhD Thesis*, University HP Nancy 1, Epinal, France (2007).

Melamine–Formaldehyde Resins without Urea for Wood Panels

E. Pendlebury [a,b], **H. Lei** [a,c], **M.-L. Antoine** [a] **and A. Pizzi** [a,*]

[a] ENSTIB-LERMAB, Nancy University, 27 rue du Merle Blanc, BP 1041, 88051 Epinal, France
[b] Department of Chemistry, University of Strathclyde, Thomas Grohen Bldg., 295 Cathedral Street, Glasgow, G1 IXC, UK
[c] Southwest Forestry University, Bai Long Si Road, 650224 Kunming, Yunnan, P. R. China

Abstract

Pure melamine–formaldehyde (MF) resins as adhesives for wood panels in which urea was not added during manufacture gave good bonding results for wood panels. The best performing formulation was found to be the one with a relatively low formaldehyde/melamine molar ratio while still being a fast-curing resin. This was confirmed by thermomechanical analysis scans of the resins and by pressing and testing laboratory wood particleboard. Its performance was explained by comparative gel permeation chromatography analysis of the resins, and by density profile analysis of the panels prepared with the resin.

Keywords

Formulations, adhesives, melamine–formaldehyde, resins, thermomechanical analysis, gel permeation chromatography, density profile, wood panels

1. Introduction

Melamine–urea–formaldehyde (MUF) resins are among the most commonly used resins for application as binders for wood panels [1]. However, the high cost of melamine has led over decades to a progressive decrease of melamine content in these resins while maintaining the high performance of the adhesive by improved resin formulation [2–7].

Recently, however, the United States Green Building Council has promoted in the USA the use of composite products manufactured without any urea–formaldehyde [8]. While the addition of urea as a potential formaldehyde scavenger in the glue-mix of a finished pure melamine–formaldehyde (MF) resin is tolerated, the use of even the best copolymerized MUF resins today on the market is not ac-

---

* To whom correspondence should be addressed. Tel.: (+33) 329296117; Fax: (+33) 329296138; e-mail: antonio.pizzi@enstib.uhp-nancy.fr

Wood Adhesives
© Koninklijke Brill NV, Leiden, 2010

ceptable to this organisation. It is for this reason that the need to return to pure MF resins and to develop adequate modern formulations has been felt by some wood panel manufactuers [8]. Thus this paper deals with the development and testing of a pure MF adhesive formulation in which urea was not used at all. Thus, three pure MF resins derived from formulations originally developed for other types of applications [9, 10] were tried and evaluated.

2. Experimental

2.1. Resin Preparation

The following procedures were used for the production of 3 different types of resins. The resins were all melamine–formaldehyde (MF) resins with differing molar ratios.

Type A Resin — Molar Ratio = 1.9 [9, 10]
70 g of paraformaldehyde were dissolved in 200 g of water and the pH adjusted to approximately 11.9 using 33% NaOH solution and 30% acetic acid. 157 g of melamine and 12 g of dimethylformamide were then added and heated at 90–96°C with constant stirring. When all the melamine had dissolved the pH was lowered to between 9.6 and 10. From this point on the pH and temperature were monitored closely until the water tolerance of 180% was reached and then the reaction was halted by cooling in cold water. The average solids content measured for the finished resin was 48.5%.

Type B Resin — Molar Ratio = 2.07 [9, 10]
25 ml of water were added to 120 ml of 37% formalin and the pH was adjusted to 9.7 using 33% sodium hydroxide solution and 30% acetic acid. 90 g melamine were added and the pH brought back to 9.7. Then it was heated at 92–96°C with constant stirring. This temperature was maintained until the turbidity point was reached. At this point the pH was brought to 10.6 and an extra 10% of melamine (9 g) (2nd addition) was added. When all the melamine had dissolved the water tolerance was tested, and when it reached 160% the reaction was halted by quenching the reaction vessel in cold water. The average solids content measured for the finished resin was 42.7%.

Type C Resin — Molar Ratio = 1.58 [9, 10]
101 g of 37% formalin were diluted with 50 ml of water and the pH was adjusted to 9.7 using 33% sodium hydroxide and 30% acetic acid. 99 g of melamine were then added along with a 2nd addition of water (20 ml). The pH was then brought to 9.8 and the reaction mixture heated at 85–90°C with constant stirring. The water tolerance was tested and when it was found to be 200–220% the reaction was halted by quenching the reaction vessel in cold water. The pH was then checked again and adjusted to 9.8 before being stored. The average solids content measured for the finished resin was 50.9%.

2.2. TMA Analysis

The three resins above were tested by thermomechanical analysis (TMA) on a Mettler 40 apparatus. Triplicate samples of beech wood alone, and of two beech wood plys each 0.6 mm thick bonded with each resin system were tested. Sample dimensions were 21 mm × 6 mm × 1.2 mm. The samples were tested in non-isothermal mode from 40°C to 220°C at heating rates of 10°C/min, 20°C/min and 40°C/min with a Mettler 40 TMA apparatus in three-point bending on a span of 18 mm. A continuous force cycling between 0.1 N and 0.5 N and back to 0.1 N was applied on the specimens with each force cycle duration being 12 s. The classical mechanics relation between force and deflection $E = [L^3/(4bh^3)][\Delta F/(\Delta f)]$ (where L is the sample length, ΔF the force variation applied and Δf the resulting deflection, b the width and h the thickness of the sample) allows calculation of the modulus of elasticity E for each case tested and to follow its rise as functions of both temperature and time. The deflections Δf obtained and the values of E obtained from them proved to be constant and reproducible.

2.3. Gel Permeation Chromatography (GPC)

Samples of resin A and resin C which were shown by TMA to be the two best MF resins were analysed by gel permeation chromatography (GPC). A Waters 515 HPLC pump and GPC system were used and the resins analysed through a Styragel HR1 column (for determination of M_w between 100 and 5000) at an elution rate of 1 ml/min, after the column was calibrated with poly(ethylene glycol) (PEG). The PEG samples used for calibration had M_w of 200, 300, 400, 600, 1000, 2000, 3400, 8000, 10 000. Each resin sample after having been disolved in dimethylformamide was tested after filtering through a 0.45 μm filter. A Waters 410 refractometer was used as the detector.

2.4. Glue Mixes and Wood Particleboard Preparation and Testing

The glue mixes for the panels were prepared by adding to the relevant resin, on basis of resin solids, 3% urea (as a 25% solution in water) and 3% ammonium sulphate (as a 30% solution in water).

One-layer laboratory particleboards of dimensions 350 mm × 310 mm × 14 mm with a target density of 690–700 kg/m³ were then produced in duplicate using an industrial wood chips mix composed of 70% by weight of beech and 30% by weight of spruce by adding 10% total MF resin solids content on dry wood particles. The panels were pressed at a maximum pressure of 28 kg/cm² (2 min from platen contact to high pressure + maintaining high pressure) followed by a descending pressing cycle of 1 min at 12–14 kg/cm² and 2 min at 5–7 kg/cm², at 190°–195°C and for a total pressing time of 5 min. The moisture content of the resinated chips was 12%. The panels, after light surface sanding, were tested for dry internal bond (IB) strength according to European Norm EN 312 [11]. The density profile of the panels along with their thickness were tested with an X-ray density profiler Grecon Da-X (Fagus-Grecon Greten GmBH, Alfeld-Hannover, Germany).

3. Results and Discussion

The thermomechanical analysis (TMA) scans of the three MF formulations shown in Fig. 1 indicate that resin C appears to give (a) the highest ultimate strength as shown by the highest modulus of elasticity (MOE) value obtained, and also (b) the quickest setting and curing of the resin as indicated by the rise of the MOE curve occurring at lowest temperature. Resin A is second best, and resin B appears to be the slowest curing and giving the weakest cured glueline. On the basis of these TMA scans resin B was excluded from further testing as being the least promising of the three.

The TMA results are confirmed by the internal bond (IB) strengths obtained for the wood particleboard panels bonded with the two formulations A and C (Table 1). In Table 1 the difference in IB strengths of resin A- and resin C-bonded wood particleboards is statistically significant. Resin C, as in the TMA test, appears to be the best performing MF formulation for use in wood based panels. Its internal bond strength is well over the relevant standard [11]. The density profiles of the panels bonded with resins A and C were determined because this is the test that controls that no errors were made during pressing of the panel. The density profiles of these two panels are shown in Figs 2 (resin A) and 3 (resin C). Two samples of nearly the same density were chosen so as to avoid any major effect due to great disparity in densities. The panel bonded with resin A (Fig. 2) shows a minimum density in the core of the panel of 576 kg/m^3, thus considerably lower than the 628 kg/m^3 for the panel bonded with resin C (Fig. 3). This indicates that at equal average panel densities the core of the panels bonded with resin A was less densified. As the moisture content of the resinated mats before pressing was the same, and also

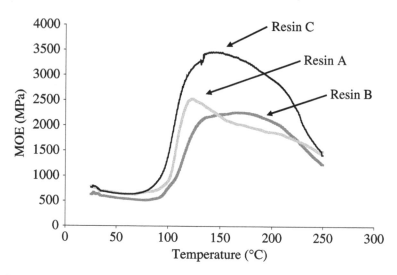

Figure 1. Results of thermomechanical analysis of resins A, B and C. Note the higher ultimate value of MOE for resin C and the increase in MOE at lower temperatures for both resins C and A, evidencing them as faster curing resins.

Table 1.
Internal bond strength results of particleboards bonded with
MF resins A and C

Resin type	Board density (kg/m^3)	Internal bond strength (MPa)
A	677 + 6	0.27 + 0.05
C	691 ± 11	0.89 ± 0.03
EN 312 requirement		⩾0.35

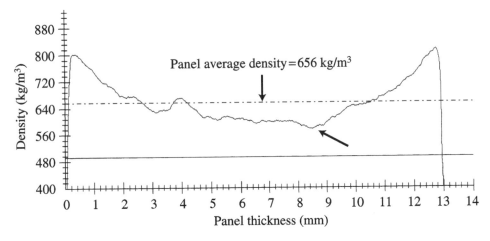

Figure 2. Density profile as a function of panel thickness for a sample from a panel bonded with resin A. Note the localized dip in the board core's density indicated by an arrow.

the pressing time, maximum pressure and pressure cycle were the same, the only reason why this difference could appear is due to lower resin flow for resin A. In short, resin A may well be more polymerised than resin C. This can be deduced from the gel permeation chromatograms of the two resins in Figs 4 and 5.

The GPC analysis helps to understand the reason for the difference in the be-haviours of the two resins, in particular why a much lower formaldehyde/melamine molar ratio (resin C) performs markedly better as a wood adhesive than resin A which has a much higher formaldehyde/melamine molar ratio. Figures 4 and 5 show the gel permeation chromatography (GPC) analysis results for the two resins. Resin A in Fig. 4 shows a more complex series of peaks. Oligomers of higher mole-cular masses are present, but the presence of a small but consistently wide GPC peak at a MW higher than 8000 indicates the presence of colloidal aggregates in the resin [12], because if oligomers of such higher MW existed they would precipitate. Resin C, in Fig. 5, shows the presence of oligomers but the lower molecular mass oligomers are more predominant than in the case of resin A. This indicated that resin A was markedly more polymerized than resin C. This means that resin A is

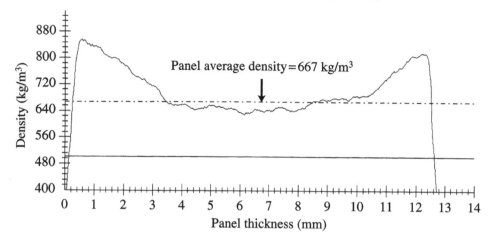

Figure 3. Density profile as a function of panel thickness for a sample from a panel bonded with resin C. Note the even density of the panel core.

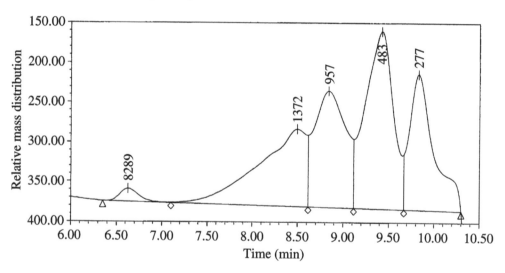

Figure 4. Gel permeation chromatogram of resin A showing both the presence of high molecular weight oligomers as well as the appearance of colloidal aggregates at around 8000 number average molar mass.

likely to present problems of either wood wetting or of flowing under heat due to its higher average molecular mass. This will cause both lower adhesion as well as lower board core density. Thus, there seems to be an optimal level of polymerization of the resin to maximize performance.

4. Conclusions

Pure MF resins as adhesives for wood panels in which urea was not added during manufacture but only as a small proportion additive in the glue-mix, according to

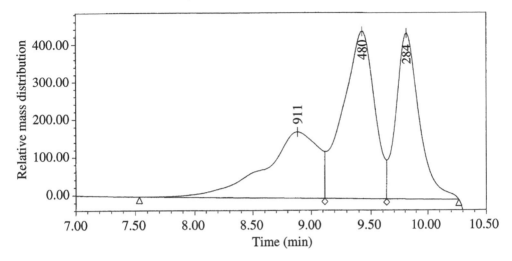

Figure 5. Gel permeation chromatogram of resin C showing the absence of high molecular weight oligomers and of colloidal aggregates of higher molar masses.

the directives to industry of the United States Green Building Council, were investigated and the following were found.

1. These adhesives gave good bonding results for wood panels.

2. The best performing formulation was found to be the one with a relatively low formaldehyde/melamine molar ratio while still being a fast-curing resin.

3. Resins that are too polymerised gave worse results due to lower flow and wetting capability during panel pressing.

Acknowledgements

The French authors gratefully acknowledge the financial support of the CPER 2007–2013 'Structuration du Pôle de Compétitivité Fibres Grand'Est' (Competitiveness Fibre Cluster), through local (Conseil Général des Vosges), regional (Région Lorraine), national (DRRT and FNADT) and European (FEDER) funds.

References

1. A. Pizzi, *Advanced Wood Adhesives Technology*. Marcel Dekker, New York, NY (1994).
2. M. Zanetti and A. Pizzi, *Holz Roh Werkstoff* **62**, 445–451 (2004).
3. A. Pizzi, M. Beaujean, C. Zhao, M. Properzi and Z. Huang, *J. Appl. Polym. Sci.* **84**, 2561–2571 (2002).
4. M. Zanetti and A. Pizzi, *J. Appl. Polym. Sci.* **88**, 287–292 (2003).
5. C. Kamoun, A. Pizzi and M. Zanetti, *J. Appl. Polym. Sci.* **90**, 203–214 (2003).
6. M. Zanetti and A. Pizzi, *J. Appl. Polym. Sci.* **90**, 215–226 (2003).
7. M. Prestifilippo, A. Pizzi, H. Norback and P. Lavisci, *Holz Roh Wekstoff* **54**, 393–398 (1996).
8. R. Lepine, private communication (2009).
9. A. T. Mercer and A. Pizzi, *J. Appl. Polym. Sci.* **61**, 1697–1702 (1996).

10. A. T. Mercer, ^{13}C NMR analysis of strength and emission of melamine and melamine–urea–formaldehyde resins for synthesis optimisation, *PhD thesis*, University of the Witwatersrand, Johannesburg, South Africa (1996).
11. European Norm EN 312, Wood particleboard — specifications (1995).
12. M. Zanetti and A. Pizzi, *J. Appl. Polym. Sci.* **91**, 2690–2699 (2004).

Bonding of Heat-Treated Spruce with Phenol-Formaldehyde Adhesive

Mirko Kariz and Milan Sernek *

Department of Wood Science and Technology, Biotechnical Faculty, University of Ljubljana,
Rozna dolina, C. VIII/34, 1001 Ljubljana, Slovenia

Abstract

The modified chemical, physical and structural properties of wood after heat treatment can affect the bonding process with adhesives. The objective of this research was to determine to what extent the degree of heat treatment influenced the bonding of treated wood with phenol-formaldehyde (PF) adhesive. Spruce (*Picea abies* Karst) lamellas were heat treated at 180°C and 220°C and then bonded. The shear strength and wood failure of the differently pretreated specimens were determined. Wettability and penetration of the adhesive were also investigated. The results showed that the shear strength of the PF adhesive bond was influenced by the heat treatment of the spruce wood, and depended on the type of pretreatment of the wood specimen prior to testing. The observed reduction in the shear strength of the PF adhesive bond was ascribed to a decrease in the wood strength itself, caused by the heat treatment, and also to other effects induced by exposure of the wood to elevated temperatures. The changes in wettability and adhesive penetration did not significantly influence the bonding process of the heat-treated wood with the PF adhesive.

Keywords

Bonding, contact angle, heat-treated wood, penetration, phenol-formaldehyde adhesive, shear strength

1. Introduction

The heat treatment of wood at elevated temperatures, ranging from 160°C to 260°C, improves the dimensional stability of wood and increases its resistance to decay [1–3]. The increased dimensional stability of heat-treated wood is mainly due to reduced hygroscopicity. Exposure of wood to high temperatures leads to chemical modifications of hemicelluloses, cellulose and lignin [4, 5], which decrease the number of hydroxyl groups, especially those in the hemicelluloses. Reduction of the quantity of free hydroxyl groups decreases the moisture uptake, and is additionally accompanied by the formation of hydrophobic substances due to cross-linking reactions of the wood polymers [6]. The improved durability of heat-treated wood is explained by the significantly modified chemical composition of the wood (it is

---

* To whom correspondence should be addressed. Fax: + 386 1 257 2297; e-mail: milan.sernek@bf.uni-lj.si

Wood Adhesives
© Koninklijke Brill NV, Leiden, 2010

assumed that fungi do not recognise this new material as a wood substrate), and by the elimination of the pentosans as a constituent of the hemicelluloses, which are the main nutrition source for fungi [7].

Heat treatment also causes some unfavourable effects, such as lower bending and tension strengths, reduced toughness and, thus, increased brittleness of wood [2, 5, 8–13]. Modified chemical, physical and structural properties of wood after heat treatment can affect the bonding process with adhesives.

Strong adhesion between the adhesive and the wood substrate is achieved by appropriate adhesive flow, penetration, wetting and curing [14]. A higher degree of thermal modification, which is indicated by a higher mass loss of wood substance, causes significant changes in the properties that affect wood bonding. The improved dimensional stability of heat-treated wood generally improves the bonding performance, because the stresses due to shrinking or swelling on the cured adhesive bond are reduced [15]. However, most of the other altered properties can reduce bond strength. The shear strength of the modified wood decreases, and this can impact the shear strength of the adhesive bond. Sometimes, it is difficult to determine whether the adhesive bond strength decreases because of poor bonding or due to the reduced strength of the heat-treated wood. Suchsland and Stevens [16] devised a useful analysis method for determining whether wood strength or bondline strength was the cause of bond failure. In this method the wood failure is plotted against plywood shear strength. A reduction in wood strength should increase wood failure and reduce shear strength. Reductions in adhesive bond strength should reduce both wood failure and wood shear strength.

Because of the less hygroscopic character of heat-treated wood [1, 8, 17], both the distribution of the adhesive on the wood surface as well as the penetration of the adhesive into the porous wood structure can be affected. The amount of adhesive penetration into a wood substrate is correlated with the bond quality [18]. Insufficient penetration causes minimal surface contact for chemical bonding or "mechanical interlocking". Overpenetration of the adhesive will create "starved" or dry bondlines. An ideal amount of adhesive penetration would repair machining damage of the wood surface, and permit more effective stress transfer between the laminates [19]. A reduction in the hydrophilic character of heat-treated wood can affect adhesive penetration into the wood structure, because most wood adhesives contain a large amount of water as a solvent. But adhesives can also overpenetrate the porous wood structure because of the limited capacity of modified wood tissue to absorb water from the curing adhesive bondline, so the adhesive stays mobile for a longer period of time [20].

The wettability of wood with water is decreased after heat treatment [21–24], mainly because the surface of the heat-treated wood is hydrophobic, less polar and significantly repellent to water. This might hinder waterborne adhesives from adequately wetting the surface and thus causes poor adhesion. Sometimes, the wettability of modified wood with adhesive can be better, if the adhesive is hydrophobic [23].

Heat-treated wood accepts less water at a slower rate, so the hardening of adhesives that cure by solvent removal is slower. In such cases, it is necessary to adapt the usual bonding process. When working with poly(vinyl acetate) (PVAc) adhesive, the water content in the adhesive should be minimized, and pressing time should be longer, because of the slower water absorption rate [25]. Swietliczny *et al.* [26] have suggested that because of the slower water absorption into heat-treated wood, the influence of external conditions (e.g., water) on the shear strength of the adhesive bond is smaller.

The reduced equilibrium moisture content of heat-treated wood can influence the hardening of adhesives that need water for the curing reaction. This is significant for 1-component polyurethane (PUR) adhesive, which cures slower when bonding heat-treated wood with a smaller content of hydroxyl groups than non-treated wood [25, 27, 28].

During the heat treatment of wood, extractives can migrate to the surface and cause inactivation of the surface or react with the adhesive [22, 29, 30]. Possible consequences of surface inactivation due to the migration of extractives can be that the adhesive fails to wet and penetrate the wood; or that the adhesive bonds to a thin layer of extractives at the wood surface; or that the extractives dissolve into the adhesive and impact the curing kinetics or make the adhesive too viscous to be able to penetrate adequately into the wood structure [12].

Heat treatment can also reduce the pH of the wood to a value of 3.5–4 [10], which might retard or accelerate the curing of adhesives, depending on the type of adhesive used for bonding. For instance, the acetic and formic acids which are present in wood after heat treatment might neutralize the alkaline hardeners used for phenolic resins and hinder the adhesive hardening. On the other hand, a low pH of the wood surface could accelerate the chemical reactions of the acid catalyzed amino resins [31].

In summary, heat treatment can cause significant changes related to adhesion to the wood. Some previous studies have considered this problem and recommended some solutions. For the bonding of heat-treated wood with the 1-component methylene diphenyl diisocyanate (MDI) system, longer open times and longer pressing times were recommended [32]. Poncsak *et al.* [33] bonded different wood species with phenol-resorcinol-formaldehyde (PRF) and PUR adhesives. The untreated samples provided better bonding strength than heat-treated samples. But the wood failure was always high, so the lower adhesive bond strength was the consequence of the lower strength of the heat-treated wood. Similar results were found when bonding heat-treated spruce with waterborne melamine-urea-formaldehyde (MUF) and PRF adhesives [15, 34]. The difference in adhesive bond strengths between non-treated and treated wood samples depended on the type of adhesive. For non-waterborne PUR adhesive, no obvious difference was found.

The objective of this research was to determine whether the degree of heat treatment of the wood influenced the bonding of treated wood with phenol-formaldehyde adhesive. The shear strength of the adhesive bond, the percentage of wood failure, surface wettability and penetration of adhesive were examined. Since the successful bonding of heat-treated wood can significantly increase the range of applications for this material, a practical objective was to try to determine whether the results for adhesive bond strength met the requirements for solid wood panels for different conditions as specified by standard EN 13353:2008 [35].

2. Materials and Methods

2.1. Heat Treatment of Spruce

Spruce (*Picea abies* Karst) lamellas were used to investigate the influence of the heat treatment of wood on the shear strength of the phenol-formaldehyde adhesive bond. Lamellas were heat treated at two temperatures: 180°C and 220°C. The process of heat treatment in a vacuum, developed by Rep and co-workers [36], was used. Prior to heat treatment, all lamellas were planed to dimensions of 350 mm × 100 mm × 18 mm, and oven dried at 103°C. The treatment was performed in a vacuum chamber (Kambič, Laboratory Equipment d.o.o., Semič, Slovenia), where an absolute pressure of 5 kPa was achieved. The lamellas were heated to the desired temperatures, which took about 1 h, and then treated for 3 h at a constant temperature. The lamellas were then left to slowly cool down to room temperature. The lamellas were oven dried before and after heat treatment to determine mass loss. Mass loss (ML) after heat treatment was estimated according to the formula:

$$\text{ML}\,(\%) = 100 \times \frac{m_0 - m_1}{m_0}, \tag{1}$$

where m_0 is the initial oven dried mass of the wood sample before heat treatment, and m_1 is oven dried mass of the same sample after heat treatment [37].

In spite of using only two temperatures for heat treatment (180°C and 220°C), several different mass losses of the heat-treated wood at the same set temperature were achieved (Table 1). The reasons for this were imprecise regulation of the temperature in the vacuum chamber and variability of the wood. The temperature was not completely constant, because during modification some exothermal processes started in the wood and the temperature rose rapidly. It was, therefore, difficult to precisely control the temperature in the vacuum chamber. The heat-treated (HT) lamellas were classified according to mass loss into five groups. Untreated wood lamellas were also prepared as a control group (C).

2.2. Bonding Process

All the prepared lamellas were stored in a standard climate chamber at 20°C and 65% relative humidity (RH) for 6 months. After this time the moisture content (MC) of the specimens was different, depending on mass loss (Table 1). Groups of lamellas were then placed into different climates in order to reach a similar equilibrium MC within all the groups. The control samples were stored in a dry climate at 20°C and 44% RH, whereas the samples treated at 180°C were exposed to a standard

Table 1.

Treatment temperatures, mass loss and moisture content in a standard climate, and moisture content prior to bonding for different group of specimens

Group of specimens	Heat treatment temperature	Average mass loss (%)	Average moisture content (%) in standard climate 20°C, 65% RH	Average moisture content (%) prior to bonding
C	Control-untreated	0.0	10.9	9.4 (dry climate)
HT-0.47	180°C	0.47	10.1	10.0 (standard climate)
HT-0.75	180°C	0.75	9.6	9.6 (standard climate)
HT-2.20	180°C	2.20	8.6	8.6 (standard climate)
HT-7.20	220°C	7.20	6.7	9.5 (wet climate)
HT-8.66	220°C	8.66	6.0	8.5 (wet climate)

climate, and the samples treated at 220°C to a wet climate at 20°C and 87% RH. The lamellas reached an equilibrium MC in 4 weeks.

Prior to bonding, all lamellas were planed (removal of wood was less than 1 mm) in order to ensure a smooth and flat surface. Two lamellas were then bonded together with commercial phenol-formaldehyde (PF) adhesive for plywood production, supplied by Fenolit d.d., Borovnica, Slovenia. The adhesive was applied with a roller at an application amount of 180 g/m^2. The press temperature was 160°C, the pressure was 0.8 MPa, and the press time was 40 min.

2.3. Testing the Shear Strength of the PF Adhesive Bond

Shear strength of the PF bond was tested according to standard EN 13354:2008 [38]. Bonded samples were conditioned for 1 week in the standard climate (20°C, 65% RH) and then cut into specimens. The dimensions and shape of a typical specimen are shown in Fig. 1. Specimens from each mass loss group were divided into four sub-groups for different pretreatments, prior to testing according to the standard. The first subgroup of specimens (the dry specimens) was tested in the dry state after conditioning in a standard climate; the second subgroup of specimens (pretreatment 1) was soaked in cold water (20°C) for 24 h (for intended use in dry conditions); the third subgroup of specimens (pretreatment 2) was boiled for 6 h, and then cooled in cold water for 1 h (for intended use in humid conditions); and the fourth subgroup of specimens (pretreatment 3) was boiled for 4 h, dried for 16 h, boiled for 4 h and cooled in water for 1 h (for intended use in exterior conditions). The shear tests were carried out on a ZWICK/Z100 universal testing machine immediately after the pretreatment.

2.4. Determination of Wettability

The contact angle was measured on planed lamellas before bonding in order to determine the wettability of the wood surface with water and with the PF adhesive. The sessile drop method was used to measure the contact angle of a drop deposited

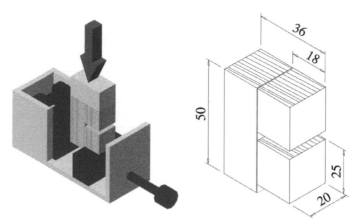

Figure 1. Mounting and loading of the test specimens for the determination of the shear strength of the PF adhesive bondline (left), and the shape and dimensions (in mm) of the typical specimen (right).

on the wood surface. A digital photo camera and a microscope were used to capture images of the drop during its spreading over the wood surface. The images were transferred to a computer and the height (h) and width (D) of the drop were measured by digital image analysis software (ImagePro). The contact angle (θ) was calculated from the equation:

$$\theta = 2 \cdot \arctan \frac{2h}{D}. \tag{2}$$

The contact angle was calculated for 0, 1, 2, 3, 5, 10, 20, 30, 45, 60 s after the drop had been deposited. 5 repeat measurements were made on each lamella.

2.5. Determination of the PF Adhesive Penetration

The measurements of the PF adhesive penetration were performed on microscopic slide sections, which were cut from end of each specimen, exposing the bondline with the cross-sectional surface. For each group of HT and C samples, 5 specimens were prepared. For each specimen, 4 randomly selected images of the PF adhesive bondline were analyzed. PF adhesive penetration was measured using a microscope and a digital image analysis technique. Digital images were captured with a Nikon Eclipse E800 microscope, a Nikon DS-Fi1 digital camera and the NIS Elements BR 3.0 software. Each image represented 1161 µm of the width of the PF adhesive bondline. The computer program (ImagePro) was used to measure the area of the PF adhesive spots (objects) in each image. The effective penetration (EP) of the adhesive was determined. The EP is the total area of the adhesive detected in the region of the bondline (represented by the sum of the all the adhesive objects — A_i) divided by the width of the maximum rectangle defining the measurement area (x_0) of the bondline [39]:

$$EP = \frac{\sum_1^n A_i}{x_0}. \tag{3}$$

Table 2.
Average shear strength, 5-percentile shear strength and average wood failure of the PF adhesive bonds with regard to the method of preparation of the specimens prior to testing

Group of specimens	Mass loss (%)	Dry specimens			Pretreatment 1			Pretreatment 2			Pretreatment 3		
		AWF (%)	5-perc. (MPa)	Bond strength (MPa)	AWF (%)	5-perc. (MPa)	Bond strength (MPa)	AWF (%)	5-perc. (MPa)	Bond strength (MPa)	AWF (%)	5-perc. (MPa)	Bond strength (MPa)
C	0.00	76	7.00	8.55	86	2.98	3.87	99	3.40	3.69	97	3.51	3.73
HT-0.47	0.47	73	3.89	6.76	44	3.87	4.36	36	3.46	3.72	69	3.63	4.00
HT-0.75	0.75	89	6.04	7.94	43	3.36	4.11	63	3.32	3.67	55	2.00	3.19
HT-2.20	2.20	45	6.68	8.35	15	3.30	4.44	14	3.21	3.61	35	1.66	3.05
HT-7.20	7.20	65	5.38	6.16	9	2.43	3.93	26	2.97	3.75	18	1.03	2.31
HT-8.66	8.66	70	4.00	4.92	30	2.24	3.16	34	2.40	3.10	23	0.51	1.91

AWF = average wood failure (%), 5-perc. = lower 5-percentile shear strength (MPa), bond strength = average shear strength (MPa).

3. Results and Discussion

3.1. Shear Strength of the PF Adhesive Bond

The average and lower 5-percentile shear strengths of the PF adhesive bond, and the average wood failure for all the specimens, are shown in Table 2. Most of the heat-treated specimens exhibited lower shear strength of the PF adhesive bond than the untreated control specimens. The percentage of average wood failure of the PF adhesive bond decreased with the severity of the heat treatment (indicated by higher mass loss). This was especially evident for all the specimens which were pretreated according to the EN 13354:2008 standard (pretreatments 1, 2 and 3). With regard to the requirements for the use of heat-treated wood for solid wood panels, only mild heat treatment of wood is appropriate (e.g., HT-0.47 and HT-0.75). The EN 13353:2008 standard requires the lower 5-percentile shear strength to be higher than 2.5 MPa, and the average wood failure to be more than 40% for each pretreatment, in order to fulfill the requirements for different uses (i.e., dry conditions — pretreatment 1; humid conditions — pretreatment 2; and exterior conditions — pretreatment 3).

3.1.1. Dry Specimens (Standard Climate)

The highest shear strength of the PF adhesive bond was observed for the untreated control specimens, where the average shear strength was 8.55 MPa. The lowest average shear strength of 4.92 MPa was observed for the most heat-treated specimens bonded with PF adhesive. A similar decrease in the shear strength of the adhesive bond, due to heat treatment, was previously found [15, 33, 34].

The heat-treated specimens showed a decreasing trend in the shear strength of the PF adhesive bond with increasing severity of the treatment, which is indicated by higher mass loss. Statistically significant differences (Fisher's LSD procedure) in the shear strengths of the PF adhesive bond were found between the specimens from the control group (C) and the specimens from the two most heat-treated groups (HT-7.20 and HT-8.66). Except for the dry specimens from HT-2.20, where 45% of wood failure was observed, the percentage of average wood failure was high and quite similar (65%–89%) for the all groups of dry specimens. This indicates that, for dry specimens, the reduction in the shear strength of the adhesive bond may be partially caused by the reduced strength of wood due to heat treatment [33]. This drop in the shear strength was expected because the mass loss increased with the severity of the heat treatment (ML was 8.66% for the most severely heat-treated wood).

3.1.2. Pretreatment 1 (24 h Soaking in Water)
The shear strength of the PF adhesive bond of the specimens which were soaked in water for 24 h was much less than that corresponding to the dry specimens. This strength decreased to approximately half of the "dry" strength, but there were differences between different groups of specimens. The control specimens lost more strength after soaking than the heat-treated specimens. For instance, the control group of specimens lost almost 55% of its shear strength after soaking in water, whereas the most severely heat-treated group of specimens lost only 35% of its "dry" strength. The reason for this could be the lower absorption of water and better dimensional stability of heat-treated wood. These cause less damage to the wood-adhesive bond, due to less swelling [26].

For the pretreatment 1, a significantly different shear strength of the PF adhesive bond was observed only between the most severely heat-treated group of specimens and the other groups of specimens. The control group had the highest percentage of wood failure (86%). All the heat-treated specimens exhibited a significantly lower percentage of wood failure, which, in general, decreased with the severity of the heat treatment. This showed that not only wood tissue, but also wood–adhesive interactions and cohesion of the adhesive developed during cure may be impacted by the heat treatment.

3.1.3. Pretreatment 2 (Boiling for 6 h, Soaking in Cold Water for 1 h)
The shear strength of the PF adhesive bond after pretreatment 2 decreased in comparison with the previous pretreatment. This was expected due to the greater severity of pretreatment 2. However, within the scope of this pretreatment, the average shear strength was almost the same for all the groups of specimens (around 3.7 MPa), except for the most severely treated sample (HT-8.66), whose shear strength was significantly lower (3.1 MPa).

The specimens from the control group had the highest percentage of wood failure (99%), whereas this percentage decreased significantly for all the heat-treated specimens. This finding again indicates that the bonding process with PF adhesive was in some way affected by the heat treatment.

3.1.4. Pretreatment 3 (Boiling for 4 h, Drying for 16 h, Boiling Again for 4 h, Soaking in Cold Water for 1 h)

The shear strength of the PF adhesive bond was similar to or smaller than the values obtained for the specimens after pretreatment 2. There was a statistically significant difference in the shear strengths and in the wood failures when comparing the less severely treated (C and HT-0.47) groups with the more severely treated (HT-7.20 and HT-8.66) groups of specimens. Shear strength as well as the percentage of wood failure decreased with the severity of the heat treatment.

3.2. Wood Failure vs. Shear Strength

A plot of the percentage of wood failure *versus* the shear strength of the PF adhesive bond for the dry specimens (Fig. 2) shows that average wood failure is high irrespective of the shear strength obtained with different heat treatment severities. For the specimens that were dry prior to testing, it is clear that the heat-treated specimens failed at a lower shear strength than the control specimens, but practically all of them (the control specimens and the heat-treated specimens) showed a similar percentage of wood failure. Thus, the decrease in the shear strength of the PF adhesive bond for the dry specimens is related to the decrease in the shear strength of the wood tissue due to the heat treatment. It has been found previously that the mechanical properties of heat-treated wood are reduced in comparison with those of untreated wood [2, 5, 8–13].

On the other hand, a plot of the percentage of wood failure *versus* the shear strength of the PF adhesive bond for the wet specimens — pretreatment 3 (Fig. 3) — shows that the average wood failure increases with increasing shear strength. The more severe the heat treatment, the lower are the shear strength and wood failure. This indicates that not only the wood tissue, but also the wood–adhesive

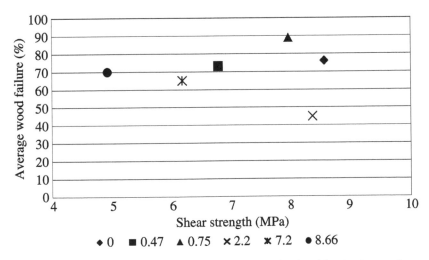

Figure 2. Wood failure *vs.* shear strength of the PF adhesive bond for the dry specimens.

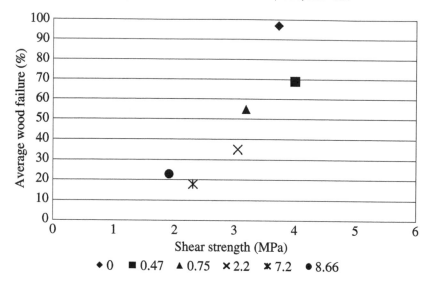

Figure 3. Wood failure *vs.* shear strength of the PF adhesive bond for the wet specimens — pretreatment 3.

interactions and the cohesion of the PF adhesive, developed during cure, may be impacted by the heat treatment.

3.3. Wettability

Table 3 shows the contact angle as a function of time for the different wood surfaces (control and heat-treated). It was found that the control specimens exhibited a smaller water contact angle than the heat-treated specimens. This means that wettability of the wood surface with water decreases with the heat treatment, which, of course, is one of the goals of thermal modification. The wettability of heat-treated wood with water decreases mainly because the surface of the heat-treated wood is hydrophobic and less polar [21–24]. This may hinder waterborne adhesives from adequately wetting the surface. However, the results concerning the contact angle of the PF adhesive on the heat-treated wood surfaces show that wettability with PF adhesive is not affected by the heat treatment (Table 4). Just the opposite, the contact angle of the PF adhesive decreased with the severity of the heat treatment of the wood. A statistically significant difference in the initial contact angle was obtained between the C and HT-2.20, HT-4.20, HT-8.66 specimens; and also between the HT-0.75 and HT-2.20, HT-4.20, HT-8.66 specimens. Improved wettability of PF adhesive on heat-treated wood was an interesting finding, but it was not within the scope of this research to investigate the phenomena behind it. An on-going study will focus on this aspect.

3.4. Penetration of the PF Adhesive

The average effective penetration of the PF adhesive in the control group of specimens was 20.9 μm, whereas it was 26.6 μm in the most severely heat-treated group

Table 3.
Time dependence of the water contact angle (°) for control and heat-treated wood specimens

Group of specimens	Mass loss (%)	Time (s)									
		0	1	2	3	5	10	20	30	45	60
C	0.00	56.7	55.7	52.8	50.6	47.3	45.3	41.7	39.5	39.1	37.8
HT-0.75	0.75	61.9	60.6	58.5	57.5	56.5	54.7	53.5	52.6	51.7	51.1
HT-2.20	2.20	76.6	74.9	73.6	72.8	71.1	70.3	68.9	67.8	67.7	66.7
HT-7.20	7.20	53.4	52.6	52.1	51.6	50.7	49.2	47.9	46.6	45.7	44.3
HT-8.66	8.66	72.6	71.4	70.8	69.9	68.9	67.7	66.5	65.8	65.0	63.9

Table 4.
Time dependence of the PF adhesive contact angle (°) for the control and heat-treated wood specimens

Group of specimens	Mass loss (%)	Time (s)									
		0	1	2	3	5	10	20	30	45	60
C	0.00	114.0	112.3	108.0	104.0	98.9	92.5	87.5	85.7	84.5	83.0
HT-0.47	0.47	114.5	113.7	109.1	105.7	100.2	94.3	89.6	87.6	85.9	84.9
HT-0.75	0.75	106.9	105.4	100.5	96.5	92.4	88.0	84.8	83.4	82.3	81.5
HT-2.20	2.20	108.3	107.3	103.5	100.0	94.5	88.9	85.1	82.5	81.0	80.2
HT-7.20	7.20	104.1	102.7	99.6	96.1	91.9	87.4	83.2	81.3	79.4	78.0
HT-8.66	8.66	108.6	107.6	104.1	100.4	95.9	90.2	86.8	84.8	82.9	81.9

of specimens. There was a slight increase in the effective penetration of the PF adhesive into the wood with the increasing mass loss (Fig 4). However, only a few statistically significant differences were identified among the different groups of specimens. It was not possible to clearly conclude that the penetration of the PF adhesive was significantly affected by the severity of the heat treatment of the wood.

4. Conclusions

This study revealed that the shear strength of the PF adhesive bond was influenced by the heat treatment of the spruce wood, and depended on the pretreatment of the specimens prior to testing. For the dry specimens, the highest shear strength of the PF adhesive bond was observed for untreated control specimens (8.55 MPa), whereas the lowest shear strength was observed for the most severely heat-treated specimens (4.92 MPa). The percentage of average wood failure was high and quite similar (65%–89%) for all groups of dry specimens (except HT-2.20). The high percentage of wood failure of the PF adhesive bond in the dry specimens suggests that the reduction in shear strength may be partially caused by the reduced strength of the wood due to the heat treatment.

The shear strength of the PF adhesive bond of the specimens which were pretreated according to EN 13354:2008 (soaking in water or a combination of boiling,

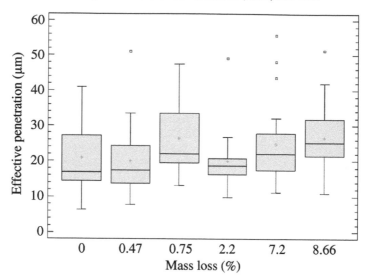

Figure 4. Effective penetration of the PF adhesive in the control and heat-treated wood with respect to mass loss.

drying and soaking) was much reduced when compared to the dry specimens. The percentage of wood failure for these specimens decreased significantly due to the increased severity of the heat treatment of the wood. This finding indicated that not only the wood strength itself but also the bonding process with the PF adhesive was in some way affected by the heat treatment.

The results on wettability showed that water wettability was lower on the heat-treated wood, but wettability with the PF adhesive improved with the heat treatment of the wood. It was found that the difference in the effective penetration of PF adhesive due to the heat treatment did not have crucial effect on the shear strength of the bonded specimens.

Acknowledgement

The authors would like to gratefully acknowledge the financial support of the Slovenian Research Agency within the framework of the research program P4-0015-0481.

References

1. C. A. S. Hill, *Wood Modification: Chemical, Thermal and Other Processes*. John Wiley & Sons, Chichester (2006).
2. S. Jämsä and P. Viitaniemi, in: *Proceedings of Special Seminar Review on Heat Treatments of Wood Held in Antibes*, France, pp. 21–27 (2001).
3. H. Militz, Thermal treatment of wood: European processes and their background. Report of International Research Group on Wood Preservation, IRG/WP 02-40241 (2002).
4. D. Fengel and G. Wegener, *Wood: Chemistry, Ultrastructure, Reactions*. Walter de Gruyter, Berlin (1989).

5. B. F. Tjeerdsma, M. Boonstra and H. Militz, Thermal modification of nondurable wood species. Part 2, Improved wood properties of thermally treated wood. Report of International Research Group on Wood Preservation, IRG/WP 98-40124 (1998).

6. M. Ohlmeyer, in: *Proceedings of the 5th COST E34 International Workshop*, Bled, Slovenia, pp. 21–29 (2007).

7. J. J. Weiland and R. Guyonnet, *Holz Roh Werkst* **61**, 216–220 (2003).

8. M. J. Boonstra, B. F. Tjeerdsma and H. A. C. Groeneveld, Thermal modification of non-durable wood species. Part 1, The Plato technology: thermal modification of wood. Report of International Research Group on Wood Preservation, IRG/WP 98-40123 (1998).

9. S. Yildiz, E. D. Gezer and U. C. Yildiz, *Building Environ.* **41**, 1762–1766 (2006).

10. M. J. Boonstra, J. Van Acker, B. F. Tjeerdsma and E. Kegel, *Annals for Sci.* **64**, 679–690 (2007).

11. A. W. Christiansen, *Wood Fiber Sci.* **22**, 441–459 (1990).

12. A. W. Christiansen, *Wood Fiber Sci.* **23**, 69–84 (1991).

13. W. J. Homan and A. J. M. Jorissen, *Heron* **49**, 361–386 (2004).

14. A. A. Marra, *Technology of Wood Bonding: Principles and Practice*. Van Nostrand Reinhold, New York (1992).

15. M. Sernek, M. Boonstra, A. Pizzi, A. Despres and P. Gerardin, *Holz Roh Werkst* **66**, 173–180 (2008).

16. O. Suchsland and R. R. Stevens, *Forest Products J.* **18**, 38–42 (1968).

17. W. Paul, M. Ohlmeyer and H. Leithoff, *Holz Roh Werkst* **65**, 57–63 (2007).

18. F. A. Kamke and J. N. Lee, *Wood Fiber Sci.* **39**, 205–220 (2007).

19. J. G. Chandler, R. L. Brandon and C. R. Frihart, in: *Proceedings of Adhesives and Sealants Council Spring 2005 Convention and Exposition*, Columbus, OH, pp. 1–10 (2005).

20. C. B. Vick and R. M. Rowell, *Int. J. Adhesion Adhesives* **10**, 263–272 (1990).

21. M. Pétrissans, P. Gérardin, I. El Bakali and M. Serraj, *Holzforschung* **57**, 301–307 (2003).

22. M. Sernek, F. A. Kamke and W. G. Glasser, *Holzforschung* **58**, 22–31 (2004).

23. J. Follrich, U. Muller and W. Gindl, *Holz Roh Werkst* **64**, 373–376 (2006).

24. P. Gerardin, M. Petrič, M. Petrissans, J. Lambert and J. J. Ehrhrardt, *Polym. Degrad. Stabil.* **92**, 653–657 (2007).

25. D. Mayes and O. Oksanen, *ThermoWood Handbook*. Finnish Thermowood Association, Helsinki (2003).

26. M. Swietliczny, M. Jabonski and P. Mankowski, *Annals of Warsaw Agricultural University, Forestry and Wood Technology* **53**, 347–350 (2003).

27. G. He and N. Yan, *Int. J. Adhesion Adhesives* **25**, 450–455 (2005).

28. F. Beaud, P. Niemz and A. Pizzi, *J. Appl. Polym. Sci.* **101**, 4181–4192 (2006).

29. M. Nuopponen, T. Vuorinen and S. Jamsa, *Wood Sci. Technol.* **37**, 109–115 (2003).

30. C. Y. Hse and M. Kuo, *Forest Products J.* **38**, 52–56 (1988).

31. A. Pizzi (Ed.), *Wood Adhesives: Chemistry and Technology*. Marcel Dekker, New York (1983).

32. Adhesive bonding of Plato wood (2006). Plato International BV, Arnhem, The Netherlands, http://www.platowood.nl/ENG07/Adhesivebondingmay06.pdf accessed January 2008

33. S. Poncsák, S. Q. Shi and D. Kocaefe, *J. Adhesion Sci. Technol.* **21**, 745–754 (2007).

34. M. Sernek, M. Humar, M. Kumer and F. Pohleven, in: *Proceedings of the 5th COST E34 International Workshop*, Bled, Slovenia, pp. 31–37 (2007).

35. EN 13353:2008 Solid Wood Panels (SWP) — Requirements.

36. G. Rep, F. Pohleven and B. Bučar, Characteristics of thermally modified wood in vacuum. Report of International Research Group on Wood Preservation, IRG/WP 04-40287 (2004).

37. M. Hakkou, M. Petrissans and A. Zoulalian, *Polym. Degrad. Stabil.* **89**, 1–5 (2005).

38. EN 13354:2008 Solid Wood Panels — Bonding quality — Test method.
39. M. Sernek, J. Resnik and F. A. Kamke, *Wood Fiber Sci.* **31**, 41–48 (1999).

Influence of Nanoclay on Phenol-Formaldehyde and Phenol-Urea-Formaldehyde Resins for Wood Adhesives

H. Lei [a,b,c], **G. Du** [a], **A. Pizzi** [c,*], **A. Celzard** [c] **and Q. Fang** [a]

[a] South West Forestry College, Bailong Si Road, 650224 Kunming, Yunnan, P. R. China
[b] Nanjing Forestry University, 159 Longpan Road, 210037 Nanjing, Jiangsu, P. R. China
[c] ENSTIB, Nancy University, 27 Rue du Merle Blanc, B.P. 1041, 88051 Epinal, France

Abstract
The addition of small percentages of Na^+-montmorillonite (NaMMT) nanoclay does not appear to improve much the performance of thermosetting phenol-formaldehyde (PF) and phenol-urea-formaldehyde (PUF) resins used as adhesives for plywood and for wood particleboard. X-ray diffraction (XRD) studies indicated that NaMMT does not become completely exfoliated when mixed in small proportions to PF resins, contrary to that observed for acid-setting urea-formaldehyde (UF) resins. Differential scanning calorimetry (DSC) indicated that NaMMT has no accelerating effect on the curing of alkaline PF resins, contrary to that observed for UF resins.

Keywords
Nanoclay, phenol-formaldehyde resins, phenol-urea-formaldehyde resins, thermosetting resins, adhesives

1. Introduction

The wood panels industry heavily relies on the use of synthetic resins and adhesives, as adhesively bonded products of one kind or another constitute about 80% of the wood products on the market today. In short, without adhesives and resins this industry would not exist [1, 2]. Among these products a certain proportion of wood panels are manufactured for exterior, weather resistant application. Phenol-formaldehyde (PF) resins, and more recently, phenol-urea-formaldehyde (PUF) resins [3–8] are the most commonly used resins among the leading adhesives for exterior grade wood panels.

Impressive enhancement of material properties achieved with the inclusion of submicrometer-size fillers in plastics and elastomers has stimulated considerable research on upgrading the performance of thermoplastic resins by addition of such

---

* To whom correspondence should be addressed. Tel.: (+33) 329296117; Fax: (+33) 329296138; e-mail: antonio.pizzi@entib.uhp-nancy.fr

materials [9]. Clay nanocomposites of different types yield a marked increase in a number of properties of thermoplastic and other resins and composites [10–15].

Choi and Chung [16] were the first to prepare phenolic resin/layered silicate nanocomposites with intercalated or exfoliated nanostructures by melt interaction using linear novolac and examined their mechanical properties and thermal stability. Lee and Giannelis [10] reported a melt interaction method for phenolic resin/clay nanocomposites, too. Although PF resin is a widely used polymer, there are not many research reports on PF resin/montmorillonite nanocomposites, and most of the research investigations have concentrated on linear novolac resins. Up to now, only limited research studies on resole-type phenolic resin/layered silicate nanocomposites have been published [17–19] and there is still no report on the influence of nano-montmorillonite on phenolic resin as wood adhesive. Normally H-montmorillonite (HMMT) has been used as an acid catalyst for the preparation of novolac/layered silicate nanocomposites. Resole resins can be prepared by condensation reaction catalyzed by alkaline NaMMT, just as what HMMT has done for novolac resins.

On the basis of PF resin, PUF resin has been developed to decrease the cost and to accelerate curing. Similar to the standard PF resin, PUF resin is prepared under alkaline conditions. Application of nanoclays to urea-formaldehyde (UF) resins has recently shown that nanoclays can improve the properties of thermosetting resins for interior-grade wood adhesives [20], and urea in PF resin could possibly be useful for the preparation of PUF resin/MMT composites, too.

This work thus deals with the addition of different types of montmorillonite nanoclays, especially NaMMT, to thermosetting PF and PUF resins to study: (1) the effect of the PF and PUF resins on the level of intercalation or exfoliation of the montmorillonite nanoclays; (2) the influence of resin structure on intercalation; and (3) the improvement in performance of the resin by analysing the performance of wood panels/composites bonded with PF/montmorillonite nanoclays.

2. Experimental

2.1. PF Resin Preparation

The phenol-formaldehyde (PF) resin was prepared at P:F = 1:1.76 molar ratio. The preparation procedure was as follows: 1.5 mol of phenol were mixed with 2.65 mol of formaldehyde (as a 37% formalin solution) in a glass reactor equipped with a mechanical stirrer, a thermometer and a reflux condenser. NaMMT was then added at room temperature and the whole mixture stirred overnight (12 h). The temperature was then increased to reflux (95°C) in 30 min under continuous mechanical stirring. Once 95°C was reached, 0.25 mol of NaOH (as a 30% aqueous solution) were added in 5 lots, each lot at 10 min interval. The mixture was maintained at reflux until the resin reached a viscosity of 400–500 mPa s, measured at 25°C. The resin was then cooled and stored.

2.2. PUF Resin Preparation

PF resin with coreacted urea was prepared at a F:P molar ratio of 1:1.7. The preparation procedure used was as follows for the resin containing 24% molar proportion of urea based on phenol coreacted in the resin (thus F:[P + U] molar ratio = 1.37).

One mole of phenol was mixed with 0.35 mol of NaOH as a 30% aqueous solution and 1.2 mol of formaldehyde (as a 37% formalin solution) were added in a reactor equipped with a mechanical stirrer, heating facility, and a reflux condenser. After stirring for 10 min at 30°C, 0.24 mol of urea were added and the temperature was slowly increased to reflux (94°C) over a period of 30 min and under continuous mechanical stirring and kept at reflux for further 30 min. 0.5 mol of formaldehyde (as a 37% formalin solution) were then added. The reaction mixture was now at pH 11 and the reaction continued at reflux until the resin achieved a viscosity (measured at 25°C) between 500 and 800 mPa s. The resin was then cooled and stored. Resin characteristics were pH = 11, and resin solids content = $50 \pm 1\%$.

Different proportions of NaMMT were added to the resin afterwards before its application to the veneers to fabricate plywood panels.

2.3. Testing

Wide angle X-ray diffraction (XRD) was carried out to investigate the effectiveness of the clay intercalation and to see if any changes in the crystalline structure of the PF and PUF resins had occurred. Samples of PF and PUF resins hardened at 103°C in an oven after mixing with 0%, 4% and 9% NaMMT were powdered and mounted on a Philips XRD powder diffractometer (sealed tube Cu source, K_α radiation ($\lambda = 1.54056$ Å)) for analysis. 2θ range from 2° to 100° in the reflection mode was scanned at 2°/min. A computer controlled wide angle goniometer was used. The Cu K_α radiation was filtered electronically with a thin Ni filter. The interlayer was calculated when possible from the (0 0 1) lattice plane diffraction peak using Bragg's equation [10]. Some information on the crystallinity levels of UF resins with and without NaMMT was obtained from the XRD investigation.

Differential scanning calorimetry (DSC) was performed on PF resins to which 0% and 4% NaMMT nanoclay was added to study the influence of the nanoclay on the hardening rate of the PF resin. A Perkin-Elmer DSC calorimeter was used for the analysis.

The resins were tested dynamically by thermomechanical analysis (TMA) on a Mettler apparatus. Triplicate samples of beech wood alone, and of two beech wood plies each 0.6 mm thick bonded with each resin system, for sample dimensions of $21 \times 6 \times 1.2$ mm were tested in non-isothermal mode between 40°C and 220°C at a heating rate of 10°C/min with a Mettler 40 TMA apparatus in three-point bending on a span of 18 mm. A force varying continuosly between 0.1 N, 0.5 N and back to 0.1 N was applied on the specimens with each force cycle of 12 s (6 s/6 s). The classical mechanics relation between force and deflection $E = [L^3/(4bh^3)][\Delta F/(\Delta f)]$ (where L is the sample length, b and h are the sample width and thickness, ΔF is the variation of the force applied, and Δf is the deflection obtained) allows the calculation of the modulus of elasticity (MOE) E for each case tested.

Duplicate three-ply laboratory plywood panels of 250 mm × 250 mm × 6 mm were prepared using 3 mm thick pine (*Pinus sylvestris*) veneers. The liquid glue mix used on the veneers was 380 g/m^2 double glueline. The plywood was pressed for 60 min after panel assembly. The press temperature was 155°C, pressure was 1 MPa, and the pressing time was 7 min. The samples were tested for tensile strength both dry and after immersion for 3 h in boiling water, i.e., wet tested.

3. Results and Discussion

The XRD patterns of NaMMT and of the PF resin with and without nanoclay are shown in Fig. 1. The peak appearing at 7° corresponds to NaMMT. In the case of the pure NaMMT the strong 2θ peak at 7° corresponds to a d-spacing of 1.26 nm according to the Bragg equation. For the PF/NaMMT system the (0 0 1) peak is still present but it does shift to a lower 2θ value (6°), i.e., a d-spacing of 1.47 nm according to the Bragg equation. This decrease in the value of 2θ indicates an increase, although relatively small, in the distance between the nanoclay planes. It appears that the PF resin intercalates to some degree between the NaMMT layers but that there is no total exfoliation.

The PF mixed with NaMMT shows some difference from that observed in UF resins where NaMMT is completely exfoliated. The different structures of the resin polymers is probably responsible for it. There are some hydrophilic groups existing in UF resins, especially amino groups. The main characteristic of MMT is that the metal ions adsorbed on the surface of MMT by electrical force can be replaced by an organic intercalator, among which alkyl ammonium salt is one of the most

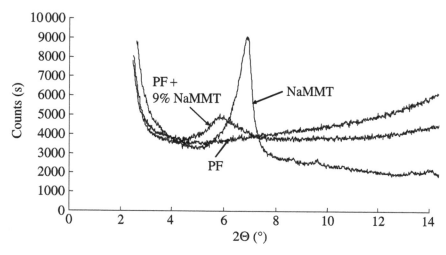

Figure 1. X-ray diffraction (XRD) patterns for NaMMT, PF resin, and PF + 9% NaMMT in the 2θ range 2°–15°.

commonly used. The basic group for the replacement reaction is $-NH_3^+$ at the end of the intercalator chain. The similarity in the structure between UF resin and alkyl ammonium salt will possibly be helpful for intercalation.

At the same time, considerable research has indicated that it is very difficult for a thermosetting resin to be intercalated into the silicate gallery as a result of its three-dimensional structure and rigidity. The PF resin used in this research has a three-dimensional structure even when uncured. Some polymer/layered silicate nanocomposites have been obtained actually for some thermosetting resins such as epoxy, unsaturated polyester, and polyurethane. But only Organo-MMT (OMMT) was used. While in this research, considering the cost and the hydrophilic groups existing in the adhesives, NaMMT was chosen. Compared with OMMT, NaMMT is much difficult to be intercalated.

Because of the irreversible nature of the hardening of thermosetting resins, addition of the nanofiller and its dispersion will occur as the resin cures. Therefore, it is better to add the nanofiller during resin synthesis to prepare thermosetting resins/nanocomposites. In this research, a resole was prepared with a higher molar ratio F/P = 1.76, which could be cured faster than a novolac resin with lower molar ratio. The resin may cure outside the nanofiller layers when the time available for intercalation is not long enough, thus limiting intercalation of MMT by the resin [21].

The results on laboratory-prepared plywood samples in Table 1 indicate that NaMMT had no obvious effect on the dry shear strength, while it improved tensile strength after boiling. Higher addition percentages, although giving higher results compared with the control, show a decrease from the maximum value obtained at 1% addition. These results indicate that addition of NaMMT does not contribute much to dry performance but improves the wet performance of PF adhesive, which is similar to what has been found in UF adhesives. Equally, the results obtained by thermomechanical analysis do not show an improvement with progressively increasing proportions of NaMMT in the PF resin (Fig. 2). Even the higher result with 5% NaMMT addition was found not to be significant.

Table 1.
Results of plywood bonded with a PF resin containing different percentages of NaMMT

NaMMT (wt%)	Dry tensile strength (MPa)	Tensile strength after 3 h boiling (MPa)
0	1.43 (22)	1.30 (30)
1	1.42 (0)	1.65 (20)
3	1.40 (5)	1.42 (10)
5	1.44 (20)	1.37 (20)

Number in parentheses = percent wood failure.

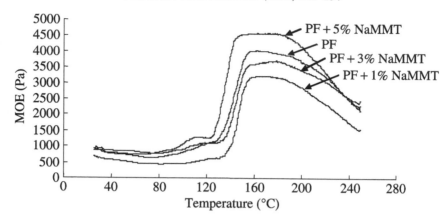

Figure 2. Thermomechanical analysis (TMA): curves of modulus of elasticity (MOE) as a function of temperature for PF resin with different percentages of NaMMT added.

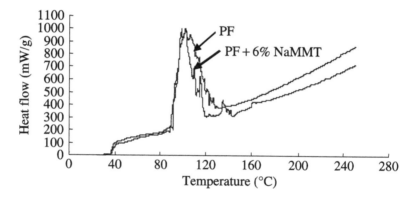

Figure 3. Differential scanning calorimetry (DSC) traces of the PF resin alone and of the PF resin + 6% NaMMT.

The differential scanning calorimetry results (Fig. 3) confirm that the addition of NaMMT does not alter the curing characteristics of the resin much. In the curves of neat PF resin and PF resin with 6% addition of NaMMT, there is only some difference in the number of peaks being shown. The neat PF resin exhibits a wide exothermic peak which is caused by the condensation either between methylol groups and phenol reactive sites to form methylene bridges or between two methylol groups to form dimethylene ether bridges. There are two differences between the PF resins with and without NaMMT. One is the pH, namely, 12.6 for the PF resin with 6% addition of NaMMT and 10 for the neat PF resin. The other is the intercalation state which is confirmed by the XRD results. Both of these could be the reasons for the peaks splitting. The difference in pH has some effect on the activation energy of the reactive groups of PF resin [22]. The difference in the curing rate caused by the intercalation will cause the observed split in the exothermic peaks of the pure and nanofilled resin (Fig. 3). Since the splitting is quite obvious, the latter reason seems more probable.

Table 2.
Results of plywood bonded with a PF resin containing 5% on resin solids of different types of MMT

MMT type	Dry tensile strength (MPa)	Tensile strength after 3 h boiling (MPa)
–	1.43 (22)	1.30 (30)
MMT	1.38 (5)	1.28 (7)
NaMMT, raw	1.44 (20)	1.37 (20)
NaMMT purified	1.51 (0)	1.29 (0)
OMMT	1.78(0)	1.81 (10)

Number in parentheses = percent wood failure.

Table 2 reports the results on PF-bonded plywood to which 5% based on resin solids of different types of montmorillonite nanopowders have been added. Thus MMT, NaMMT of two different grades, and OMMT, an organo-montmorillonite prepared by an ion exchange reaction between Na^+-montmorillonite and $CH_3(CH_2)_{15}$–$N(CH_3)_3Br$, were tested. Again, from the results in Table 2, there is no significant difference between PF-bonded plywood and type of inorganic MMT nanoclay, indicating that all types of inorganic MMTs with 5% addition amount do not contribute to improve the performance of PF adhesive resins. The only exception is OMMT where an improvement in both dry and after boiling tensile strengths can be noticed.

To ascertain if it is the alkaline environment of the resin that denies the effect observed in acid setting resins rather than just the absence of urea, a phenol-urea-formaldehyde (PUF) resin was prepared. PUF resin is alkaline setting but contains urea copolymerized with the phenol through formaldehyde [7, 8]. The results of plywood prepared with these resins are shown in Table 3. In this case the results improve up to 3% addition of NaMMT. The reason for the performance decrease once the addition percentage is higher than 3% may be that it is more difficult to obtain a uniform mixture between a higher amount of NaMMT and resin. It is worth noting that all PF and PUF resin/nanoclay systems showed increase of water resistance with the addition of NaMMT.

The results obtained by TMA support what was found on the panels. Thus in Fig. 4 there is no difference in the maximum MOE value if NaMMT is added. However, an interesting difference is noticeable: curing is faster as the percentage of NaMMT increases progressively up to 3%, as can be seen by the curing temperature shifting to lower temperatures. For more than 3%, this advantage is lost (Fig. 4).

The XRD patterns of the NaMMT nanoclay and of the PUF resin with 3% of nanoclay are shown in Fig. 5. For the PUF/NaMMT system the (0 0 1) peak is still present but it does shift to a lower 2θ value (6.35°) hence to a d-spacing of 1.39 nm according to the Bragg equation. This is even lower than that of the PF resin. Furthermore, while in the PF resin the d-spacing increases even at lower percentages

Table 3.
Results of plywood with a PUF resin containing different percentages of NaMMT

NaMMT (wt%)	Dry tensile strength (MPa)	Tensile strength 3 h boiling (MPa)
0	1.35 (23)	1.51 (11)
1	1.45 (31)	1.55 (39)
3	1.55 (18)	1.62 (23)
4	1.31 (26)	1.40 (10)
6	1.28 (26)	1.41 (43)

Number in parentheses = percent wood failure.

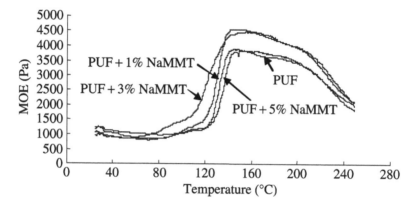

Figure 4. Thermomechanical analysis (TMA): curves of modulus of elasticity (MOE) as a function of temperature for PUF resin with different percentages of NaMMT added.

of nanoclay, this is not the case for the PUF resin where low amounts of NaMMT do not shift the 2θ value. The reason for the poor performance of inorganic MMTs in PF resins is that under alkaline conditions exfoliation does not occur or is low. In an acid environment inorganic MMTs are completely exfoliated [20], while in an alkaline environment the increase in d-spacing due to nanoclay exfoliation is indeed very limited.

4. Conclusions

The following conclusions could be drawn from the results on phenol-formaldehyde (PF) and phenol-urea-formaldehyde (PUF) resins investigated in this study.

(1) XRD results showed that although Na-montmorillonite (NaMMT) nanoclay had some degree of intercalation after being mixed with PF and PUF resins, it was difficult for resole-type phenolic resins with three-dimensional structure to be effectively mixed with nanoclay at the molecular level.

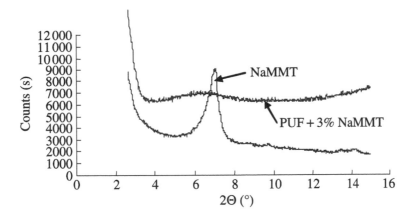

Figure 5. X-ray diffraction (XRD) patterns for NaMMT and PUF + 3% NaMMT in the 2θ range 0°–15°.

(2) When mixed in small proportions in PF and PUF resins, NaMMT improved the wet performance of both PF and PUF resins, while it did not improve much the dry performance. To increase the amount of NaMMT in the resins had no obvious effects on the performance of laboratory-prepared plywood samples.

(3) When OMMT with larger interlayer distance was used with the addition proportion up to 5% based on PF resin solid, an improvement in both dry and after boiling tensile strengths was found.

Acknowledgements

The French authors gratefully acknowledge the financial support of the CPER 2007-2013 "Structuration du Pôle de Compétitivité Fibres Grand'Est" (Competitiveness Fibre Cluster), through local (Conseil Général des Vosges), regional (Région Lorraine), national (DRRT and FNADT) and European (FEDER) funds.

The Chinese authors gratefully thank the financial support from the Key Research Foundation of Southwest Forestry University grant 2005088, Yunnan, China.

References

1. A. Pizzi, in: *Wood Adhesives: Chemistry and Technology*, A. Pizzi (Ed.), Chapter 2. Marcel Dekker, New York, NY (1983).
2. A. Pizzi, L. Lipschitz and J. Valenzuela, *Holzforschung* **48**, 254–261 (1994).
3. Y. Lei, Q. Wu, C. M. Clemons, F. Yao and Y. Xu, *J. Appl. Polym. Sci.* **106**, 3958–3966 (2007).
4. K. Yano, A. Usuki, A. Okada, T. Kurauchi and O. Kamigaito, *J. Polym. Sci. Part A: Polym. Chem.* **31**, 2493–2498 (1993).
5. A. Usuki, Y. Kojima, M. Kawasumi, A. Okada, Y. Fukushima, T. Kurauchi and O. J. Kamigaito, *Mater. Res.* **8**, 1179–1184 (1993).
6. P. B. Messersmith and E. P. Giannelis, *Chem. Mater.* **10**, 1719–1725 (1994).
7. J. W. Gilman, *Appl. Clay Sci.* **15**, 31–49 (1999).

8. R. A. Vaia, G. Price, P. N. Ruth, H. T. Nguyen and J. Lichtehan, *Appl. Clay Sci.* **15**, 67–92 (1999).
9. R. K. Bharadwaj, *Macromolecules* **34**, 9189–9192 (2001).
10. J. Lee and E. P. Giannelis, *ACS, Polym. Preprints, Div. Polym. Chem.* **38**, 688–689 (1997).
11. M. H. Choi and I. J. Chung, *Chem. Mater.* **12**, 2977–2983 (2000).
12. H. Wang, T. Zhao, L. Zhi, Y. Yan and Y. Yu, *Macromol. Rapid Commun.* **23**, 44–48 (2002).
13. Z. Wu, C. Zhou and R. Qi, *Polym. Composites* **23**, 634–646 (2002).
14. H. Wang, T. Zhao, Y. Yan and Y. Yu, *J. Appl. Polym. Sci.* **92**, 791–797 (2004).
15. W. Jiang, S. H. Chen and Y. Chen, *J. Appl. Polym. Sci.* **102**, 5336–5343 (2006).
16. M. H. Choi and I. J. Chung, *J. Appl. Polym. Sci.* **90**, 2316–2321 (2003).
17. L. B. Manfredi, D. Puglia, J. M. Kenny and A. Vazquez, *J. Appl. Polym. Sci.* **104**, 3082–3089 (2007).
18. M. Lopez, M. Blanco, J. A. Ramos, A. Arbelaiz, N. Gabilondo, J. M. Echeverria and I. Mondragon, *J. Appl. Polym. Sci.* **106**, 2800–2807 (2007).
19. L. J. Dai, *China Chemistry and Adhesion* **28**, 79–81 (2006).
20. H. Lei, G. Du, A. Pizzi and A. Celzard, *J. Appl. Polym. Sci.* **109**, 2442–2451 (2008).
21. W. H. Lv, Preparation of wood/montmorillonite (MMT), *PhD Thesis*, Beijing Forestry University, Beijing (2004).
22. G. He and B. Riedl, *J. Polym. Sci. Part B: Polym. Phys.* **41**, 1929–1938 (2003).

Emulsion Polymer Isocyanates as Wood Adhesive: A Review

K. Grøstad * and A. Pedersen

DYNEA AS, Svellevn. 33, 2000 Lillestrøm, Norway

Abstract

Emulsion Polymer Isocyanate (EPI) adhesives were introduced in the Japanese market for gluing of wood-based products approximately 30 years ago. The water-based emulsion adhesives with isocyanate as cross-linker are used in many parts of the world for production of different types of wood-based products such as: solid wood panels of different types, parquet, window frames, furniture parts, plywood, finger joints and load-bearing constructions like glulam beams and I-beams. The curing characteristics of EPI adhesives are quite complex and include film formation of the emulsion adhesive as well as chemical reactions of the highly reactive isocyanate towards water, hydroxy-, amines- and carboxy-groups. The advantages obtained by the use of EPI adhesives are fast setting speed, cold curing, light colored glueline, low creep of the glueline and high moisture resistance.

Keywords

EPI, Emulsion Polymer Isocyanate, MDI, pMDI, wood adhesives, solid wood panel, parquet, window frame, furniture, plywood, finger joint, glulam beam, I-beam, cross-laminated timber

1. Introduction

Emulsion Polymer Isocyanate (EPI) adhesives are two-component adhesive systems consisting of water-based emulsions cured with an isocyanate cross-linker. The first EPI adhesives were developed in Japan in the early 1970's and this adhesive type has been extensively used in Japan since then [1–3].

EPI adhesives can be formulated to provide a wide range of properties. They have very good adhesion properties, give good bonding to most wood species and can also glue wood to metal, for example aluminum to wood [2, 4, 5]. EPI adhesives are cold curing but may also be cured by heat or radiofrequency. They are fast setting even at room temperature and give light colored gluelines and possess low creep as well as high heat and moisture resistance. Due to the wide temperature range for curing, high moisture resistance and relatively short pressing times, EPI adhesives are used for a wide range of applications. Typical uses are production of solid wood panels of different types, parquet, window frames, furniture parts, plywood, finger

---

* To whom correspondence should be addressed. E-mail: kristin.grostad@dynea.com

joints and production of load-bearing constructions (glulam beams, I-beams and finger joints for constructions) [1–6].

This paper provides a review of EPI adhesives for gluing of wood-based products. The following topics regarding the use of EPI adhesives are discussed: chemical aspects, gluing properties, different applications, and handling and safety.[1]

2. History

Water-based Polymer Isocyanate (JIS name [7]) or Emulsion Polymer Isocyanate (EPI) adhesives were developed in Japan by Kuraray Co., Ltd, Koyo Sango Co., Ltd and Asahi Plywood Co., Ltd in the early 1970's [8]. The driving forces were to establish new markets for poly(vinyl alcohol) (PVA) and to develop alternatives to formaldehyde-based wood adhesives. The technology was patented [9], and thereafter licensed to different companies worldwide [8]. The licenses led to introduction of EPI adhesives into Europe, North America and Oceania. Since the EPI adhesives are of Japanese origin, their popularity is naturally concentrated in the Asian markets. Non-Asian consumption is growing, but the market for this adhesive type is still limited.

Originally the adhesive system consisted of an aqueous poly(vinyl alcohol) solution with an isocyanate cross-linker. At this stage the common name was Aqueous Polymer Isocyanate (API) [3]. Further developments including use of different types of polymer emulsions, like poly(vinyl alcohol) (PVA), ethyl(vinyl acetate) (EVAc), styrene butadiene rubber (SBR) or acrylic-styrene (AcSt) emulsion, led to adhesives systems with improved performance, hence today the common abbreviation is EPI [4, 8]. The isocyanate cross-linking agent has also been further developed to improve the compatibility and the reactivity with the water-based component, thus several different types are now available for use in EPI adhesives.

EPI adhesive technology has been described for adhesion of wood in the Japanese Industrial Standard JIS K6806 since 1985 [7]. In this standard the adhesive system is described as an aqueous polymer and/or emulsion as a base adhesive and an isocyanate compound as cross-linker. In Japan EPI adhesives are approved according to JIS K6806 as a supplement to phenol-resorcinol-formaldehyde (PRF) adhesives (approved according to JIS K6802) for production of small and medium glulam beams (load-bearing) posts for indoor use [7, 10]. The Japanese market has been a relatively important market for European beam producers during the last 10–15 years; hence the European demand for EPI adhesives initially came from these producers in the late 1990's. The driving force for this demand was to reduce the pressing time, to obtain light colored gluelines and to use formaldehyde-free adhesive systems. Initially EPI adhesive production was based on the Japanese licenses.

---

[1] Due to the European origin of the authors, the information in this article is based on literature and patents available in English as well as experience gained in the European market. Information from patents and other literature available only in Japanese or other Asian languages is not included.

Lately, new adhesives have been developed to satisfy the requirements for different applications as these vary significantly from one part of the world to another.

3. Chemical Composition

EPI adhesives are two-component systems. In the following the emulsion/polymer based component (the adhesive component) and the isocyanate based cross-linker component will be discussed separately.

3.1. Emulsion Polymer Component

The water-based component is the main adhesive component in EPI adhesives. Generally, it consists of water, poly(vinyl alcohol) (PVA), one or more water-based emulsions, filler(s) and a number of additives such as defoamers, dispersing agents and biocides [1, 4, 5, 8, 9]. As with traditional thermosetting and thermoplastic wood adhesives, properties such as viscosity and solids content vary with the intended application. In the European market the typical viscosities are 2000–8000 mPa s at 25°C and the solids content is normally 50% or more. The adhesives are normally neutral with a pH in the range of 6–8 [3]. The storage stability of the EPI adhesive component is typically half a year when stored at a temperature between 10 and 30°C.

As already mentioned most EPI adhesives contain the water-soluble polymer poly(vinyl alcohol) (PVA). The PVA has multiple roles in the EPI adhesive formulation: it takes part in cross-linking reactions with the isocyanate during curing, it is used to adjust viscosity/rheology of the adhesive and also to prevent sedimentation of the filler. Although PVA is the most commonly used thickener, other hydroxyl functional polymers, such as hydroxyl ethyl cellulose (HEC) and starch, are also used [4]. Even if most EPI adhesives contain PVA or other hydroxyl functional thickeners, it has also been reported that it is beneficial to limit the hydroxyl functionality to the latex exclusively. This is to ensure high degree of cross-linking in the latex film [6] as this may improve the water resistance of the glueline.

PVA is produced by hydrolysis of poly(vinyl acetate) (PVAc). Both modified and unmodified PVA grades with different molecular weights (M_w) and differing degrees of hydrolysis are commercially available. Fully hydrolyzed PVA, where all the acetate groups have been hydrolyzed, is more moisture resistant than partially hydrolyzed PVA due to the lower probability of further hydrolysis of the acetate groups. Fully hydrolyzed PVA has higher surface tension than partially hydrolyzed PVA, however, fully hydrolyzed PVA modified with ethylene groups in the polymer structure has lower surface tension (see the section on foaming for the effect of surface tension on the behavior of the EPI). According to Schilling [11] partially hydrolyzed PVA may offer an advantage in that fewer hydroxyl groups are available for reaction with isocyanate [11] to enhance the cross-linking reactions between the latex polymers (cf. previous section). However, a high degree of hydrolysis is reported to be preferable to obtain a more water resistant glueline [4, 11].

Different types of water-based emulsions are used in EPI adhesives. The most common are poly(vinyl acetate) (PVAc) emulsion, ethylene vinyl acetate (EVAc) emulsion, vinyl acetate-acrylate copolymerized (VAAC) emulsion, acrylic-styrene (AcSt) emulsion or styrene-butadiene rubber (SBR) latex or modified versions of these emulsion types [1–4, 8, 9]. It has also been reported that tri- or ter-polymer emulsions like vinyl acetate-butyl acrylate-hydroxypropyl methacrylate or emulsions with different combinations of block copolymers can be used [4]. Emulsion polymers containing cross-linking functional groups are especially well suited [4, 6, 9]. The choice of emulsion(s) will, to a large extent, influence the adhesive properties such as setting time, bond quality, heat resistance, and moisture resistance. EPI adhesive systems are, however, very complex and the total composition (including the choice of cross-linker) and the interaction between the different components will determine the properties of the adhesive. Due to this it is difficult to describe in detail the effect of choosing one type of emulsion over the other.

Fillers in EPI adhesive formulations are used to reduce the cost of the adhesive systems, to increase the solids content of the adhesive and to improve the gap filling properties and heat resistance of the cured glueline. The most commonly used filler is calcium carbonate ($CaCO_3$). Other fillers include organic and inorganic materials such as wood fibers, shell flours, clays, silica and talc [4, 6, 9, 11]. The particle size of the filler is important to prevent settling of the filler during storage and to ensure good quality of the cured glueline. It should preferably be in the same range as the emulsion particles to obtain a smooth and stable adhesive [6]. The hardness of the filler is also important to minimize the tool wear during planing of the glued samples. The hardness of the filler is often a question of purity and geometry of the inorganic particles. The amount of filler in the adhesive varies both with the type of filler and the intended use of the adhesive, and ranges from 0–50% [4, 6].

It is common to add a dispersing agent in the formulation to obtain good and stable dispersion of the filler and to prevent sedimentation. Other surface active ingredients such as defoamers and/or air release agents are usually added to prevent foam and/or air bubbles in the adhesive. Biocides/fungicides are also added to prevent microbial growth.

3.2. Cross-linking Agent

Isocyanates are used as a cross-linker in EPI adhesive systems. Theoretically, any isocyanate with two or more NCO groups would be suitable. In practice two main parameters are important for the choice of isocyanate: The volatility and reactivity of the isocyanate [1, 12]. The use of isocyanates with low volatility, low vapor pressure, is preferred to minimize the health risks concerned with the use and handling of the isocyanate. Isocyanates that are typically used, or have been used, as cross-linkers in EPI adhesive systems are based on the following isocyanate monomers: toluene diisocyanate (TDI), diphenylmethane diisocyanate (MDI), hexamethylene diisocyanate (HMDI), 3-isocyanatomethyl-3,5,5-trimethylcyclohexyl isocyanate (IPDI) and triphenylmethane-triisocyanate (TTI)

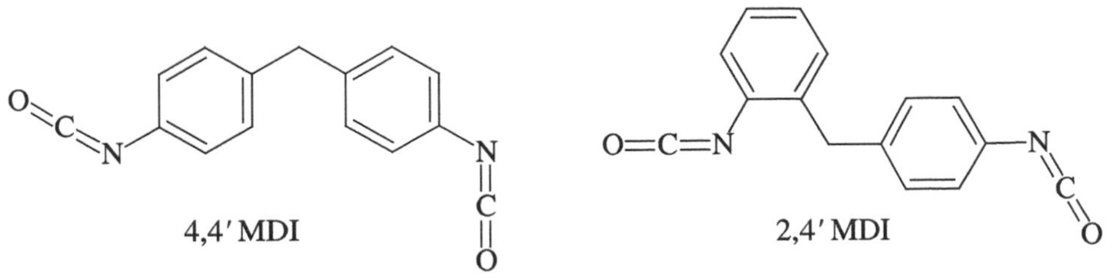

Figure 1. Structures of 4,4′ and 2,4′ isomers of MDI.

[2–4, 6]. These are polyfunctional isocyanates with two or more NCO groups which may cross-link the adhesive system. MDI is the preferred isocyanate due to its high reactivity and low vapor pressure. HMDI and IPDI are aliphatic and cyclic isocyanates with very low reactivity due to low solubility in the water and the absence of the aromatic structure found in TDI and MDI. TDI, on the other hand, has quite high vapor pressure; hence increased health risks involved with the use of this isocyanate are significant [12].

MDI is produced from the reaction between aniline and formaldehyde with hydrochloric acid as a catalyst. The mixture of polyamines obtained is phosgenated to a mixture of polyisocyanates [13]. This is separated into polymeric MDI (pMDI) and pure MDI. Pure MDI is a mixture of 4,4′ and 2,4′ isomers (Fig. 1). The 4,4′ isomer is approximately three times more reactive than the 2,4′ isomer. The 2,4′ isomer is not available in a pure state. The 4,4′ isomer, with a melting point of 38°C, can be isolated from a mixture of 4,4′ and 2,4′ isomers. Due to the high melting point, storage temperature is an important issue for the 4,4′ isomer. Commercial products with up to 50% of the 2,4′ isomer are available. The 50/50 mixture of the 2,4′ and 4,4′ isomers has, with a melting point of 18°C, a good balance between reactivity and easy storage and handling [13].

The most usual form of MDI used in EPI adhesives is polymeric MDI (pMDI) (Fig. 2). Polymeric MDI is generally a composition of approximately 40–50% pure MDI dimer (Fig. 1), 20–30% trimer, 10% tetramer and the rest is polymer. All components are mixtures of 2,4′ and 4,4′ isomers. The average functionality is in the range 2.5–2.7 [14].

The NCO groups in polymeric MDI, or especially in pure MDI, may undergo a condensation reaction and form carbodiimide and uretonimine (Fig. 3) which give low functionality isocyanates (2.0–2.1). Uretonimine is liquid at room temperature but crystallizes at approximately 15°C and care should be taken during storage of MDI with uretonimine to prevent crystallization. Uretonimine crystals dissolve by heating above 40°C but this is not recommended on industrial scale and uretonimine-containing pMDI should be stored at temperatures of 20°C or higher [13].

MDI polymerized with a polyol can also be used as a cross-linker for EPI adhesives (Fig. 4). This is referred to as modified-MDI or prepolymer of MDI and contains moieties with urethane linkages [12, 14, 16]. The urethane groups may

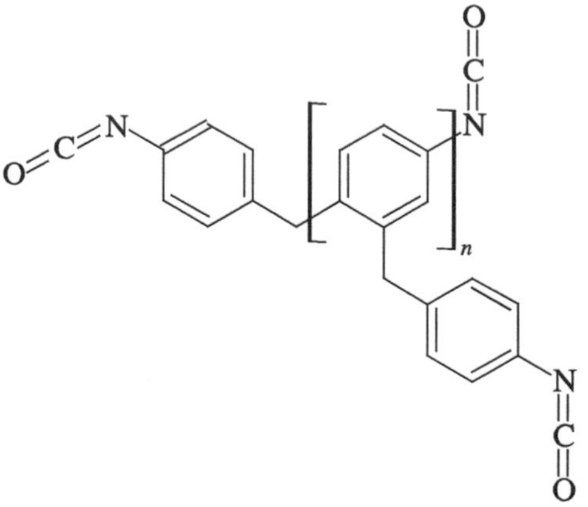

Figure 2. Structure of polymeric MDI (pMDI).

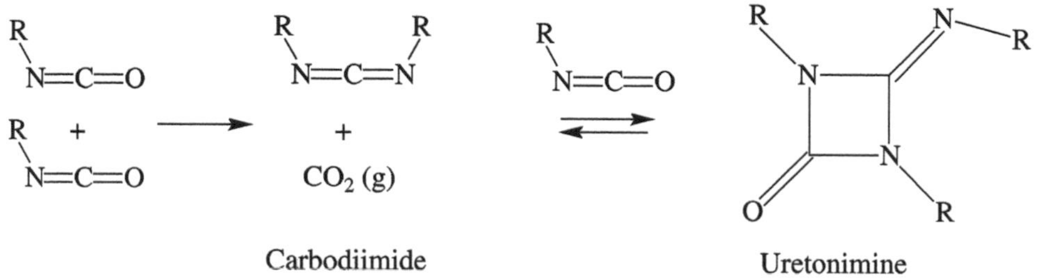

Carbodiimide   Uretonimine

Figure 3. Self-condensation of isocyanate *via* carbodiimide to uretonimine.

$$n\text{R(NCO)}_2 + n\text{HOC}_2\text{H}_4\text{OH} \longrightarrow \left[-\text{CONHRNHCOOC}_2\text{H}_4\text{O}- \right]_n$$

Figure 4. Reaction of pMDI with polyol to form polyurethane.

form internal hydrogen bonds and as a result higher viscosity. Compared to a standard pMDI the number of NCO groups available will thus be reduced and this will result in less cross-linking in the final glueline.

Commercial pMDI is stable at room temperature and is available in different grades with respect to functionality, NCO content, ratio between isomers, chemical modification, viscosity and storage stability. pMDI reacts readily with water, thus the cross-linker should be stored dry without contact with humid air or water. Typical for all of these grades is that the non-polar MDI grades are difficult to disperse as small particles into the polar water-based adhesive component. Originally organic solvents like toluene, ketones, xylene or dibutyl phthalate (DBP) were included in the EPI adhesive, either in the water-based emulsion component or in the pMDI component (depending on the choice of solvent), for easy dispersion of the pMDI into the adhesive [3, 4, 8, 9]. This gave unwanted volatile organic compounds (VOCs) in the system, and if the solvent was mixed into the MDI, it could influence the storage stability of the MDI. It also influences properties such as pot-life, foaming, and moisture resistance of the EPI adhesive (this will be discussed in more detail in later sections). The difficulty in dispersing pMDI led to the development of 'aqueous emulsified isocyanate' where the pMDI is modified with non-ionic surfactants to facilitate the dispersion of the pMDI into the adhesive [3].

It is possible to protect the isocyanates from reaction with water by reacting the –NCO groups with mono hydroxyl- or amino-compounds. Primarily solid TDI dimer (2,4-toluene diisocyanate) is used for this application and the –NCO groups on the surface of the solid isocyanate particles are blocked [5, 6, 15]. Such blocked isocyanates will not react with water or any other compounds until they are 'unblocked'. Heat is normally required to unblock the isocyanate groups. The necessary temperature will depend on the chemical group used to block the isocyanate. To the knowledge of the authors, isocyanate grades that 'unblock' at temperatures as low as 60–70°C are available. Since the blocked isocyanates do not react with water they can be dispersed in water and such dispersions are commercially available. The use of dispersions with blocked isocyanate ensures good distribution of the isocyanate in the adhesive. Comparison of the moisture resistance of EPI adhesives cured with standard polymeric isocyanate and the blocked version shows that the standard isocyanate is more efficient and gives better water resistance to the final glueline [6].

Aziridine compounds having two or more pendant aziridine rings have been described as alternative cross-linkers for EPI adhesives, offering approximately the same reactivity as isocyanates [4]. To the best of our knowledge these compounds are not frequently used.

4. Chemical Reactions in EPI Adhesives

The reactivity of different isocyanates varies widely, and the most reactive NCO groups can react with almost any compound that contains an active hydrogen [1, 2, 16]. The reactivity of the nucleophilic groups also varies; primary amines are more reactive towards NCO than primary alcohols, followed by water, secondary and tertiary alcohols, other urethanes, carboxylic acids, and carboxylic acid amides in that order [16]. The isocyanate will, of course, react with the water present in the EPI formulation to form amines followed by further reactions producing urea and biuret. The mechanism of this reaction is shown in Fig. 5. As can be seen from the reaction mechanism CO_2 is a byproduct of this reaction.

The isocyanate will also react with hydroxyl groups in the EPI adhesive as shown in Fig. 6. The reaction between isocyanate and hydroxyl gives urethane linkage.

The main source of hydroxyl groups in the adhesive is usually the PVA, but hydroxyl and other functional groups such as carboxyl groups and amines on the emulsion polymer chain can also react with the isocyanate, provided that they are physically available. In addition, the hydroxyl groups in the wood may take part in

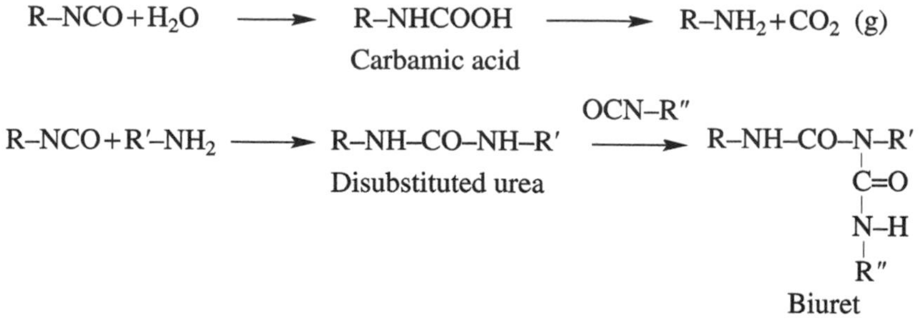

$$R\text{–}NCO + H_2O \longrightarrow R\text{–}NHCOOH \longrightarrow R\text{–}NH_2 + CO_2 \ (g)$$

Carbamic acid

$$R\text{–}NCO + R'\text{–}NH_2 \longrightarrow R\text{–}NH\text{–}CO\text{–}NH\text{–}R' \xrightarrow{\ OCN\text{–}R''\ } R\text{–}NH\text{–}CO\text{–}N\text{–}R'$$

Disubstituted urea

$$\begin{aligned} &\overset{|}{C}{=}O \\ &\overset{|}{N}\text{–}H \\ &\overset{|}{R''} \end{aligned}$$

Biuret

Figure 5. Reaction of isocyanate (NCO) with water, and reaction of amine with isocyanate to form disubstituted urea and biuret.

$$R\text{–}NCO + R'\text{–}OH \longrightarrow R\text{–}NH\text{–}COO\text{–}R'$$

Figure 6. Reactions of isocyanate (NCO) with hydroxyl to form urethane.

the reaction with isocyanate to some degree [1, 2, 12]. Most of the reactions taking place in the EPI adhesive mixture are irreversible.

The reaction mechanism of isocyanate in EPI glue mixes has been studied by Raman spectroscopy. This showed that the reaction with water to give urea was the most dominant reaction [17]. However, biuret and urethane were formed in higher amounts than expected. The reaction rate over time was investigated by FT-IR analysis and during the first 12 h of reaction large amounts of urea and biuret were formed. In this step there was still sufficient water in the system to enable the NCO to react with water instead of with hydroxyl. After 12 h the glueline was solid and the samples had constant weight. Monitoring of the reaction showed that the extent of reaction between isocyanate and hydroxyl groups to form urethane groups increased as seen by a gradual reduction of the amount of NCO groups.

Due to the large number of competing reactions between components of widely different concentrations, it is difficult to predict to what extent the different reactions occur in an EPI adhesive mixture. Due to the large amount of water in the adhesive mix, it is clear, however, that there is extensive reaction between the isocyanate and water.

5. Gluing Process and Glueline Properties of EPI Adhesives

Since the EPI adhesives are emulsion based but cross-linked with isocyanate, they share characteristics with both thermosetting and thermoplastic adhesives. The adhesives are multi-phase systems comprising emulsion particles, polymer solution, cross-linker droplets and filler particles. Just as for other emulsion adhesives, the coalescence of the emulsion particles [18, 19] and the distribution of these in the glue film is important for the bond quality. The cross-linking in the adhesive film is also of great importance for the bond quality as well as for the moisture resistance and heat resistance of the adhesive.

During gluing with an EPI adhesive a number of potential reactions and processes can occur simultaneously. These are:

- Water transport away from the glueline;

- Coalescence of the emulsion particles to form a glue film;

- Reaction of –NCO groups with water;

- Reaction of –NCO groups with other –NCO groups;

- Reaction of –NCO groups with –OH groups present in PVA or with other hydroxyl-functional groups in the water phase;

- Reaction of the –NCO groups with functional groups available in/on the emulsion polymer;

- Reaction of –NCO groups with hydroxyl groups on the wood cell walls.

These processes/reactions occur in parallel to each other and at different rates. The rate and extent of each of these depend on the composition of the EPI adhesive and the actual gluing conditions (temperature, gluespread, wood moisture content (MC), etc.). Hence, it is virtually impossible to accurately describe what is occurring at any given time during the gluing process. An attempt to discuss the individual processes and what may affect them will, however, be made.

The initial film formation to obtain a glue film may take from minutes up to one hour, depending on factors such as temperature, emulsion type(s), solids content, gluespread, wood species and wood moisture content. The coalescence of the emulsion particles may, however, continue for days after the gluing is performed [16]. Although the film formation process is relatively fast even at room temperature, higher temperature will increase the water transport away from the glueline and result in faster film formation. The effect of the increasing temperature is, however, small in this physical process compared to the effect the increased temperature has on the reactivity and chemical processes of thermosetting adhesives. How fast the initial strength of the glueline builds up depends, to a large extent, on the time taken for removal of water from the glueline and how quickly the coalescence process proceeds [2, 6].

The chemical cross-linking reactions between the isocyanate and the adhesive components occur before, during and after the film formation process and may continue for days after the gluing has been performed [16]. The choice of type(s) and amount of PVA polymer is important for the isocyanate cross-linking reactions occurring in the water phase. If the EPI is not formulated correctly with respect to this, the cross-link density will suffer, resulting in inferior moisture resistance. Reaction with hydroxyl functional groups in or on the emulsion polymer is also possible, provided that the functional groups are physically and sterically available for reaction. Even reaction between the isocyanate and the hydroxyl groups in the wood substrate has been reported [1, 2, 12] although recent research indicates that this is still open for debate [20, 21]. The contents of water and hydroxyl groups are very high in EPI adhesive gluelines, hence the probability of reaction between isocyanate and hydroxyl groups in the wood is most likely very low. The adhesion between the glue and the wood will, however, be improved by ionic interactions and mechanical and chemical bonds, e.g., hydrogen bonds [4, 6].

The glueline will have very good mechanical properties if the coalescence of the emulsion particles is optimal. However, if the isocyanate reaction proceeds too far before the glue film is formed, i.e., the glue mix is too old before it is used (cf. 'pot-life' Section 6.1), adequate coalescence will not be possible and the glueline will have a more rubbery consistency [2]. Hence, a good balance between these processes is essential.

By cross-linking with isocyanate [3] the glueline properties change from those of a typical thermoplastic to those similar to thermosetting adhesives. The EPI adhesives are unique in that the emulsion-based component can be formulated to have low glass transition temperature (T_g) and minimum film forming temperature (MFFT) to facilitate formation of a good glue film at room temperature. The final glueline will still have low creep, high heat and moisture resistance, and high T_g due to the isocyanate cross-linking reactions. Unfortunately, there is little information available on how the T_g of the different emulsions influences the T_g of the final glueline [2]. Qiao *et al.* describe the variation in T_g for EPI adhesives based on PVAc emulsions and different cross-linkers [2].

The water resistance of a newly formed glueline is generally poor, but increases gradually as the cross-linking reactions proceed. How quickly full water resistance can be obtained varies from adhesive to adhesive, but most adhesives will have very good moisture resistance 1–2 days after gluing. This time might be reduced if the gluing process is done at elevated temperatures. The final water resistance of the glueline depends on the choice of cross-linker, the degree of cross-linking, the amount and type(s) of PVA, the composition of the water-based emulsion component, and the mixing ratio between the two adhesive components.

A proper mixing of the adhesive emulsion component and the isocyanate cross-linker is very important for the resulting bond quality. Poor emulsification of the non-polar pMDI grades into the polar water-based adhesive component results in partial phase separation during drying/curing and results in non-uniform cross-linking in the glueline. Due to this, the best moisture resistance is obtained by using aqueous emulsifiable isocyanates which are easier to disperse into the water-based emulsion component [3]. As already mentioned, solvents can also be used to improve the dispersability of the isocyanate. There is, however, a risk that the solvent will act as a plasticizer for the emulsion(s), which may speed up the film forming process but give softer gluelines.

The amount of isocyanate used in the EPI adhesive will significantly influence the adhesive properties. Krystiofiak *et al.* [22] showed that both the hardness of the glue film and the moisture- and heat-resistance of the glueline increased with increasing amounts of cross-linker in a series of tests where 10–20 parts cross-linker to 100 parts emulsion component were used. Hu *et al.* [3] concluded that the resistance to warm and boiling water increases with increasing loading of cross-linker. Testing of up to 20% cross-linker in the adhesive system showed that 15–20% was required to obtain a boil resistant glueline with a well-formulated emulsion component. The best performance was obtained with an emulsion component based on tri-copolymer latex and 20% dosage of aqueous emulsifiable isocyanate.

6. Use of EPI Adhesives

When using EPI adhesives the same considerations as for other adhesives have to be taken. In the following subsections properties such as pot-life, assembly time and setting time will be discussed. In addition, the foaming caused by formation of CO_2 in the reaction of water with isocyanate will be discussed. The mixing ratio of the water-based emulsion component and the isocyanate is commonly 100 parts by weight (pbw) adhesive to 10–20 pbw isocyanate, although dosages down to 5 pbw of isocyanate are also used.

6.1. Pot-Life

The term 'pot-life' is commonly used for two-component adhesives to describe the time from mixing the two components until the glue-mix has too high viscosity to be used. The pot-life is commonly determined by measuring the viscosity increase with time. The pot-life of an EPI adhesive system is, however, not necessarily so easy to define. First of all, the CO_2 released from the reaction between isocyanate and water foams the glue-mix and influences the viscosity measurement. In practice the foaming may also make the glue-mixture difficult to handle and, depending on the application equipment, may cause the glue-mix to be usable for only a limited length of time. Secondly, some EPI adhesives will form a semi-gel after reaction with the isocyanate, whereas others will only increase in viscosity to a given level. This latter type of adhesive may still run well in the application equipment a long time after mixing, but the bond quality of the glued product will be poor because the cross-linking reactions have proceeded too far for proper coalescence of the glue film and/or because too much of the isocyanate has been consumed. Finally, for applications where a certain moisture resistance is important, it is crucial that the amount of unreacted isocyanate is high enough to give the required cross-linking in the glueline. Hence in most cases the pot-life is really determined by how long the glue-mix will give the required bond quality and/or moisture resistance. Due to this it is common to determine the 'pot-life' of EPI adhesives by gluing tests with glue-mixes of different ages, followed by testing the bond quality and/or moisture resistance of the resulting glueline. The pot-life is determined as the time after mixing when the adhesive no longer gives the required bond quality/moisture resistance. For example, for adhesives for production of laminated beams it is common to determine the pot-life by carrying out delamination tests according to EN 302-2 at different times after preparation of the glue-mix.

The mixing ratio between the water-based adhesive and the isocyanate influences a number of properties of the adhesive-mix. Krystofiak *et al.* [22] show that increased dosage of cross-linker gives higher initial glue-mix viscosity and that the viscosity of the glue-mix increases faster with higher dosages. Due to this glue-mixes with low dosage of cross-linker may be easier to handle than glue mixes with high dosage. On the other hand, glue-mixes with low amounts of isocyanate will quickly reach the point where the amount of unreacted isocyanate is too low for adhesive applications where high moisture resistance is required. Hence, the pot-life will be shorter than if larger amounts of isocyanate were used.

6.2. Assembly Time

Assembly time is important in most gluing operations. The assembly time is defined as the time from glue application until clamping pressure is applied. Exceeding the maximum assembly time will lead to poor bonding.

For EPI adhesives the assembly time is too long if the coalescence of the emulsion particles proceeds too far before pressure is applied. Hence, the maximum assembly time will vary from adhesive to adhesive as it depends on the adhesive composition. For a given adhesive, the assembly time depends on factors such as glue spread (amount of glue mix on a specific glueline area), temperature, wood moisture content, the relative humidity in the air (RH), and the velocity of the air flow (ventilation, draught). Addition of small amounts of organic solvents may increase the assembly time. The film forming process may be similar or sligthly faster with solvent but the solvent may 'protect' the isocyanate so that the immediate reaction between isocyanate and water is delayed [3]. Unfortunately, this will most likely increase the setting time as well.

6.3. Foaming and Viscosity Increase

One property that distinguishes isocyanate cross-linking adhesives from other types of wood adhesives is the CO_2 that is formed during the reaction of isocyanate with water. The release of CO_2 causes the glue-mix to foam over time. The amount of CO_2 formed, how fast it is formed, and how easily it is released from the adhesive-mix depend on the formulation of the water-based adhesive component as well as on the type and amount of isocyanate used for cross-linking. Glue mixes with high amounts of free isocyanate will foam more than glue mixes with lower amounts. Also the distribution of the isocyanate in the water-based adhesive component will influence the foaming. The better the isocyanate is dispersed, the greater the surface area of the isocyanate droplets, the faster it reacts, and the more foam is formed. It is also reported that EPI adhesives containing EVAc may have a higher degree of foaming in the glue-mixture than when SBR emulsions are used [23]. The observation is well documented but an explanation is not given. Foaming is not only a practical problem when handling an EPI glue-mixture, it may also cause micro-defects in the glue film, which will reduce the mechanical strength of the glueline.

Hori *et al.* [23] observed that increasing the amount of EVAc in the formulation resulted in reduced setting time but increased the degree of foaming. These authors also discuss the influence of the ethylene/vinyl acetate (E/VAc) ratio on the foaming due to the reaction between isocyanate and water. A high E/VAc ratio results in less gas generation during curing of an EPI adhesive film, which indicates that the E/VAc ratio influences the initial curing process. However, after storage the gluelines with different E/VAc ratios had similar cross-linked structures and performance properties.

6.4. Setting Time — Reactivity

EPI adhesives can be cured in a wide temperature range and one of the attractive properties of this adhesive type is the fast setting time even at room temperature. As described earlier, warm conditions will speed up both the film formation process and the chemical reactions in the glueline. Since neither the film formation nor the cross-linking of the isocyanate is pH-dependent, the pH value of the different wood species does not influence the setting time of the adhesive. Coalescence of the emulsion particles to form a continuous glue film occurs as the water is transported away from the glueline. Factors that influence this water transport positively or negatively will, therefore, influence the setting time the same way. Gluing of dense wood species and materials with poor water absorbency will, for example, require longer setting times than gluing of low density wood species and other materials with good water absorbency. High glue spread will give longer setting times than low glue spreads.

The adhesive composition will also influence the setting time. Adhesives with higher solids content will result in shorter setting time than adhesives with lower solids content when comparing similar formulations. The emulsion type(s) in the adhesive will also influence the setting time. For example, use of significant amounts of EVAc in the formulation will generally result in fast setting adhesives whereas use of significant amounts of SBR or AcSt emulsions will increase the setting time. The amount of isocyanate in the glue mix will also influence the setting time. Higher amounts of the cross-linker may increase the setting time due to slower film formation.

7. EPI Adhesives in Different Wood Applications

A wide range of EPI adhesive systems are used for the production of different wood-based products and composite materials where wood is one of the components. A comprehensive review covering all the applications in all markets is outside the scope of this article. Hence the focus will be on the most common applications in Europe with input from articles discussing markets and applications outside Europe.

Adhesives for wood applications traditionally have been poly(vinyl acetate) (PVAc), urea-formaldehyde (UF), melamine-urea-formaldehyde (MUF), phenol-formaldehyde (PF) and phenol-resorcinol-formaldehyde (PRF) adhesives. Due to the thermoplastic properties during the gluing process, EPI adhesives can easily be used as replacement for thermoplastic PVAc adhesives in a number of applications. As the final glueline also has thermosetting characteristics, EPI adhesives can also replace thermosetting adhesives for products where normal emulsion-based adhesives, such as PVAc, do not fulfill the requirements [3].

EPI adhesives can be cured at room temperature and even adhesives which cure at temperatures down to 5°C are available [24]. Thus, production processes with low energy consumption can, as described in earlier sections, be combined with high glueline performance. For many applications the same production equipment and adhesive application equipment may be used for EPI adhesives as for traditional adhesive systems.

The main challenge with EPI adhesives is the relatively short assembly time of most of the systems. This is a problem for applications where the surface area for glue application is large and/or the number of wood layers is high. One possible way of solving this problem would be to use blocked isocyanates. EPI adhesives with up to 8 h pot-life and long assembly times are reported using this technology [5, 6]. Although heat is required to 'unblock' the isocyanate groups, the necessary temperature is still significantly lower than the temperatures used with MUF and PF adhesives.

EPI adhesives have very good adhesion properties to the wood surface and thus can glue different wood species very well. Since this adhesive type also has very good adhesion to materials like metals, plastics and foams, it is very well suited for gluing composite materials where wood is combined with other materials [2, 4, 5]. The excellent adhesion of EPI adhesives to metals has to be taken into account in the production process since the glued pieces may adhere to the press plates if no surface treatment of the press plates is used. Release agents are available and are in use.

In the following sections the different applications where EPI adhesives are used will be discussed. Where it is considered to be of interest, the requirements, standards and regulations for the application will be included in the discussion.

7.1. Laminated Solid Wood Panels

Laminated solid wood panels are produced by joining wood lamellas using an adhesive. The panels are used for different types of products such as table tops, furniture, stairs, garden furniture, cutting boards, etc. (Fig. 7).

Depending on the end use and the requirement for moisture and heat resistance, the laminated solid wood panels are produced with UF, MUF, PVAc or EPI adhesives. EPI adhesives are viable and price competitive alternatives to MUF adhesives for products where high heat and moisture resistance is important (e.g., table tops for use in kitchens, garden furniture, etc.) and are, to a certain degree, in use for this purpose. EPI adhesives are, in general, too expensive to compete with PVAc and UF adhesives for products where moisture resistance is of less importance.

Figure 7. Laminated solid wood panels.

7.2. Window Frames

Window frames are exposed to all types of weather, hence they have to be produced with adhesives with high heat and moisture resistance. Although MUF adhesives are used to some extent, the European production mainly utilizes EPI adhesives and PVAc adhesives with high heat and moisture resistance (adhesives classified as D4 according to EN 204/205). The dispersion based adhesives are preferred due to the relatively fast curing rate at low temperatures. Since EPI adhesives give gluelines with higher heat and moisture resistance than what can be obtained with D4 PVAc adhesives, the trend is that more and more window frames are produced with EPI adhesives.

7.3. Flooring/Parquet

In Europe the most common types of wood based flooring are two- and three-layer sandwich parquet. In these constructions a decorative surface layer is glued perpendicular to a core consisting of pine- or spruce-ribs or different board materials. Veneer is used as backing material for the three-layer construction to stabilize the construction (Fig. 8). Two-layer parquet is mainly produced with PVAc. Three-layer sandwich parquet is produced with both UF adhesives and PVAc adhesives, but the use of UF adhesives is predominant.

EPI adhesives are competitive with UF adhesives regarding bond quality for gluing of parquet. With respect to pressing times, EPI adhesives can normally compete with high and medium reactivity UF systems used in single- and multi-opening hot presses. The current EPI technology is, on the other hand, unable to match the pressing times of fast setting UF systems used in radiofrequency presses with short cycle times. Very short pressing times are, nevertheless, obtained in a cold press

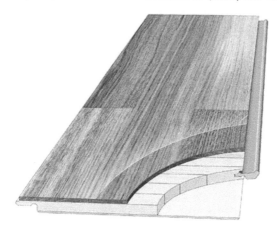

Figure 8. Three-layer parquet.

line designed by the German company Bürkle GmbH. In this line an EPI adhesive mix with high solids is applied on the surface layer and then fed through an IR zone for evaporation of water from the glueline. The parquet is assembled and pressed immediately after the IR zone. In this line the pressing times can, with a well formulated adhesive, be in the range of 30–45 s, depending on the wood materials used in the construction.

Currently the volume of EPI adhesives used for the production of parquet in Europe is very low. The volume is expected to increase as the demand for parquet produced with non-formaldehyde adhesives increases. Due to their high moisture- and heat-resistance EPI adhesives give more robust parquet than PVAc adhesives, especially if the parquet is used with floor heating.

7.4. Veneered Boards

Board materials resembling solid wood are obtained by gluing veneer onto board materials, such as particleboard. Traditionally, this is done with UF or PVAc adhesives, but EPI adhesives can also be used. The technical advantages of using EPI adhesives over PVAc adhesives are minor and this adhesive type is not competitive with respect to price. Due to this only a few producers are using EPI adhesives for this gluing process in Europe.

7.5. Plywood

Traditionally, UF, MUF and PF adhesives have been the preferred adhesive systems for production of plywood worldwide. In the European market these adhesive types are still predominant. There are, however, a few producers using EPI adhesives for products where metal is included in the construction.

Plywood is included in the Japanese standard for EPI, JIS K 6806 [7] and the impression from the literature is that it is more common to use EPI adhesives for production of plywood in Asia and USA [1, 5, 6].

Figure 9. Cross-laminated timber.

7.6. Multilayer Boards

In Europe the term multilayer board includes cross-laminated timber elements used for construction parts in buildings and shuttering boards. Both these types of boards are mainly produced with MUF adhesives in gluing processes with relatively long pressing times or in processes with high temperatures. Polyurethane (PUR) adhesives are already in use as an alternative to MUF adhesives for production of building elements whereas EPI adhesives are in the introduction phase.

Building elements (cross-laminated timber) require moisture resistant gluelines. The elements are large and consist of many layers of wood (Fig. 9). Thus, it is very energy consuming to produce these constructions by hot pressing. The elements also contain a large amount of adhesive and the formaldehyde emission has to be taken into account when MUF adhesives are in use. This is no concern when EPI and PUR adhesives are used.

The use of EPI adhesives for production of building elements offers a much simpler and less expensive process than when MUF adhesives are used, since the curing can be done without application of heat. Building elements glued with EPI adhesive systems are price competitive and are, due to the short pressing times at ambient temperature, good alternatives to MUF adhesives. The flexible gluelines obtained with EPI adhesives are also favorable for the building elements.

Shuttering boards are thin cross-glued boards. Since the volume of wood is low, hot pressing is less energy consuming than for building elements. Thus, EPI adhesives are not frequently used for this application due to the higher cost of the EPI adhesives compared to the MUF adhesives used for these products.

7.7. Glulam Beams

Glulam beams (laminated beams) are wood lamellas glued together for use as structural parts in buildings (Fig. 10). The glue types and methods of production of

Figure 10. Glulam beam for structural applications.

laminated beams are strongly regulated. Different test institutes in different countries are approved to perform tests both on the resins and on the final beam to ensure safe use. The test methods are described in the standards discussed below.

Since an EPI adhesive was first developed in Japan, the Japanese were the first to include EPI adhesives in their standards. The Japanese Industrial Standard (JIS) K 6806-1985 [7] deals with 'Water Based Polymer — Isocyanate Adhesives for Wood'. The Standard describes the quality of the adhesive, the performance properties and the requirements which should be fulfilled to be classified as Class 1 or 2 or used for laminate timber, furniture, decorative veneer, or plywood.

In North America the standards ASTM D 2559 and D 3535 give the requirements for adhesive systems used for structural purposes. EPI adhesives formulated for structural purposes meet the requirements in these standards and have the same performance properties as other structural adhesives [1, 5].

In Europe a standard for EPI adhesives used for glulam beams is under development, but for the time being the approvals are regulated by national authorities. The test methods are based on EN 301/302 with additional tests, such as creep test (ASTM D 3535). The approved adhesive systems for glulam beams, I-beams, and finger joints are shown in adhesive lists from the authorized organizations performing these types of tests. Authorized organizations in Europe include the Norwegian Institute of Wood Technology [25] and the Materialprüfungsanstalt Universität Stuttgart [26].

A wide range of EPI adhesive systems are approved for production of posts for the Japanese market. In North America EPI adhesives for structural applications are mainly used for I-beams and finger joints. In Europe EPI adhesives are used for finger joints, I-beams, and glulam beams. There is only one EPI adhesive approved for production of glulam beams in Europe [27].

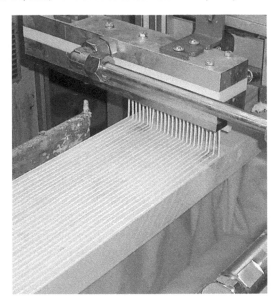

Figure 11. Ribbon spreader.

EPI adhesive systems used for the production of laminated beams are used in a ratio of 100 parts adhesive component to 15 parts pMDI (cross-linker) calculated on weight basis. This gives good performance properties for the final glueline and reasonable pot-life and assembly time.

Equipment used to apply EPI adhesives is mainly a ribbon spreader (Fig. 11) but roller coaters can also be used. Proper mixing of the adhesive component and the cross-linker is extremely important to ensure good glueline properties. New equipment for mixing adhesive and hardener has been developed for EPI adhesives. For production of laminated beams, use of the aqueous emulsifiable pMDI is an advantage to obtain a good distribution of the cross-linker in the adhesive and to obtain good moisture and boil resistant properties of the final glueline. The necessary glue spread depends on the wood species, planing quality and the press used. Pressing time increases with increased glue spread and moisture content in the wood.

The fast setting time at ambient temperature makes cold curing using EPI adhesives advantageous. The systems may also be cured in radiofrequency (RF) presses or hot presses. In most new production lines the benefits of fast cure at ambient temperature are utilized. An EPI adhesive can be specially formulated for RF curing. The formulation can be optimized by adding salts to increase conductivity. This gives a faster temperature increase in the glueline and thus shorter pressing time. Due to the thermoplastic behavior of EPI adhesives during the initial stage of curing, it is common to combine pressing time with RF followed by a pressing time without RF to ensure that the temperature in the glueline is reduced when the glued sample leaves the press. This ensures safe handling.

Assembly time is critical in a production process and especially when short pressing times are required. One method to increase the assembly time is to spray water onto the wood surface after the adhesive mix is applied and before the lamellas are joined. This method is patented [28]. The assembly time may be increased by up to 50% by spraying 10–15% water calculated on the amount of adhesive already applied to the wood surface.

7.8. Finger Joints

EPI adhesives are well suited and their use is widespread for finger joints. The EPI adhesive systems give short pressing times and high strength. In production units where finger jointing of timber is the only gluing process EPI adhesives have become very popular. The reason for this is the good gluing properties at low temperatures. Figure 12 shows the time required from gluing the finger joint until a water resistant glueline is obtained at different temperatures. Good gluelines are obtained even with timber having temperatures as low as 5°C. EPI adhesive works very well in contactless application equipment for finger joints due to the high viscosity of the glue mix (Fig. 13).

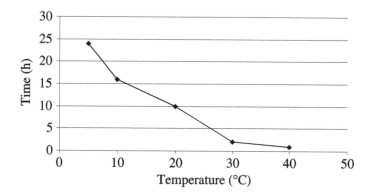

Figure 12. The time required to obtain a water-resistant glueline for a commercial EPI adhesive at different temperatures.

Figure 13. Contactless application of EPI for gluing of finger joints. Adhesive ribbons on the tips of the fingers are applied without contact between the application head and the wood fingers.

Pre-heating of the wood before gluing is used to obtain a faster water removal and thus shorter time to water resistance. Troughton [29] has shown that preheating of spruce, pine and fir maintained the same good strength and glueline properties in the finger joint as by cold curing with EPI adhesives.

7.9. I-Beams

I-beams are glued very similarly to finger joints but with different joint profiles. The most common wood type for the upper and lower parts in the I-beam (flench) is spruce or laminated veneer lumber (LVL) or Douglas fir (Fig. 14). In the middle part (web) oriented strain board (OSB) is most common; medium density fiberboard (MDF) can also be used.

Fire and heat resistance for elements in constructions has been the focus during recent years. In glulam beams the fire develops quite slowly due to the solid wood behavior. I-beams have to be insulated, due to the thin web and flench, to ensure adequate fire and heat resistance.

In the USA both standard EPI adhesives, with emulsion adhesive and pMDI, and polyurethane emulsion polymers (PEPs) are used for production of I-beams. These adhesive systems fulfill all the requirements for a structural adhesive [30].

Figure 14. I-beam.

7.10. Other Applications

The applications known to the authors are presented above. However, there is plenty of literature related to the use of EPI adhesives in applications not mentioned thus far. This section will give a short summary of some of the different applications with references.

According to literature from Japan and other Asian countries, EPI adhesives can be used for production of particleboards (chipboards) [3]. The advantage is the possibility to use different wood materials, higher moisture content in the wood and lower pressing temperatures. The disadvantage is the price.

Structural panels with outer facings, skins, fire retardants, and cores with foam are glued with EPI adhesives [1, 4]. Well formulated EPI adhesives have shown very good wetting and adhesion properties to metals and due to the good processing properties EPI adhesives have taken over some of the epoxy, urethane and cross-linked PVAc markets [1], especially in the USA.

EPI adhesives are very well suited for gluing of different wood species and treated wood species. This area is huge and outside the scope of this article and only a few references will be given.

Thermally and chemically modified wood is dimensionally more stable and has lower moisture content than untreated wood. The resistance to moisture or biological degradation can also be improved. The treatment may affect the wood's ability to absorb water, due to changes in the hydrophilic nature of the wood, and thus the setting time, the performance of the glueline, and the bond quality of the EPI adhesive may be influenced.

Gluing of acetylated wood was tested by Vick *et al.* [31]. The gluelines with EPI adhesives showed low shear strength in wet conditions, but had very high strength in dry conditions, regardless of whether they had gone through a water saturation/ drying cycle or not. This behavior is normal for EPI adhesives; the adhesion and the strength of the gluelines are reduced in wet conditions, but gluelines of good quality EPI adhesives will re-gain their strength if the glueline is dried.

Studies on modification of yellow poplar sapwood glued with phenol-formaldehyde, melamine-formaldehyde, epoxy and EPI adhesives show that EPI adhesives give extremely low adhesion strengths when the wood is acetylated [32] compared to the untreated wood. The butylene oxide modified wood gave the same strength as the untreated wood whilst propylene oxide modification gave lower strength. The authors' hypothesis that the additional hydroxyl groups in butylene oxide would give improved adhesion was not confirmed because of reduced strength of the wood itself. The reason for the low strength and adhesion obtained for EPI adhesives in acetylated wood, both in dry and wet conditions, may be related to the formulation of the EPI adhesive when we compare this with the work of Vick *et al.* [31].

Henriksson *et al.* [33] show that gluing of thermally modified wood by EPI adhesives gives the same strength as for other structural adhesives. The study did, however, show that thermal treatment of posts glued before treatment reduced the strength of the glueline for EPI adhesives significantly. All thermally-treated posts or posts glued with thermally-treated wood produced lower shear strength than untreated wood, independent of the adhesive system used for gluing. The results in this study show why it is not recommended to use thermally-treated wood for load-bearing constructions.

8. Health and Safety Aspects

EPI adhesives are viable alternatives to formaldehyde-based adhesives and the fact that EPI adhesives are formaldehyde free is often used to promote this adhesive type. EPI adhesives are considered to be environmentally friendly due to the absence of volatile hazardous chemicals. Nevertheless, both the water-based component and the isocyanate cross-linker in the EPI adhesive pose some chemical hazards. As for all types of emulsion-based adhesives, some people may suffer from allergic reactions when they are in contact with these adhesives. Thus, contact with EPI adhesives carries the risk of allergic reaction. Isocyanates, in general, are known to be skin sensitizers and should be handled with care. Some people may be especially sensitive to isocyanate and this effect may be aggravated by repeated exposure to this chemical. The risk of inhalation is low in low-temperature gluing processes provided that isocyanates with low vapor pressure are used. If the gluing process is performed at high temperatures, precautions should be taken and good ventilation is required. All in all, normal industrial practice for handling and use of chemicals should be followed.

ISOPA (European Diisocyanate & Polyol Producers Association) discusses the environmental, health and safety aspects of handling of MDI. Their recommendations should be followed [34].

Provided that the EPI adhesive systems are solvent free and that the emulsions used in the adhesive are 'VOC free', there will be no VOC emission from fully cured gluelines.

9. Conclusion

EPI adhesives are thermoplastic polymer dispersion adhesives combined with isocyanate cross-linking that gives a degree of thermosetting characteristics and properties. The adhesives are complex, multiphase systems with a number of physical processes and chemical reactions occurring in parallel during the gluing process.

The main advantages of EPI adhesives are short pressing times, very good cold cure properties, light colored gluelines, and high heat and moisture resistance. EPI adhesives have very good adhesion to different materials such as metals and plastics. Most wood species can be glued with EPI adhesives.

EPI adhesives have secured foothold in all regions of the world and are used for a number of applications.

References

1. H. F. Pagel and E. R. Luckmann, *Appl. Polym. Symp.* **40**, 191–202 (1984).
2. L. Qiao, A. J. Easteal, C. J. Bolt, P. K. Coveny and R. A. Franich, *Pigment Resin Technol.* **29** (4), 229–237 (2000).
3. H. Hu, H. Liu, J. Zhao and J. Li, *J. Adhesion* **82**, 93–114 (2006).
4. J. Zheng, E. Payton, C. Reed and G. Nieckarz, *PCT WO* 2007/056357 (2007).
5. H. F. Pagel and E. R. Luckman, *Adhesives for Wood: Research Applications, and Needs*, H. Gillespie (Ed.), pp. 139–149. USDA Forest Serv., Forest Prod. Lab., Madison, WI (1984).
6. N. J. Gruber and C. E. Powell, US Patent No. 4,609,690 (1986).
7. Japanese Industrial Standard JIS K6806 (1985), revised (1995).
8. D. Asai, *Jushi Gosei Kogyo* **38** (10), 150–153 (1991).
9. S. Sakurada, H. Miyazaki, T. Hattori, M. Shiraishi and T. Inoue, US Patent 3,931,088 (1976).
10. Japanese Industrial Standard, JIS 6802 (1986), revised (1995).
11. B. Schilling, *Adhesion - Kleben & Dichten* **48** (5), 20–24 (2004).
12. F. F. P. Kollmann, E. W. Kuenzi and A. J. Stamm, *Principles of Wood Science and Technology II, Wood Based Materials*, pp. 86–88. Springer Verlag (1975).
13. T. Gurke, *Proc. Eurocat 2002*. Barcelona, Spain (2002).
14. Huntsman International LLC, Huntsman Polyurethanes, Product Data Sheet, Suprasec® MDI products (02/2003).
15. O. R. Ganster and J. Buechner, *Adhesives Age* **45** (8), 36–40 (August 2002).
16. H. Warson, *The Applications of Synthetic Resin Emulsions*. Ernest Benn Limited, London (1972).
17. N. Hori, A. Takemura and H. Ono, in: *Proc. Wood Adhesives Conference*, South Lake Tahoe, Nevada, USA, pp. 167–168 (2000).
18. D. Urban and K. Takamura, *Polymer Dispersions and Their Industrial Applications*. Wiley-VCH (2005).
19. *Surface Coatings, Vol. 1, Raw Materials and Their Usage*, pp. 175–183. Chapman and Hall, London (1983).
20. D. J. Yelle, J. E. Jakes, J. Ralph and C. R. Frihart, in: *Proc. Wood Adhesives Conference* (2009).
21. J. E. Jakes, D. J. Yelle, J. F. Beecher, D. S. Stone and C. R. Frihart, in: *Proc. Wood Adhesives Conference*, Lake Tahoe, Nevada, USA (2009).
22. T. Krystofiak, S. Proszyk and M. Jozwiak, *Forestry Wood Technol.* **53**, 214–217 (2003).
23. N. Hori, K. Asai and A. Takemura, *J. Wood Sci.* **54**, 294–299 (2008).
24. DYNEA AS, Technical Data Sheet for two component EPI, Prefere 6151 (May 2009), available at: www.dynea.com
25. Available at: www.treteknisk.no
26. Available at: http://www.mpa.unistuttgart.de/organisation/fb_1/abt_12/listen_und_verzeichnisse/listen_und_verzeichnisse.html
27. J. Horacek, *Holzkurier* **47**, 24–25 (2005).
28. E. Moerland, M. Keiser and H. Michael, *PCT WO* 2008/056992 (2008).
29. G. E. Troughton, *Forest Products J.* **36**, 59–63 (1986).
30. Ashland Inc., Speciality Polymers and Adhesives, Columbus, OH 43216, Technical Data Sheet, Isoset Polyurethane/Emulsion Polymer Adhesive (2007).
31. C. B. Vick, P. Ch. Larsson, R. L. Mahlberg, R. Simonson and R. M. Rowell, *Intl. J. Adhesion Adhesives* **13**, 139–149 (1993).
32. R. Brandon and I. R. Ibach, in: *Proc. Wood Adhesives Conference*, San Diego, California, USA, pp. 111–114 (2005).

33. M. Henriksson, M. Sterley and J. Danvind, in: *Proc. European Conference on Wood Modification*, Stockholm, Sweden, pp. 577–584 (2009).
34. European Diisocyanate & Polyol Producers Association, available at: www.isopa.org

Adhesives for On-Site Rehabilitation of Timber Structures

Helena Cruz [a,*] and João Custódio [b]

[a] Laboratório Nacional de Engenharia Civil — LNEC, Structures Department, Av. Brasil 101,
1700-066 Lisboa, Portugal
[b] Laboratório Nacional de Engenharia Civil — LNEC, Materials Department, Av. Brasil 101,
1700-066 Lisboa, Portugal

Abstract

The use of adhesives to produce assembled structural joints in the building industry is increasing, particularly in the context of on-site rehabilitation of timber structures. On their own or together with steel or fibre reinforced polymer composite connecting materials, adhesives can provide low intrusive, fast, versatile and effective on-site repair or reinforcement interventions to timber structures. Most common applications involve sealing and repair of cracks, drying fissures and delamination of glued laminated members; replacement of decayed beam ends; strengthening of timber members; and repair and strengthening of mechanical timber joints. The performance of bonded joints highly depends on their design and detailing, surfaces preparation, selection and application of adhesives, and full compliance with their cure schedule. Therefore, the work should be carried out by well-informed, trained and certified operators following a Quality Assurance Program to ensure satisfactory end-product strength and durability.

Despite some recent developments, the exploitation of the full potential for on-site bonded joints is mainly restrained at present by the lack of structural design guidance, standards for durability assessment and on-site acceptance testing.

This article discusses briefly the use of adhesives on the construction site in the context of structural repair and reinforcement; the requirements and practical difficulties in the work on site with regards to the strength and durability of the rehabilitated timber structure; and the consequent need for quality control. It also highlights the characteristics and requirements that must be fulfilled by structural adhesives and reinforcing materials; factors affecting performance and durability of bonded joints; and ways to improve adhesion and durability. Finally, it points out some research needs and future developments identified by the authors.

Keywords

Timber structures, rehabilitation, on-site polymerized adhesives, limitations, requirements, performance, durability, quality control

---

* To whom correspondence should be addressed. Tel.: 00351 218443295; Fax: 00351 218443071; e-mail: helenacruz@lnec.pt

Wood Adhesives

1. Introduction

Although most common applications of adhesives in the wood industry are related to the manufacture of engineered wood products, their use to produce assembled structural joints is becoming more and more important in the building industry, particularly in the context of structural repair and reinforcement.

The potential of structural adhesives to produce efficient joints is enormous due to their ability to distribute the applied load over the entire bonded joint area, resulting in a more uniform distribution of stress, compared to mechanical connections, and to the use of the whole timber cross-section, without the normal reduction caused by holes (as in bolted connections). Bonded joints introduce little or no damage to the adherends also by eliminating the problem of wood splitting caused by driving fasteners like nails or screws without pre-drilling, and they add very little weight to the structure. They are suitable for joining dissimilar materials, and can also be invisible, which in certain cases has great aesthetical advantages. Structural bonded joints have inherently a much higher stiffness than mechanical joints. This results from avoiding the high deformation of fasteners like nails, bolts or screws, their embedding in wood due to stress concentration, and the initial joint slip associated with the fastener's hole clearance.

One key area in which bonded structural joints have proved their great potential is in the on-site repair and reinforcement of existing timber structures, where adhesives may be used on their own or together with steel or fibre reinforced polymer (FRP) composite materials. Typical examples include the use of plates or rods bonded into slots or drilled holes either to connect two timber sections or to improve strength and stiffness of timber members, and bonded-in rods inserted across the grain through a single timber section to repair or prevent delamination of glued laminated timber, drying fissures or cracks in joints. Such interventions have minimal intrusion into the original structure and the normal functioning of the buildings by avoiding extensive displacement of materials. The repair is often invisible and can even be encased within the timber or shaped to match the appearance of the original structure. These often constitute the less invasive and less time consuming, and even less expensive repair approaches to timber structures.

One often raised argument against the use of synthetic adhesives in the repair of historic buildings is the irreversibility of the intervention. However, although the adhesive cannot be removed from timber, such repairs enable retaining an appreciable amount of the original timber, compared to a repair done by traditional carpentry techniques. If required later, the whole repair system may be removed, still enabling traditional repair as if the adhesive based repair system had not been applied.

Another frequent criticism is the relatively little experience in the use of synthetic adhesives and PRF composites and consequent lack of confidence in their long-term performance. But one should acknowledge that similar adhesives have been successfully used for more than fifty years (in fact, for as long as the expected service life for a common building) and bonding design is normally very conservative.

2. Structural Adhesives for On-Site Applications

Adhesives used for on-site rehabilitation of timber structures must have good adhesion and produce strong and durable bonds to several different materials. They should produce negligible dimensional variation during curing, have relatively long open assembly time, be able to cure without pressure applied and at room temperature, and be only slightly sensitive to bondline thickness variation. Depending on the job, gap-filling properties and/or thixotropy might also be required. Typically, 2 mm thick bondlines are used in on-site rehabilitation work [1, 2].

In practice, epoxy, polyester, acrylic and polyurethane adhesives have been used. However, despite their limitations, epoxy adhesives have been used for bonding timber for more than thirty years and are presently still the best option for on-site structural timber repair. Epoxies are synthetic thermosetting adhesives, in the form of one- or two-parts and include a wide range of formulations and products with distinct properties. They may exhibit good gap-filling properties; excellent tensile/shear strength; high dry and wet strength; excellent moisture, chemicals and solvent resistance; although they have poor peel strength and may delaminate with repeated wetting and drying. They are, therefore, regarded as structural adhesives for limited exterior service environments [3], corresponding to Service Classes (SC) 1 and 2 defined in EN 1995-1-1 [4] where the average moisture content of most softwoods will not exceed 20%.

A structural adhesive is (according to EN 923 [5]) 'able of forming bonds capable of sustaining in a structure a specified strength for a defined long period of time', and according to The Adhesive and Sealant Council [6] 'an adhesive of proven reliability in engineering structural applications in which the bond can be stressed to a high proportion of its maximum failing load for long periods without failure'. Semi-structural adhesives generally do not withstand long-term static loading without deformation and non-structural adhesives are suitable only for low strength or temporary fastening.

Toughened acrylic adhesives are also classified as structural adhesives for limited exterior service environments, while polyurethane and polyesters (unsaturated) are considered as semi-structural adhesives for limited exterior service environments [3].

Limited exterior service environments include heated and ventilated buildings, as well as exterior protected from weather or exposed to weather only for short periods, situations for which both adhesive types I and II defined in EN 301 [7] are acceptable.

Thixotropic epoxies are convenient when the adhesive has to be applied from below or on vertical surfaces (Fig. 1, left). Epoxy grouts are employed to fill large volumes and should, therefore, be able to eliminate trapped air bubbles, should not stratify, should exhibit a low cure exotherm and be self-levelling (Fig. 1, right).

Figure 1. Application of a thixotropic adhesive to be used in conjunction with GFRP rods to strengthen a bridge stringer (left) and an epoxy grout to repair a carrier beam in a roof structure (right) [8] (reprinted with permission from Rotafix Ltd.).

3. Reinforcing Materials

Fibre reinforced polymers (FRPs) are composed of a reinforcement material (glass, aramid or carbon fibres) surrounded and retained by a (thermoplastic or thermosetting) polymer matrix (unsaturated polyester, epoxy, vinyl ester, or polyurethane). FRPs were first used in the rehabilitation of reinforced or pre-stressed concrete, but they have also been widely used in the reinforcement of timber structures.

Glass fibres are the most frequently used, having moderate cost and good mechanical properties. They exist in the form of pultruded profiles (FRPs), fabrics (tissues) or mats. Aramid fibres (e.g., Kevlar) originate from aromatic polyamides, produced by extrusion and drawing, and are available in various forms, including sheets, fabrics, ropes and ribbons. Carbon fibres are mainly used in the form of pultruded profiles of solid, open or hollow cross-sectional shapes. Advantages of FRPs are their high strength and rigidity at low weight, and the fact they are corrosion resistant materials. Composite profiles are regarded as lighter, easier to handle, cut, clean and use on-site than steel connecting materials. Their major disadvantage is still their high price. Surface preparation is of paramount importance and it should be performed prior to bonding. Normally, it includes mechanical abrasion followed by solvent wipe.

Steel rods or plates should be protected against corrosion, especially when used with acidic timbers like oak. Stainless steel or hot dip zinc coated steel are frequently used. Stainless steel may give poor adhesion and, therefore, it is normally surface coated for improved roughness and adhesion. If hot dip zinc coated steel rods or bars are used, the application of a priming product to improve adhesion is normally required. Threaded rods or ribbed bars and textured plates may be used instead, and in this case the mechanical anchorage will also contribute to the bonding strength. Surface preparation is particularly critical in uncoated steel and it should preferably include grit blasting and cleaning with an adequate solvent to remove oil, grease, salts, dirt or other contaminants [9–11].

Fabrics made of glass or carbon fibres may also be used to make fibre reinforced composites on site. Normally a layer of adhesive is applied to the timber surface, one or more layers of the fabric are then pressed into the adhesive and a new layer of adhesive is applied to it, resulting, after cure, in a fibre reinforced composite. The principal direction of the fabric is normally oriented parallel to the wood fibres to improve bending strength and stiffness.

The use of this method to wrap around timber members is not recommended, because the resulting sharp angles of the composite (in the case of rectangular cross-sections) will create high stress concentration that may lead to a premature failure.

4. Use of Adhesives on the Construction Site

4.1. Sealing of Fissures and Delamination

On their own, adhesives have been used as an attempt to repair minor delamination of glued laminated timber or drying fissures (Fig. 2). Besides safety considerations, fissures and delamination may contribute to increased risk of biological attack, therefore, it may be advisable to have them sealed for the sake of durability.

These interventions generally require cutting wider the fissure or delamination, to refresh its edges for improved adhesion, to remove any loose fibres or old adhesive, provide a more regular bondline thickness and enable the adhesive to reach as deep as possible.

Normally a higher viscosity or thixotropic adhesive, a sealant, or a thin wood fascia is first applied to the surface to seal the fissures. Small diameter plastic tubes are left in place at regular intervals to be used later to inject in a lower viscosity adhesive, that will hopefully penetrate the whole depth of the fissure, and let the air out. Injection starts from the lower injection tubes until the adhesive starts coming out from the next lower tube, and moves up. Plastic tubes will be cut and disguised after cure. This is nevertheless a tricky operation as the right balance between gap filling and penetration ability of the adhesive must be found.

Figure 2. Sealing of fissures by adhesive injection.

These solutions are not recommended for exposed structures since subsequent delamination will further contribute to water intake. The elastic modulus of the adhesive should match that of the timber to avoid it acting as a wedge should the timber swell and try to close the fissure. Another inconvenience of such interventions is that the sealant will hide the fissures, preventing periodical inspection to check its possible progress.

Nowadays, in order to create a higher mechanical and visual compatibility between timber member and filling material, a matching timber wedge is preferably used to fill the gap, thus reducing to a minimum the amount of adhesive used to fix it in place (Fig. 3).

However, if safety is at stake, these approaches are not at all suitable due to the high variability in strength and durability regain provided by such an approach [13]. Not only will the adhesive most likely fail to reach the (closed) very tip of the fissure, but also this type of bonded joint itself is very prone to delamination due to timber dimensional variations.

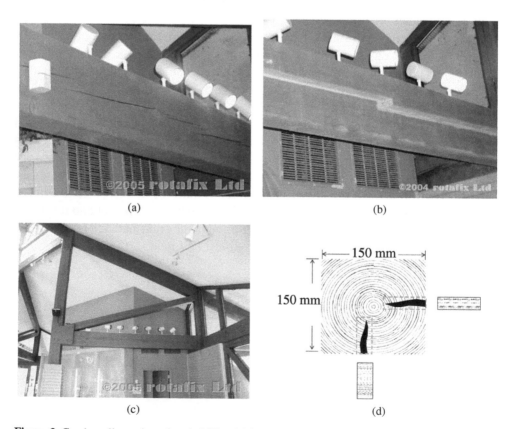

(a)

(b)

(c)

(d)

Figure 3. Crack sealing using a bonded fillet: (a) beam before repair; (b) beam after routing; (c) beam after repair; (d) schematic of a typical repair [12] (reprinted with permission from Rotafix Ltd.).

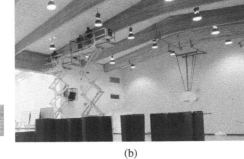
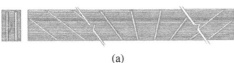

(a)   (b)

Figure 4. Fissures or delamination repair by bonded-in rods: (a) schematic solution using GFRP bars bonded with a thixotropic structural adhesive into pre-drilled blind holes made from the underside of the beam; (b) glulam beam after repair and after cosmetic finish with wooden plugs (reprinted with permission from Rotafix Ltd.).

4.2. Fissures and Delamination Repair

Fissured or delaminated (in the case of glulam) members may require proper strengthening. The strength increase may be obtained by bonding in steel or FRP rods inserted approximately across the grain (Fig. 4). Glass fibre mats bonded to the sides of the beams or sandwiched between two beam parts may also be used. By avoiding premature failure in tension perpendicular to grain, bending strength may fully develop, thus contributing to a significant beam reinforcement.

4.3. Repair of Decayed Beam Ends

Structural adhesives have a high potential to repair the decayed ends of beams or trusses in building floors and roof systems. These are frequently exposed to high moisture content due to water intake from exterior walls and roofs and are, therefore, prone to biodegradation by fungi and insects.

Possible approaches to deal with this problem include partial reconstruction of timber structural members by replacing the degraded timber with epoxy grout cast on-site into a permanent timber formwork, or, preferably, replacing decayed material with a new timber splice; in both cases, the load is transferred to the remaining sound timber by FRP or steel rods or plates (Fig. 5).

These techniques are more effective and less invasive than traditional repair techniques since they maintain the structural functioning and keep most of the original material. Besides, they have a reduced visual impact as both timber formwork or timber splices may be chosen to match the original timber.

The correct identification of the decayed volume, proper design and detailing, choice of materials, workmanship and quality control are all essential factors for the efficiency and durability of the intervention.

Several configurations for the intervention may be adopted, depending on the access to the beam end and working conditions (Fig. 6).

Except for possible minor variations, the following general steps are required:

Figure 5. Bonded-in steel rod connector for rotten beam-end repairs (reprinted with permission from Rotafix Ltd.).

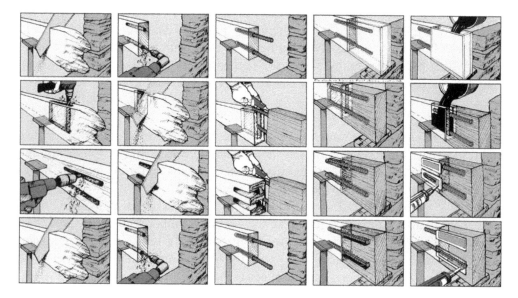

Figure 6. Typical repair configurations [14] (reprinted with permission from Rotafix Ltd.).

1. Propping the beams and scaffolding installation;

2. Cutting off the deteriorated beam part to reach sound timber;

3. Fabrication of the permanent timber formwork or the splice;

4. Drilling holes or cutting slots in the remaining sound timber to insert the connecting rods or plates. No more time than necessary should be allowed to elapse

between final surface preparation and bonding, in any case less than 24 h before bonding;

5. Careful cleaning of the surfaces to be bonded from dust and debris, by vacuum cleaning or by compressed air;

6. Partial injection of the adhesive into holes or slots;

7. Cleaning up and insertion of rods or plates;

8. Placing splice or formwork and checking their alignment with the remaining beam;

9. Pouring the bonding grout in the formwork or injection of the adhesive into holes or slots;

10. Disguising of slots if necessary (with timber or timber sawdust);

11. Removal of the temporary supports after complete cure of the adhesive products.

The timber splice should normally match the original timber member's strength, stiffness, durability, dimensional stability and visual appearance. If the durability of the original timber species is clearly insufficient to face the risk of biological attack, a more durable species or preservative treated timber might have to be specified. Timber moisture content is normally required to stay below 16–20% but not too low, to minimize subsequent dimensional variations.

4.4. Repair of Damaged Joints

Adhesive injection has been extensively used in the past to consolidate loose or saggy mechanical timber joints. Sealing of the joint area prior to adhesive injection currently includes the application of timber boards, structural plywood or steel plates bonded across the whole joint area, affecting the various members that meet in the joint.

Although improving the joint strength and stiffness, major criticisms to this practice are the irreversibility of the intervention, the prevention of future inspection, and the modification of the joint behaviour. Joints become highly rigid, which may change the overall stress distribution in the structure with possible negative effects.

In the case of decayed timber (Fig. 7(a)), the whole joint area may be replaced by a new solid structural node made with cast-in epoxy grout, connected with rods to sound timber parts (Fig. 7(b)). A much better alternative, although requiring more time and means and more skilled operators, is the individual repair of the members meeting in the joint, thus maintaining the original joint behaviour (Fig. 7(c)).

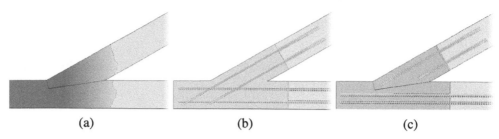

Figure 7. Repair of decayed joint (a) by integral node casting (b) and by individual members' repair (c).

4.5. Strengthening of Timber Members

Strengthening of beams or truss members might be needed to overcome insufficient strength or stiffness, namely as a result of design errors, of increasing load or safety requirements, and of ageing or biological degradation of timber.

Overall strengthening normally involves most or the whole length of the timber member (resulting in a composite member); a local deficiency may be dealt with a local intervention, but care must be taken to ensure that stresses flow from sound to sound timber and the transition between strengthened and un-strengthened lengths is progressive and/or does not occur in a highly loaded cross-section to avoid breaking the timber in the transition area.

Metal or FRP reinforcing and connecting materials may be bolted, or, preferably, bonded to the member surfaces or, alternatively, inserted into the timber members bonded with structural adhesives. Although requiring higher skill and careful workmanship, the use of structural adhesives to fix reinforcing materials is preferable to mechanical fasteners due to the gap-filling ability of adhesives and the more rigid connection and thus more effective composite member behaviour (Fig. 8).

The application of profiles to the surface is far easier, the disadvantages being their higher visual impact, the higher exposure of the adhesives to high temperature and fire, and the need for fire protection of the profiles and protection against corrosion in the case of steel plates. On the other hand, insertion of rods or bars in the timber members requires more working space and specific tools to drill long holes or slots and it is more difficult to control the adhesive thickness.

Selection of adhesive, surface preparation, cleaning and bonding should follow the same principles as applicable for the replacement of decayed beam ends. This technique has low intrusiveness and visual impact, and although not reversible, it allows keeping in place the original element that would otherwise have to be replaced.

4.6. Perpendicular to Grain Reinforcement of Mechanical Joints

Joints are often the weakest part of the structure, normally susceptible to premature risk of splitting near the mechanical fasteners. By improving the tension perpendicular to the grain of the timber in this location, failure will not occur in the wood but

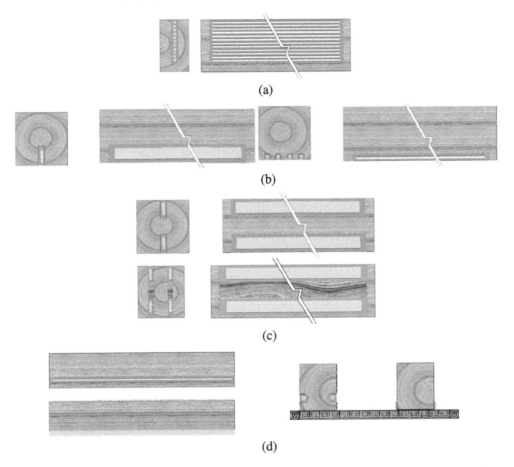

(a)

(b)

(c)

(d)

Figure 8. Principal strengthening methods currently used: (a) modified flitch upgrading; (b) tension zone upgrading; (c) compression and tension zones upgrading; (d) reinforcement of beams or joints above decorative ceilings [15] (reprinted with permission from Rotafix Ltd.).

it will be controlled by the type, number and quality of the mechanical fasteners, increasing the joint strength and ductility.

This effect may be achieved by using fibreglass fabric bonded separately to each of the contacting faces of the members in the joint area, in directions of 45° or 90° with respect to the wood grain direction. For members with dowels or bolts loaded perpendicular to the grain, reinforcement can ensure full load carrying capacity even for very small end-distances. Local reinforcement for tension perpendicular to the grain of the timber can also be very effective for end-notched beams.

Similar effects may be achieved by bonding strong material plates to the joint's inner surfaces, like structural plywood, densified veneer wood, or even FRP plates. Not only these will avoid wood splitting but they will also increase the embedding strength of the resulting member loaded by the fasteners, thus improving the joint's load carrying capacity.

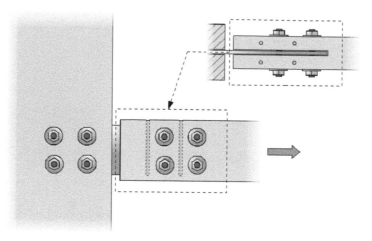

Figure 9. Bonded-in rods driven across the grain for joint strengthening.

The above reinforcements can be made almost invisible, but this can only be done prior to the joint assembly or when disassembly is possible.

Alternatively, bonded steel or fibre reinforced rods can be inserted from the sides of members in the joint area, in the direction perpendicular to the grain, thus stitching the timber fibres together and avoiding splitting (Fig. 9). Whenever possible, long rods should be driven all the way across the timber member width, or shorter ones should be driven from both sides.

The main advantage of bonded-in rods in this case is that they can be inserted also during the service life of the structure, as a preventive or even remedial measure, although complex geometry or tri-dimensional joints may hinder access for the proper positioning of rods.

5. Requirements and Practical Difficulties in the Work on Site

Despite its great potential, the use of structural adhesive bonding of timber is still limited by the lack of suitable design guidance (and guidelines) together with a general lack of experienced personnel, whether engineers, designers, contractors or clients. Structural bonding on site aimed at conservation has recently attracted a great deal of interest from the European scientific community, namely within COST Action E34. Relevant information can be found in the Timber Engineering STEP manual [16, 17], in the COST E34 Core Document [18] and website (http://www1.uni-hamburg.de/cost/e34/index.htm), and in the Low Intrusion Conservation Systems for Timber Structures Project website (www.licons.org).

Nevertheless, design guidance provided by EN 1995-1-1 is limited to the requirement that 'Adhesives for structural purposes shall produce joints of such strength and durability that the integrity of the bond is maintained in the assigned service class throughout the expected life of the structure' and that adhesives should comply with specifications defined in EN 301 for the relevant service class. In relation to structural detailing and control, EN 1995-1-1 specifies that the adhesive manufacturer's recommendations relevant to the proper use of the adhesive should be followed, full strength development of bonded joints prior to loading should be achieved, and that 'where bond strength is a requirement for ultimate limit state design, the manufacture of glued joints should be subject to quality control'. More restrictive is EN 1995-1-2 (Structural fire design) [19], which specifies that 'Adhesives for structural purposes shall produce joints of such strength and durability that the integrity of the bond is maintained in the assigned fire resistance period' and acknowledges that 'for some adhesives, the softening temperature is considerably below the charring temperature of the wood'. Although not listing acceptable/unacceptable adhesives, this is clearly an indication that epoxy adhesives and others alike do not fulfill structural design requirements in the case of fire.

The great diversity of on-site bonding repair situations, timber species and exposure conditions to deal with, structural adhesives and composites available are also a source of problems and challenges.

Proper intervention planning and design, workmanship and quality control are essential since the non-reversibility of adhesion does not allow inspection and correction after cure. Compared to currently used mechanical timber joints, bonding requires more skilled and well-trained operators, and better organisation and record keeping for future traceability of possible problems and defective joints.

Poor joint performance frequently results from poor surface preparation (surface inactivated by too long preparation time, debris, grease or dust contamination), incorrect adhesive mixing (mistake between proportions by weight or volume), altered adhesive components (leftovers being used), incomplete mixing or adhesive contamination due to the use of dirty containers or tools, and incomplete curing (early removal of supports or holding devices, or early surface finishing).

Selection of improper adhesive for the intended use is one major source of poor performance. It should uniformly cover the wood surface and ideally penetrate the wood surface, filling the small voids caused by pores, checks, and other anatomical features. The right balance between gap-filling and penetration ability will, therefore, depend on the specific job. Adhesives with suitable strength and stiffness similar to those of timber are normally recommended, but good adhesion to the various materials involved, durability, small creep, and good thermal stability should also deserve consideration. The last is particularly important since most adhesives suitable for bonding on-site have very limited elevated temperature resistance and may be affected by service environments. Besides, overheating of bonded joints in service should be prevented by design, through shading and ventilation.

The working life characteristics of individually manufactured products in terms of polymerization and exothermic rate will depend on room temperature, material temperature and volume of installed material. Data covering realistic working temperatures are not always clearly stated in the Product Data Sheets nor clearly understood by the operators. Very cold environments (below 5°C) may require local heating to allow polymerization, whilst hot climates may require early morning application, smaller portions of adhesives to be handled and pre-cooling of the ad-hesive components, all in all more difficult to control.

6. Performance and Durability

The ability of a structural joint to maintain satisfactory long-term performance, of-ten in severe environments, is an important requirement of a structural adhesive joint, as the joint should be able to support design loads, under service conditions, for the planned lifetime of the structure. A number of factors determining the dura-bility of structural adhesive joints have been identified and can be grouped in three categories: environment, materials and stresses. All of which are discussed next.

6.1. Environment

The environment to which joints are exposed plays an important role in their dura-bility. Moisture and temperature are the most important factors in determining the strength loss of a joint exposed to the service environment.

6.1.1. Moisture
Water is often regarded as one of the most distressing agents that can affect the bonded joint performance and durability. Most bonded structures when exposed to water or humidity will lose strength over a period of time and in rare cases they may collapse, although this effect is limited to extreme conditions.

Water ingress into a structural adhesive joint can decrease the bond performance by several reversible and irreversible mechanisms. The effect of water, at least ini-tially, can be reversible provided that any bond degradation has not proceeded too far. When the joint dries out, the bond can regain some of its lost strength. How-ever, with time, the various irreversible processes that can occur become a serious threat to the long-term durability of the joint. Figure 10 shows the effect of water on the stiffness of four commercial epoxy adhesives. It can be seen that each epoxy adhesive formulation will behave differently for the same immersion period. For instance, while for adhesives A and B the immersion in water at 20°C for a period of 18 months does not result in a loss in the storage modulus, for adhesives C and D it results in some degradation as their storage modulus after emersion and drying does not recover to the initial value.

Generally, the amount of moisture in the wood will greatly affect wetting, flow, penetration and cure of water-borne wood adhesives. However, research studies car-ried out in recent years [21, 22] have found that in the case of on-site polymerized adhesives, like epoxies and polyurethanes, this is not always the case and they are able to bond timber with moisture content up to 22% without any significant deteri-oration in bond strength. However, the major inconvenience in bonding timber with a high moisture content will be the stress originated due to the subsequent drying and shrinkage of the timber.

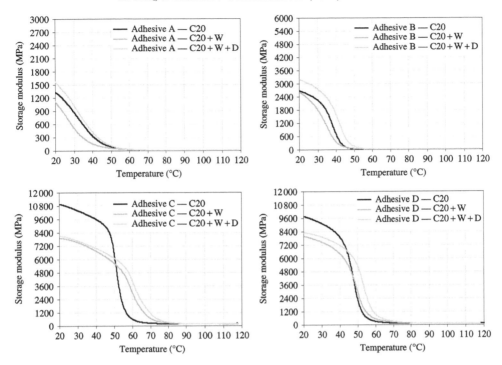

Figure 10. Comparison of the storage modulus curves obtained for four epoxy adhesives before immersion (C20), after emersion (C20 + W), and after emersion and drying at 20°C and 65% RH for one month (C20 + W + D) [20].

6.1.2. Temperature

As moisture, temperature is also an important factor in the durability of structural adhesive joints, since it can affect the creep, fatigue and fire performance of bonded joints.

Well-designed and well-made joints should retain their mechanical properties indefinitely if the wood moisture content stays within reasonable limits (i.e., <15%) and if the temperature remains within the range of human comfort. However, when bonded joints are exposed either intermittently or continuously to abnormally high or low temperatures for long periods they will eventually deteriorate [23].

The effect of temperature variation on the strength of adhesive-repaired structures can be divided in two categories. One category considers the effect of temperature changes due to natural environmental causes. In this category, the temperature varies from −18 to 65°C, a reasonable expected variation. The second major effect to be considered is fire, where extreme temperatures are reached [23].

Today the prevailing belief of practitioners is that since the bondlines in a structural joint are hidden in the interior of the timber element, they experience, due to the low thermal conductivity and high specific heat of wood, considerably lower temperatures compared to ambient conditions. However, the authors [24–27] found that the service temperature to which timber elements are exposed dictates the tem-

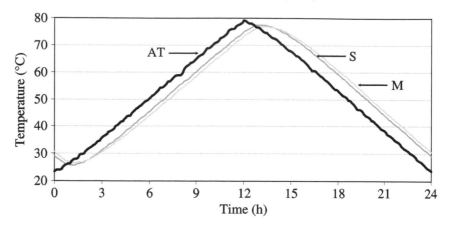

Figure 11. Temperatures attained at the bondline when bonded-in rod timber specimens are subjected to a temperature variation cycle, AT. The graph depicts the temperature variations obtained for timber specimens made of spruce, S (density 440 kg/m^3) and mucoso, M (density 1090 kg/m^3), both specimens having a cross-sectional area of 7.6 × 7.6 cm^2 (corresponding to a bondline timber cover of 3 cm) [24].

perature reached at the bondline placed well within the bonded elements, despite the insulation provided by the timber cover that produces only a certain delay and damping as compared to the ambient climate (Fig. 11).

The shear strength of a bonded joint is also a function of the time for which a given temperature is sustained or has been sustained. This is important when considering the case of an epoxy-repaired timber structure exposed to fire, as compared to situations where the bondline would be subjected to prolonged or repeated exposure to hot environments, e.g., in roof trusses in countries with hot summers [20].

Little information is available on epoxy-repaired timber structures exposed to fire, although there has been some investigation about the effects of fire on wood structures, in general.

The temperature range between 80°C and 100°C that has been recommended [28] as safe surface temperatures for wood exposed for long periods is not valid for all adhesive types. Figure 12 shows the thermal behaviour of eleven typical adhesives used in the rehabilitation of timber structures. It can be seen that all adhesives exhibit a pronounced decrease in their stiffness with the increase in temperature. If one considers the temperatures these adhesives will have to withstand in service (usually up to 50°C), a very careful selection of the adhesive is necessary. Thus, the temperature of the adhesive joint in service must be below the safe working temperature precribed by the adhesive manufacturer, which should correspond to 10–20°C below the adhesive glass transition temperature (taken from the peak of the tan delta curve, which is the most prevalent criterion appearing in the literature and the most used by adhesive manufacturers) [29].

Because of the sensitivity of these adhesives to temperatures in the range of 30–80°C (depending on the adhesive formulation), the fire resistance of an adhesive-

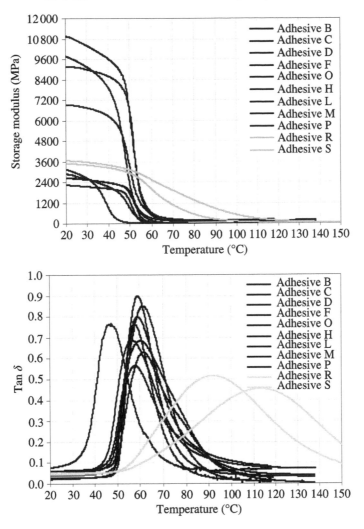

Figure 12. Viscoelastic properties determined by Dynamic Mechanical Analysis at different temperatures for two-component epoxy (black lines: B, C, D, F, H, L, M, O and P) and polyurethane (gray lines: R and S) adhesives.

repaired joint would depend primarily on the distance of the bondline from the surface [24, 25, 30]. Also, adhesives used in these applications can display significantly different viscoelastic responses over the temperature ranges attained normally in service. Thus temperature-induced creep is a risk factor that needs to be considered cautiously when selecting the adhesive for a specific application.

In summary, the rehabilitation and repairing techniques for timber structures involving structural adhesives should always take into consideration the effect of service temperature on the adhesive performance, thus care should be taken regarding the structural joint design and the adhesive selection.

A large number of commercial Product Data Sheets on adhesives do not give detailed information about their thermal stability. If this is the case then it is necessary to obtain that information from the manufacturer or to determine it experimentally. In addition, adhesives showing a similar glass transition temperature can have significantly different strength and thermal behaviour [29]. Thus, in our opinion, besides information about the glass transition temperature, the products data sheet should also contain information about the magnitude of the strength decrease with increasing temperature.

6.2. Materials

Besides the environmental factors mentioned above, the materials involved in a structural joint also influence bond strength and durability. The factors in the material category include: the adherends; the adhesive; the design of the joint; absence of surface contamination (including contamination with wood extractives); stability of the adherend surface; the ability of the adhesive to wet the surface, and entrapment of air/volatiles. Thus, the condition of the adhesive/adherend interface becomes a decisive factor affecting the initial bond strength as well as the long-term durability of the bonded joint [31].

6.2.1. Surface Preparation
Bonding timber is not normally difficult and it is generally possible to obtain a good bond, provided that adequate surface preparation is undertaken before bonding. The main reasons for preparing the wood surface before bonding are: (1) to produce a close fit between the adherends; (2) to produce a freshly cut or planed surface, free from machine marks and other surface irregularities, extractives and other contaminants; (3) to produce a mechanically sound surface, without crushing or burnishing it, which would inhibit adhesive wetting and penetration.

The longer a freshly machined wood is exposed to the atmosphere, the more inactivated it becomes. To prevent surface inactivation in wood two measures should be taken. First, the wood must be dried with care taken not to overdry or overheat the wood. Second, the wood should preferably be planed before bonding to remove hydrophobic and chemically active extractives and other contaminants that could interfere with bonding.

6.2.2. Age of Surface
Wood surfaces can be chemically inactivated by external contamination and self-contamination. External contamination results from airborne chemical contaminants, oxidation and pyrolysis of wood bonding surface from overdrying or exposure to high temperatures, impregnation with preservatives, fire retardants, and other chemicals. Self-contamination results from a natural surface inactivation process where the hydrophobic wood extractives might migrate with time to the wood surface and affect adhesion.

To achieve optimal adhesion conditions it is important to control the time dependence of the inactivation process. It is recommended that no more time than necessary should be allowed to elapse between final surface preparation and bonding. The prepared surfaces should be kept covered with a clean plastic sheet or other relatively inert material to maintain cleanliness prior to the bonding operation. Experimental studies have demonstrated a substantial reduction in wettability during the first 24 h after preparing the surfaces of several wood species [32]. So it is commonly accepted that wood should be surfaced or resurfaced within 24 h before bonding to remove extractives and other contaminants that interfere with bonding.

6.2.3. Wood Species

Wood is a porous, permeable, hygroscopic, and orthotropic material of extreme chemical diversity and physical complexity. Because of this, its properties vary between species, between trees within a species, and even within a tree. This variability can result in an adhesive to produce bonded joints that will perform inconsistently and with different levels of performance among the several types of timber species.

Wood extractives present in the timber are complex mixtures of chemicals such as tannins, anthocyanins, flavones, catechins, kinos, lignans and volatile hydrocarbons. Due to this diversity, countless opportunities exist for chemical reactions between extractives and the atmosphere, and between these materials and adhesives or other chemicals that may contact the materials at the wood surface. The pH and buffering capacity of the wood can be strongly affected by the type and amount of extractives. The setting or curing reactions of some adhesives have been reported to be sensitive to these factors [28]. Wood extractives are then extremely important because of their often undesirable and unpredictable effect on adhesive bonding.

Generally, hardwoods contain more extractives than softwoods. In many species there is a difference between the gluing characteristics of sapwood and heartwood, probably due to the nature and the amounts of extractives found in the heartwood. Most timbers are easy to bond and it is generally possible to obtain good bonding. Some species have characteristics which make them less easy to bond, but that does not mean that those timbers should be avoided when adhesives are used, it merely indicates that a special surface preparation or another adhesive should be used. In addition, most adhesive manufactures give advice concerning their products and they may also produce variants of the standard adhesives which have been formulated to overcome specific problems. The simplest way of minimizing the negative effect of the presence of extractives is to remove them before bonding [31].

6.2.4. Treatment of Wood

Depending on the species and the end-use, timber can be treated with chemicals to enhance its performance against biological agents, fire and weather. Wood can be protected against the attack of decay fungi, wood boring insects, or marine borers by applying chemical preservatives. Timber protection against fire can be achieved through its impregnation or painting with fire retardants. Water repellents can be used to improve timber resistance against the weather. All these treatments should be considered as 'contaminants', as far as adhesion is concerned. Types of surface chemical treatments, adhesives, conditions of joint assembly and adhesive cure can have varied and strong effects on the strength and durability of bonds. Certain combinations of these factors can lead to excellent bonds, despite the interference from chemical treatments [33–45].

6.3. Stresses

A bonded joint that is stressed during ageing will exhibit either decreased lifetime or decreased residual strength [46]. As with moisture, there may be a critical stress level below which failure does not occur or is not accelerated (depending on the moisture level) [47]. The type of stress is also important. For example, cyclical stresses degrade the bond more rapidly than constant stresses [48, 49]. Structural adhesives depend on primary and secondary chemical bonds, both within the polymer itself and across the adhesive/adherend interface. Any chemical reaction involving the destruction of these bonds would be accelerated if the bonded joint is stressed [50]. Stress can also increase the rate of transport of moisture in the adhesive, possibly *via* crazing, the formation of microcracks, or increasing the free volume of the polymer allowing for more moisture ingress [51]. Therefore, the effects of moisture and heat in combination with an applied stress have a considerable influence on the performance and durability of structural adhesive joints. The bonded joint will degrade at a rate determined by the temperature, moisture and level of stress [20].

6.4. Assessment

Because an adhesive should be able to maintain the integrity of the bonded assembly under the expected conditions of service for the life of the structure, it is important to understand the properties of an adhesive that allow it to perform this function.

Rehabilitation systems involving structural adhesives have been used for many years in on-site repair and strengthening of timber structures. However, because they exhibit excellent initial joint strength when tested in standard climate conditions, there has not been a major concern about their service durability and reliable and realistic accelerated ageing tests do not, therefore, currently exist [20].

Furthermore, the application of existing European or national test and performance standards for epoxy bonded products are much too penalising, since they merely impose severe conditions that are not verified in service, or are inadequate because they were developed originally for other adhesives, namely for phenolic and aminoplastic adhesives used in very thin bondlines [7, 52–58]. Moreover, current standard proposals developed for gap-filling adhesives focus only on the initial bond quality control [59–63]. The lack of standards in this field impedes the objective evaluation of the reliability of a bonded-in rod connection, causing engineers to avoid this type of approach altogether.

The scope for further work to enhance the knowledge and design methods employed is very large. Therefore, although these systems are becoming better understood and more usable by designers and building contractors, adhesive material selection and performance assessment still poses a major obstacle to the realisation of their full potential.

7. Ways to Improve Adhesion and Durability

High initial bond strengths are relatively easy to achieve with the type of adhesives commonly used in the rehabilitation and repair of modern and historic timber structures. However, maintaining good bond durability in some situations is comparatively more difficult with these adhesives. For instance, while adhesives such as phenolic, resorcinolic, and aminoplastic resins produce durable bonds in EN 1995-1-1 [4] service classes 1, 2 and 3, the typically used on-site polymerized adhesives do not form bonds of adequate durability in service class 3, when bonding some preservative treated timbers, in situations in which the adhesive is bonded to dense hardwoods, or when bonding wood to non-wood materials, such as FRP profiles, steel rods, etc.

The most effective method of ensuring good durability would be to change the environment to which a bond is exposed, but as this is not always feasible, design is used to protect a joint from the aggressive environment (reducing service class level) and/or to minimize the peak stress concentration in the joint. This leaves the provision of improved bonding (excluding adhesive type) to physico-chemical methods, e.g., *via* the introduction of primary and/or physical bonds that are less susceptible to degradation. These include the use of primers, coupling agents and other surface treatments. Despite the inevitable extra cost associated with these techniques, their use is of particular value where structural bonds might be subjected to repeated wetting and drying [64]. In spite of their widespread use in the aerospace, automotive, and plastics industries, where they are used to develop highly durable bonds to metals, advanced composites, ceramics and plastics, such treatments are virtually nonexistent in the wood products industry [65].

Several adhesion promoters and coupling agents exist, but, unfortunately, the majority of the reported studies refer to their application only to non-cellulosic substrates. One of the most commonly used chemical treatment is through the use of silanes, which have been available commercially for many years, and are the most commonly used coupling agents. They are used to produce highly durable bonds to glass, metals, advanced composites, ceramics, and plastics in the aerospace, automotive, plastics and composites industries [66]. More recently their value in improving the adhesion of surface coatings and adhesives has been investigated, in particular, the improvement in the 'wet' adhesion of coatings and adhesives which results from their use [65].

The only studies found in the literature regarding the treatment of lignocellulosic material with silanes refer to their use to enhance wood properties such as cell wall bulking, anti-swelling efficiency, reduced moisture uptake, and durability [67–70]. No studies were found concerning their use to improve adhesion to timber, and to explain their mechanism of interaction with wood.

Another chemical treatment that has resurged recently due to improvements made in its formulation is the hydroxymethylated resorcinol (HMR). HMR has been used successfully with several timbers and adhesives, to promote the exterior durability of their bonded joints [65, 71]. Consequently, this technique seems ready for industrial application, at least for the species and adhesives tested. Nevertheless, studies to clarify some aspects of its action mechanism are still needed [20].

Successful adhesion between non-bondable or hard-to-bond materials can also be achieved trough the combination of physical and chemical treatment methods. It seems likely that optimum combinations of treatments and adhesion promoters exist for different species and adhesive systems. The synergistic effects are very real for the investigations undertaken so far but, due to the limited research conducted until now on cellulosic adherends, more research is needed to overcome this situation and to allow the use of optimized techniques at a commercial level.

8. Quality Control

Site work should be conducted by well-informed, trained and certified operators under the supervision of the Site Manager, and should comply with local regulations, the specifications of the Quality Plan and the Health and Safety Plan [18].

Quality control of the materials should cover: timber splices (species, dimensions, moisture content), metallic or FRP rods or plates (type, dimensions, roughness and anti-corrosion protection), and adhesives or grouts (type, storage conditions, shelf-life, container condition). Adhesive products should be supplied in separate, clearly labelled closed packages in pre-weighed proportions (stoichiometric quantities), supplied by the manufacturer in batch coded containers. All materials used should require a Technical Data Sheet, as well as a Safety Data Sheet. Their handling and disposal must be properly controlled, monitored, documented and must follow the appropriate current regulations [72].

Quality control of tools and equipment used for the structure propping, timber cutting, drilling, surface cleaning, mixing and application of adhesives and grouts involves the selection of suitable tools, and guarantee that they are kept clean and in good condition.

A Quality Assurance Program should be conducted to assure satisfactory end-product strength and durability. This should (1) establish limits on bonding factors to ensure acceptable joint and end-product; (2) monitor the production processes and quality of bond in joint and end-product; and (3) enable detection of unacceptable joint and end-product, determining the cause, and correcting the problem [3].

The mixing procedure, preparation conditions and cure schedule all have a profound effect on the mechanical properties of the adhesives and, consequently, will affect their durability [20, 73]. Because of this, expeditious on-site tests should be conducted to detect possible deficiencies. A number of test methods have been proposed by the CEN/TC193/SC1/WG11) in the pre-standard 'Adhesives for on-site assembling or restoration of timber structures — on-site acceptance testing'. This proposed standard includes three parts, namely Part 1 — Sampling and measurement of the adhesive's cure schedule; Part 2 — Verification of the adhesive joint's shear strength; and Part 3 — Verification of the adhesive bond strength using tensile proof-loading. However, until now this proposed standard has not been approved.

After completion of the work, the quality of the finished product should be evaluated by the Quality Assurance Manager.

9. Research Needs and Future Developments

The widespread use of on-site repair and reinforcement of existing or new timber structures using structural adhesives has been hindered by the lack of well established design and strength calculation methods for on-site bonding processes, the absence of systematic on-site quality assurance procedures, the absence of harmonized European standards concerning adhesives for on-site assembling or restoration of timber structures, and concerns about the performance against hazards (fire) and durability of these systems in specific situations (e.g., harsh environments, bonding treated timber). However, progress has been made in recent years and some of the existing obstacles for bonding on-site have been eliminated. For instance, suggestions for strength calculation methods have been presented and discussed for the strengthening of timber beams, including pre-stressing and bonded-in rods [18]. Quality control tools for bonding on-site have been created. Standard drafts to evaluate initial bonded joint strength have been proposed to the European Committee for Standardization (CEN). Research studies focusing on the performance and durability of these rehabilitation systems have also been conducted and have contributed to the increase in confidence in the reliability of theses techniques.

Nevertheless, more research work is needed: to develop appropriate test methods for the accrual of accurate property performance data that consider realistic loading and environmental conditions; to overcome the lack of information about extreme service temperature and fire resistance of bonded connections; to develop realistic predictive tools for the long-term behaviour of bonded timber joints and connections. These are required in order to promote their wider use through the increased confidence of architects, designers and owners alike.

Due to sustainability concerns, in the next years the use of bonded timber systems will continue to increase. Also new applications for bonded wood and greener adhesives will appear, and the widespread use of on-site bonding techniques will continue to rise.

In order to allow a proper choice of adhesive, its technical data sheet should provide additional information such as the glass transition temperature of the adhesive (and how it was obtained) and the thermal stability of the adhesive (and providing magnitude of the decrease in strength with increasing temperature for the entire temperature range of the expected service conditions). Producers should put their efforts in producing adhesives especially formulated for specific situations, e.g., high shear strength adhesives for applications involving pre-stressing with improved ductile behaviour so that the full length of the bond anchorage is mobilized instead of concentrating the peak shear stresses over a short anchorage length, and less toxic adhesive formulations for sub-ambient cure applications. Finally, there is also a need for adhesives with higher glass transition temperatures but which maintain the shear strength and modulus of elasticity comparable to those of the existing commercial products that show good creep behaviour.

References

1. C. Mettem and G. Davis, *Constr. Repair* **10** (2), 23–28 (1996).
2. C. Mettem and G. Davis, *Constr. Repair* **10** (3), 43–47 (1996).
3. J. G. Broughton and J. Custódio, in: *ICE Manual of Construction Materials*, M. C. Forde (Ed.), Vol. 2, pp. 739–760. Thomas Telford Ltd, London (2009).
4. BS EN 1995-1-1:2004 + A1:2008 Eurocode 5. Design of timber structures. General. Common rules and rules for buildings.
5. BS EN 923:2005 + A1:2008 Adhesives. Terms and definitions.
6. *Standard Definitions of Terms Relating to Adhesives*. The Adhesive and Sealant Council, Bethesda, MD (2006).
7. BS EN 301:2006 Adhesives, phenolic and aminoplastic, for load-bearing timber structures. Classification and performance requirements.
8. Case Studies. The Tourand Creek Bridge, Winnipeg, Manitoba, Canada. Available at: http://www. rotafix.co.uk/case-timber-tourandimages.htm [16/10/2009].
9. BS EN ISO 12944-4:1998 Paints and varnishes. Corrosion protection of steel structures by protective paint systems. Types of surface and surface preparation.
10. BS EN ISO 8504-1:2001, BS 7079-D1:2000 Preparation of steel substrates before application of paints and related products. Surface preparation methods. General principles.
11. BS EN ISO 8504-2:2001, BS 7079-D2:2000 Preparation of steel substrates before application of paints and related products. Surface preparation methods. Abrasive blast cleaning.
12. Case Studies. Structural Timber Repair. Timber Slip and Fissure Injection Method. Zootique, Winnipeg, Manitoba, Canada. Available at: http://www.rotafix.co.uk/case-timber-zootique.htm [03/04/2009].
13. H. Cruz, J. Custódio and D. Smedley, in: *Proceedings of the 3rd International Conference on High Performance Structures and Materials — HPSM 2006*, Ostend, Belgium, pp. 539–548 (2006).
14. *Resiwood System Timber Engineering Product Data*. Rotafix Ltd., Abercraf, Swansea, UK (1997).
15. Principal Repair Methods currently used in in-situ repair/restoration. Available at: http://www. rotafix.co.uk/presentations/index.htm [03/04/2009].
16. H. J. Blass, P. Aune, B. S. Choo, R. Gorlacher, D. R. Griffith, B. O. Hilson, P. Racher and G. Steck (Eds), *Timber Engineering STEP 1: Basis of Design, Material Properties, Structural Components and Joints*. Centrum Hout, Almere, the Netherlands (1995).
17. H. J. Blass, P. Aune, B. S. Choo, R. Gorlacher, D. R. Griffith, B. O. Hilson, P. Racher and G. Steck (Eds), *Timber Engineering STEP 2: Design, Details and Structural Systems*. Centrum Hout, Almere, the Netherlands (1995).
18. M. Dunky, B. Källender, M. Properzi, K. Richter and M. V. Leemput (Eds), *Core Document of the COST Action E34 — Bonding of Timber*. University of Natural Resources and Applied Life Sciences, Vienna (2008).

19. BS EN 1995-1-2:2004 Eurocode 5. Design of timber structures. General. Structural fire design.
20. J. Custódio, Performance and durability of composite repair and reinforcement systems for timber structures, *PhD Thesis*, Department of Mechanical Engineering, School of Technology, Oxford Brookes University, Oxford (2009).
21. A. S. Wheeler and A. R. Hutchinson, *Int. J. Adhesion Adhesives* **18** (1), 1–13 (1998).
22. J. G. Broughton and A. R. Hutchinson, *Construct. Build. Mater.* **15** (1), 17–25 (2001).
23. J. Custódio, J. Broughton and H. Cruz, *Int. J. Adhesion Adhesives* **29** (2), 173–185 (2009).
24. H. Cruz, J. Custódio and J. S. Machado, *Int. J. Construlink* **3** (10), 1–8 (2005).
25. H. Cruz and J. Custódio, in: *Proceedings of the 9th World Conference on Timber Engineering – WCTE 2006*, Portland, OR, USA (2006).
26. H. Cruz and J. Custódio, in: *Proceedings of the 2° Encontro sobre Patologia e Reabilitação de Edifícios — PATORREB 2006 (2nd Symposium on Building Pathology and Rehabilitation)*, Porto, Portugal, pp. 149–158 (2006).
27. J. Custódio and H. Cruz, *Ciência & Tecnologia dos Materiais* **18** (3–4), 23–30 (2006).
28. *Wood Handbook: Wood as an Engineering Material. Gen. Tech. Rep. FPL–GTR–113.* United States Department of Agriculture, Forest Service, Forest Products Laboratory, Madison, WI (1999).
29. J. Custódio, D. Rodrigues, R. André, J. Ferreira and H. Cruz, in: *Proceedings of the COST Action E34 Final Conference on Bonding of Timber — Enhancing Bondline Performance*, Sopron, Hungary, pp. 20–30 (2008).
30. H. Cruz, J. Custódio and J. S. Machado, in: *Proceedings of the 2° Congresso Nacional da Construção, Repensar a Construção — CONSTRUÇÃO 2004 (2nd National Construction Congress, Re-think Construction)*, Porto, Portugal, pp. 907–912 (2004).
31. J. Custódio and B. Broughton, in: *Core Document of the COST Action E34 'Bonding of Timber'*, M. Dunky, B. Källender, M. Properzi, K. Richter and M. V. Leemput (Eds), pp. 56–73. University of Natural Resources and Applied Life Sciences, Vienna (2008).
32. T. Nguyen and W. E. Johns, *Wood Sci. Technol.* **13** (1), 29–40 (1979).
33. M. P. Dimri and K. S. Shukla, *J. Timber Dev. Assoc. India* **39** (2), 18–22 (1993).
34. H. B. Manbeck and K. R. Shaffer, *Wood Fiber. Sci.* **27** (3), 239–249 (1995).
35. T. Sellers and G. D. Miller, *Forest Prod. J.* **47** (10), 73–76 (1997).
36. C. B. Vick and R. C. Groot, *Forest Prod. J.* **40** (2), 16–22 (1990).
37. C. B. Vick and T. A. Kuster, *Wood Fiber. Sci.* **24** (1), 36–46 (1992).
38. C. B. Vick, P. C. Larsson, R. L. Mahlberg, R. Simonson and R. M. Rowell, *Int. J. Adhesion Adhesives* **13** (3), 139–149 (1993).
39. C. B. Vick and A. W. Christiansen, *Wood Fiber. Sci.* **25** (1), 77–89 (1993).
40. C. B. Vick, *Forest Prod. J.* **47** (7/8), 83–87 (1997).
41. C. B. Vick, *Forest Prod. J.* **45** (3), 78–84 (1995).
42. B. Herzog and B. Goodell, *Forest Prod. J.* **54** (10), 82–90 (2004).
43. W. R. Kilmer and P. R. Blankenhorn, *Wood Fiber. Sci.* **30** (2), 175–184 (1998).
44. J. H. Lisperguer and P. H. Becker, *Forest Prod. J.* **55** (12), 113–116 (2005).
45. C. Tascioglu, B. Goodell and R. Lopez-Anido, *Composites Sci. Technol.* **63** (7), 979–991 (2003).
46. A. Pizzi and K. L. Mittal (Eds), *Handbook of Adhesive Technology*, 2nd edn. Marcel Dekker, New York, NY (2003).
47. K. W. Allen (Ed.), *Adhesion*, Vol. 1. Applied Science Publishers, London (1977).
48. E. W. Thrall and R. W. Shannon (Eds), *Adhesive Bonding of Aluminum Alloys.* Marcel Dekker, New York, NY (1985).

49. W. S. Johnson (Ed.), *Adhesively Bonded Joints: Testing, Analysis and Design*. ASTM International, West Conshohocken, PA (1988).
50. A. J. Kinloch (Ed.), *Durability of Structural Adhesives*. Applied Science Publishers Ltd., Barking, UK (1983).
51. A. J. Kinloch, *Adhesion and Adhesives: Science and Technology*. Chapman & Hall, London (1987).
52. ASTM D 2919-01 Standard Test Method for Determining Durability of Adhesive Joints Stressed in Shear by Tension Loading.
53. ASTM D 3535-07a Standard Test Method for Resistance to Creep Under Static Loading for Structural Wood Laminating Adhesives Used Under Exterior Exposure Conditions.
54. ASTM D 4680-98 (2004) Standard Test Method for Creep and Time to Failure of Adhesives in Static Shear by Compression Loading (Wood-to-Wood).
55. BS EN 14292:2005 Adhesives. Wood adhesives. Determination of static load resistance with increasing temperature.
56. ASTM D 2559-04 Standard Specification for Adhesives for Structural Laminated Wood Products for Use Under Exterior (Wet Use) Exposure Conditions.
57. BS EN 302-1:2004 Adhesives for load bearing timber structures. Test methods. Determination of bond strength in longitudinal tensile shear strength.
58. BS EN 302-2:2004 Adhesives for load-bearing timber structures. Test methods. Determination of resistance to delamination.
59. Standard Proposal CEN TC193/SC1/WG11 N20 Adhesives for on-site assembling or restoration of timber structures. On-site acceptance testing. Sampling and measurement of the adhesive's cure schedule. European Committee for Standardization (CEN), Brussels (2003).
60. Standard Proposal CEN TC193/SC1/WG11 N21 Adhesives for on-site assembling or restoration of timber structures. On-site acceptance testing. Verification of the adhesive joint's shear strength. European Committee for Standardization (CEN), Brussels (2003).
61. Standard Proposal CEN TC193/SC1/WG11 N22 Adhesives for on-site assembling or restoration of timber structures. On-site acceptance testing. Verification of the adhesive bond strength using tensile proof-loading. European Committee for Standardization (CEN), Brussels (2003).
62. Standard Proposal CEN TC193/SC1/WG11 N48 Adhesives for on-site assembling or restoration of timber structures. Evaluation of the shear strength of tubular adhesive joints. European Committee for Standardization (CEN), Brussels (2005).
63. Standard Proposal CEN TC193/SC1/WG11 N260 Adhesives for on-site assembling or restoration of timber structures. Comparative evaluation of the shear strength of adhesive joints and solid wood. European Committee for Standardization (CEN), Brussels (2006).
64. J. Custódio, J. Broughton, H. Cruz and P. Winfield, *Int. J. Adhesion Adhesives* **29** (2), 167–172 (2009).
65. J. Custódio, J. Broughton, H. Cruz and A. Hutchinson, *J. Adhesion*. **84** (6), 502–529 (2008).
66. K. L. Mittal (Ed.), *Silanes and Other Coupling Agents*, Vol. 5. VSP/Brill, Leiden (2009).
67. K. I. Brebner and M. H. Schneider, *Wood Sci. Technol.* **19** (1), 75–81 (1985).
68. S. Donath, H. Militiz and C. Mai, *Holzforschung* **60** (2), 210–216 (2006).
69. M. H. Schneider and K. I. Brebner, *Wood Sci. Technol.* **19** (1), 67–73 (1985).
70. S. Donath, H. Militz and C. Mai, *Holzforschung* **60** (1), 40–46 (2006).
71. J. Custódio, H. Cruz, J. G. Broughton, A. R. Hutchinson and P. H. Winfield, in: *Proceedings of the 10th International Conference on the Science and Technology of Adhesion and Adhesives — Euradh 2008/Adhesion'08*, Oxford, UK, pp. 129–133 (2008).

72. G. Mays and A. R. Hutchinson, *Adhesives in Civil Engineering*. Cambridge University Press, Cambridge (1992).

73. J. Custódio, H. Cruz and J. G. Broughton, in: *Proceedings of the 11th International Conference on Durability of Building Materials and Components — DBMC 2008*, Istanbul, Turkey, pp. 635–642 (2008).

Part 3
Environment-Friendly Adhesives

Thermal Characterization of Kraft Lignin Phenol-Formaldehyde Resin for Paper Impregnation

Arunjunaj Raj Mahendran [a], **Günter Wuzella** [a] **and Andreas Kandelbauer** [b,c,*]

[a] WOOD Carinthian Competence Center, Kompetenzzentrum Holz GmbH,
Klagenfurterstrasse 87–89, A-9300 St. Veit/Glan, Austria
[b] Department of Wood Science and Technology, University of Natural Resources and
Applied Life Sciences, Peter Jordan Strasse 82, A-1190 Vienna, Austria
[c] Faculty of Applied Chemistry, Reutlingen University, Alteburgstrasse 150,
D-72762 Reutlingen, Germany

Abstract
Both kraft lignin phenol-formaldehyde (KLPF) and phenol-formaldehyde (PF) resins were synthesized and the curing kinetics of the resins was determined using thermal analysis data. The kinetic parameters were predicted using two popular model-free kinetic (MFK) methods: the advanced form of the Vyazovkin method and the Kissinger–Akahira–Sunose method. From the experimental and predicted values, the rate of kraft lignin phenol-formaldehyde curing was less compared to the phenol-formaldehyde resin. To increase the cure rate different cure additives were tried. Among these, only potassium carbonate (KC) showed a positive effect: An increase in additive concentration to 4% reduced the curing time to almost 50% as compared to pure KLPF resin at 160°C. Comparison of the predicted values from MFK calculations with isothermal experimental data showed that both MFK approaches were suitable to predict the curing characteristics.

Keywords
Lignin, phenol-formaldehyde resin, accelerator, isoconversional kinetic analysis, model-free kinetic analysis, laminates, paper impregnation, differential scanning calorimeter

1. Introduction

Phenol-formaldehyde resins are among the most important polymeric adhesives used in the wood based composite panel manufacturing industries [1]. Phenolic resins are prepared by the reaction of phenol or any substituted phenol with formaldehyde or other aldehydes, in the presence of acidic or basic catalyst. The price of phenol depends on the oil price and is likely to ever increase due to shortage of fossil resources. Hence, several lignin substitute products based on renewable materials derived from annual plants such as flax [2, 3] or kenaf [4], agricultural waste such as sugar cane bagasse [5] and wheat straw [6] or by-products from the

---

[*] To whom correspondence should be addressed. E-mails: andreas.kandelbauer@reutlingen-university.de, or andreas.kandelbauer@boku.ac.at

Wood Adhesives
© Koninklijke Brill NV, Leiden, 2010

wood, pulp and paper industries [3, 7, 8] have been tested as a phenol replacement for both resol and novolac resins. In this context, lignins are a very promising class of natural compounds because of their relatively low price and high availability. Lignin polymers contain phenolic hydroxyl groups and are basically natural PF resins. They are the second most abundant compounds next to cellulose [4] and are obtained from wood using different pulping methods during the paper manufacturing process. Kraft lignins are obtained as a by-product from the kraft pulping process [9]. Kraft lignins contain intermolecular linkages between phenylpropane, guaiacyl, syringyl, p-hydroxyphenyl and biphenyl nuclei. The formulation of resins with kraft lignin has been studied before to obtain binder materials for engineered wood products [10–17]. Olivares et al. obtained resins with 20% phenol replacement by kraft lignin for application as a glue in particleboard [18].

However, there is practically no research work published focussing on the preparation of kraft lignin-PF hybrid resins for the impregnation of papers intended for decorative laminates. Here, typically resol PF resins of low viscosity and rapid curing characteristics are applied. Kraft papers are impregnated with PF and stacks of such papers are cured in a press at elevated temperature to form cured panel products that are preferably used as exterior claddings and in kitchen and bathroom surfaces where high moisture resistance is required.

One major drawback of kraft lignin modified PF (KLPF) type resol resins is the lower reactivity of lignin compared to PF. Hence, the addition of lignin causes lower curing rates in the press than obtained with pure PF resin composites. To increase the cure rate and to investigate the effect on the molecular weight distribution of PF resin, cure additives such as ammonia based salts, amines, and amides were investigated by Duval et al. [19]. Carbonates as curing additives for phenol-formaldehyde resins and their curing mechanism were discussed by Pizzi and coworkers [20, 21] and Park et al. [22].

In the present contribution, the synthesis and characterization of partially substituted kraft lignin phenol-formaldehyde (KLPF) impregnation resins containing different additives for use in decorative paper laminate manufacturing are discussed based on dynamic DSC experiments.

2. Experimental

2.1. Chemicals

Phenol, formaldehyde in form of a 37% aqueous solution (formalin), and sodium hydroxide were purchased from Sigma Aldrich, Steinheim, Germany. Kraft lignin was supplied in the form of a dried black powder by Mead Westvaco Corporation, Charleston, South Carolina, USA. The salts potassium carbonate, zinc borate, and propylene carbonate were also purchased from Sigma Aldrich, Steinheim, Germany. All chemicals were analytical grade and used as such for the preparation of the resins without further purification.

2.2. Preparation of Phenol-Formaldehyde (PF) Resin

A phenolic resin with a phenol to formaldehyde molar ratio of P:F = 1:1.8 was prepared according to the procedure described in [23]. In a 3 l round bottom flask fitted with a temperature sensor and a mechanical stirrer, 208.8 g phenol were dissolved in distilled water to give a 91% (w/w) aqueous solution to which 292 ml of 37% formalin were added at room temperature. The pH of the mixture was adjusted to 8.5 by slowly adding 12 ml of 50% aqueous sodium hydroxide solution. The reaction mixture was heated to 90°C and the temperature was kept constant for approximately 180 min. The water tolerance of the reaction was tested at regular time intervals. When the water tolerance (the percentage of water that could be added to the liquid resin without precipitation) reached a value of lower than 300%, the condensation reaction was stopped by cooling the reaction mixture to room temperature. Subsequently, vacuum distillation was carried out to adjust the solids content of the resin to 60%. Resin gel time was measured by subjecting the resin to a heat treatment on a hot plate and measuring the time until solidification (hot plate test) and it was determined as 100 s. Resin viscosity expressed as the time required by the resin to flow through the hole in the bottom of a cone of defined volume was determined as 19 s using a flow-cup viscosimeter according to the DIN EN ISO 2431 standard.

2.3. Preparation of Kraft Lignin Modified Phenol-Formaldehyde (KLPF) Resin

A kraft lignin-modified phenolic resin was synthesized which was based on the same phenol to formaldehyde ratio of P:F = 1:1.8 as the pure PF-resin, the major difference being that 20 wt% of the phenol was substituted by kraft lignin of which the exact molecular weight was unknown. Kraft lignin (KL), phenol and formalin were charged in a 3-neck round bottom flask and the initial pH was adjusted to 8.5 by addition of 6 ml sodium hydroxide (50%). The reaction mixture was heated to 80°C and kept at this temperature for 2 h to introduce reactive methylol groups into the lignin framework as described earlier for a different type of lignin (lignosulfonate) by Alonso *et al.* [24]. After this activation of the lignin, 91% aqueous solution of phenol was added to the methylolated kraft lignin and the pH was adjusted to 8.2 by adding another portion of sodium hydroxide (50%). The mixture was heated to 90°C and the kraft lignin phenol-formaldehyde resin was allowed to polymerize. Water tolerance was checked at regular time intervals. When the water tolerance fell below 300%, the reaction was stopped by cooling the mixture to room temperature. The solids content was adjusted to 60% by vacuum distillation of the resin. Gel time and viscosity of the KLPF resin were 120 s and 22 s, respectively.

The thermal properties of freshly prepared pure and additive-modified PF and KLPF resins were studied with DSC. In an attempt to accelerate the curing of KLPF, the additives zinc borate, potassium carbonate, sodium carbonate, and propylene carbonate were added to the resin and the curing behaviour was compared to the curing of unmodified KLPF and PF resins. Based on dynamic DSC experiments, isoconversional kinetic analysis of resin curing was performed.

2.4. Differential Scanning Calorimetry (DSC) Analysis

All thermograms were recorded using a differential scanning calorimeter 822^e DSC equipment by Mettler Toledo (Greifensee, Switzerland).

For dynamic DSC experiments, 2.0–3.5 mg samples of PF and pure or additive-containing KLPF resins were subjected to a temperature gradient ranging from 25 to 250°C with three heating rates (5, 10 and 20°C/min). As additives for KLPF, 4 and 8% (weight per weight) of zinc borate, potassium carbonate and propylene carbonate were used. To suppress evaporation of volatiles during condensation, the samples were sealed in high pressure, gold coated stainless steel crucibles of 30 µl total volume. The enthalpy changes were recorded and analyzed for the peak maximum temperature, T_{peak}, the onset temperature, T_0, and the normalized enthalpy integral, H, using the STAR 8.10 software package (Mettler Toledo, Greifensee, Switzerland). All experiments were repeated twice.

For experimental verification of the kinetic models, validation experiments were performed using isothermal DSC. Here, 2.0–3.5 mg of each validation resin mixture was weighed into a high-pressure, gold-coated stainless steel crucible (30 µl) and thermograms were recorded at 160°C for 30 min. The validation samples were inserted into the oven which was preheated to the isothermal temperature. Thermograms were recorded after stabilization of the oven temperature which took approximately 1min. In all validation experiments, the enthalpy changes ΔH were recorded and analyzed for the peak maximum temperature, T_{peak}, the onset temperature, T_0 and the enthalpy integral, H.

2.5. Analysis of Thermochemical Data

From the temperature integral of the thermograms, both conversion, α and the change of conversion with time, $\alpha(t) = d\alpha/dt$, were determined at a specific cure time (t). The $\alpha(t)$-value was determined from dynamic runs as the ratio between the heat released until the time t and the total heat of the reaction according to equations (1) and (2):

$$\alpha(t) = \frac{H_{t_0,...,t}}{H_\infty}, \tag{1}$$

$$\frac{d\alpha(t)}{dt} = \frac{H_t}{H_\infty}, \tag{2}$$

with t_0 = time of start of curing reaction and t_∞ = time of 100% cross-linking.

Integration of the enthalpy curve from t_0 to t yielded the degree of conversion $\alpha(t)$ for any time t between t_0 and t_∞ as the ratio $H_{t_0,...,t}/H_\infty$, where H_∞ is the integral of the enthalpy curve from t_0 to t_∞ and $H_{t_0,...,t}$ is the integral of the enthalpy curve from t_0 to t. The differential $d\alpha/dt$ for each time t between t_0 and t_∞ was calculated as H_t/H_∞, where H_t is the measured enthalpy at time t.

Kinetic analysis according to the model-free approach by Kissinger–Akahira–Sunose [25, 26] was performed with the computer program Excel. For kinetic analysis according to the advanced Vyazovkin method [27], the STAR software package was used. The theory behind these isoconversional methods has recently been briefly summarized [28].

As a result of the kinetic analyses with both methods, the dependence of the activation energy on the conversion (apparent activation energy E_α) was calculated for all resins (PF, KLPF and modified KLPF). The experimental verification of the model suitability was done by comparing the calculated curves of conversion *versus* the predicted reaction times t_α at 160°C with the curve of conversion *versus* time as directly measured by isothermal DSC data at 160°C. The reaction temperature of 160°C was used because it is well within the range of industrially relevant temperatures at which hot pressing is performed.

3. Results and Discussion

3.1. Curing of Pure PF and KLPF Resins

The exothermic curing curves for the pure PF and KLPF resins are shown in Fig. 1. The PF resin displays two exothermic peaks occurring at 153°C and 211°C. From the literature it is well known that there are two major reactions taking place during phenol-formaldehyde curing: methylolation and condensation [29, 30]. According to Christiansen and Gollob [31], the first enthalpy peak is related to methylolation and usually occurs at temperatures $\leqslant$100°C, whereas the second peak at higher temperature is related to the condensation reaction. Park and Riedl [32] suggested another reason for the observed thermogram of PF resin related to the distribution of the average molecular weights in PF resins: while species of higher molecular

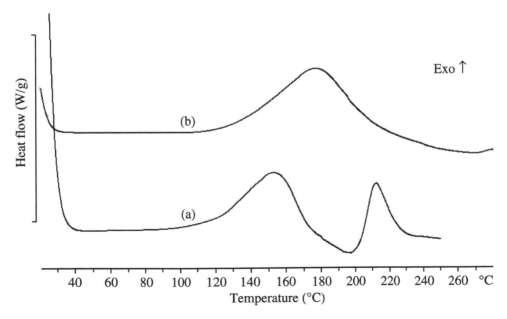

Figure 1. DSC thermograms of (a) PF resin and (b) KLPF resin at a heating rate of 10°C/min.

weight will cure at lower temperature, smaller species will require higher temperatures resulting in the observed enthalpy course. According to this, the PF resin used would contain two major fractions of distinctly different average molecular weights.

In contrast, for the KLPF resin only one broad enthalpy peak was found over the whole temperature range which is associated with the condensation reaction. Peak onset and peak maximum were at higher temperatures than for PF, indicating a lower overall reactivity of the KLPF resin.

From the thermograms of the KLPF resin that were recorded at different heating rates, the function for the time-dependence of conversion $\alpha(t)$ was calculated for each heating rate. The data from the $\alpha(t)$-curves were used for kinetic analysis to predict the conversion-dependent activation energy using the advanced model-free approach by Vyazovkin (VA) [27] and the model-free approach suggested by Kissinger–Akahira–Sunose (KAS) [25, 26] (Fig. 2a). For better visualization, representative error bars are only indicated for the dataset calculated with the KAS method to give an indication of the overall error. For verification of the analysis, the curves of conversion *versus* the predicted reaction time t_α at 160°C were compared with the curve of conversion *versus* time as obtained directly from isothermal DSC at 160°C (Fig. 2b).

In all calculations, the activation energy was not constant throughout the curing reaction. Non-constant activation energy during a chemical reaction indicates that no single reaction mechanism is valid for the whole reaction. Since the cure mechanism of the KLPF resin is similar to unmodified PF, the observed change in reaction mechanisms during curing is due to the varying contributions of the two different reaction types addition and condensation to the overall reaction. Initially, the addition of hydroxymethylol groups to free *ortho*- and *para*- positions in the kraft lignin phenoxy units takes place which is followed by condensation between two or more methylol lignins [33].

While the KAS method led to a rather uniform value for the activation energy throughout the reaction, the advanced VA method yielded a more distinctive dependence of the activation energy on the conversion. Initially, when the addition reaction is predominant, there is a higher value for the activation energy which gradually decreases as the contribution of the condensation reaction gains importance. While the activation energy remains constant at conversions from 30 to 60%, it is thought that condensation is the predominant reaction taking place. Toward the end of the reaction, especially at conversions >70%, the activation energy increases again indicating that the reaction transits from being chemically controlled to be diffusion controlled. While the values for the predicted conversion calculated with the practically constant activation energy obtained by the KAS approach differ significantly from the observed values from the isotherm at any stage of the reaction, the advanced form of the isoconversional Vyazovkin method matches the experiment perfectly well over a wide α-range and predicts well the observed overall course of conversion.

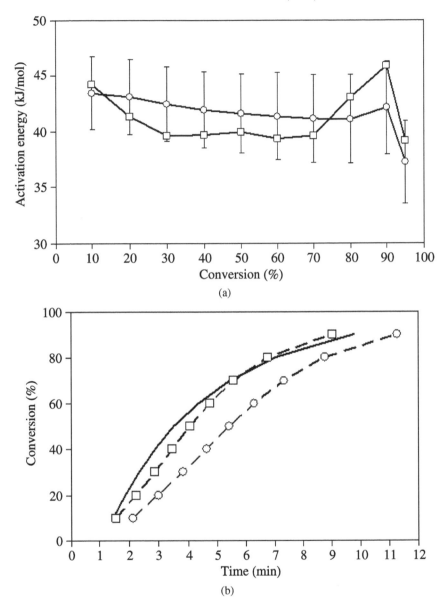

Figure 2. (a) Activation energy E_α for KLPF resin estimated by the advanced Vyazovkin method (VA, □) and by the Kissinger–Akahira–Sunose method (KAS, ○). (b) Conversion α in % over time for KLPF at 160°C. Comparison of the isothermal experiment (—) with curves calculated with the VA (□) and with the KAS (○) methods.

3.2. Curing of Modified KLPF Resins

To study the accelerating effect of some additives, in a screening experiment, the three salts zinc borate, potassium carbonate and propylene carbonate were added to the KLPF resin at a concentration of 4 wt% and thermograms were recorded

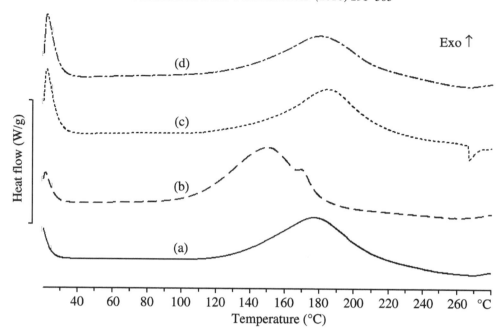

Figure 3. DSC thermograms of KLPF resin and KLPF resin with three different accelerators at a concentration of 4 wt% and at a heating rate of 10°C/min, (a) KLPF resin, (b) KLPF resin with potassium carbonate (KC), (c) KLPF resin with zinc borate (ZnB) and (d) KLPF resin with propylene carbonate (PC).

Table 1.
Normalised enthalpy and peak temperatures obtained from the thermograms of the KLPF resin with different accelerators

Resin	Additive	Enthalpy (J/g)	Peak onset temperature (°C)	Peak maximum temperature (°C)	Peak endset temperature (°C)
PF	none	130[a]	119[a]	153[a]	173[a]
		54[b]	203[b]	212[b]	226[b]
KLPF	none	205	131	177	210
KLPF	potassium carbonate	261	108	150	183
KLPF	zinc borate	210	147	185	217
KLPF	propylene carbonate	205	131	177	211

[a] First enthalpy peak in DSC thermogram.
[b] Second enthalpy peak in DSC thermogram.

using dynamic DSC (Fig. 3). Table 1 lists the values for the normalised curing enthalpies, peak onset, endset and peak maximum temperatures for the three different additives.

Zinc borate and propylene carbonate both had either no or a diminishing effect on the curing process; the peak maximum and peak onset temperatures obtained with the KLPF resin containing these salts were practically identical to those obtained with the pure KLPF resin. With potassium carbonate, however, a significant shift of the peak maximum and peak onset temperature values toward lower temperatures was observed and a pronounced reduction in curing time was obtained. While it took approximately 10 min to fully cure the pure KLPF resin, addition of 4% K_2CO_3 reduced the curing time to ca. 5 min in the isothermal experiment. When 8% of catalyst was used, the curing time was further reduced by 50% to ca. 2.5 min. Hence, by addition of potassium carbonate highly reactive lignin-modified resins may be obtained with some potential in substituting phenol as a raw material for impregnation resins. Several carbonates have been used to accelerate curing of plain phenol-formaldehyde resins earlier [20–22, 32], with propylene carbonate being the most effective accelerator in terms of gelation time reduction at 120°C and the alkali metal salts sodium and potassium carbonate having a comparable influence [32]. The differences in catalytic activity of different carbonates for PF resins have been studied by Pizzi *et al.* [20] in detail. While sodium carbonate accelerated PF curing by purely enhancing the condensation reaction, propylene carbonate increased the average functionality of the system by an additional cross-linking reaction, namely Kolbe–Schmitt type addition of carboxyl functions to reactive sites in the phenolic ring system [20]. In the present study of lignin-modified PF, propylene carbonate, however, showed no positive effect at all which might be attributed to the lower number of reactive sites available for this catalytic mechanism in the case of the bulky lignin molecules and the comparatively lower content of pure PF after 20 wt% substitution by lignin. Since only potassium carbonate accelerated the cure process of kraft lignin-modified phenol-formaldehyde resin, the cure kinetics of the system K_2CO_3/KLPF was analyzed in more detail using isoconversional methods for two additive concentrations and the power of the model to predict the curing behaviour from dynamic DSC experiments was investigated.

In Fig. 4a, the predicted courses of activation energy *versus* conversion as calculated with the KAS and VA methods are given for the curing of KLPF containing 4 wt% of K_2CO_3. Fig. 4b compares the corresponding values for conversion predicted by the two models with the measured values determined from the isothermal experiment at 160°C. Figure 5a and 5b summarizes the analogous results for the curing of KLPF resin containing 8 wt% of K_2CO_3.

Compared to the un-accelerated reaction, with both concentrations of additive the values for the activation energy were generally higher and the reactions proceeded distinctly faster. Faster reactions do not necessarily require a reduction in the values for the activation energy, since the activation energy is not the only factor governing the rate of a reaction. An increase in catalyst concentration may also increase the efficiency of molecular interactions and thereby enhance the reaction [26].

The conversion dependence of the activation energy was less pronounced than with the un-accelerated resin; in fact, both kinetic analysis approaches yielded practically identical and constant activation energy throughout the curing reaction except for a sharp rise at a conversion of 80% which was obtained with the VA

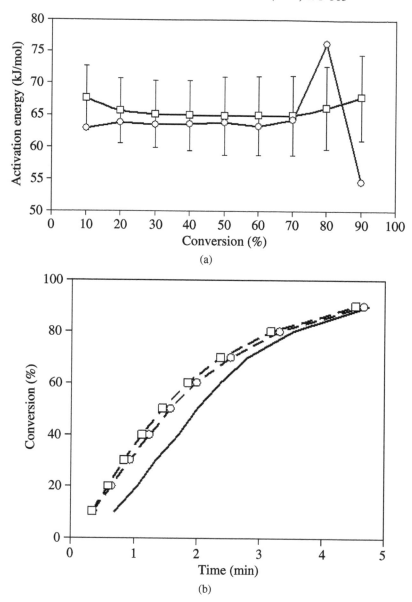

Figure 4. (a) Activation energy E_α for KLPF resin with accelerator KC at a concentration of 4 wt% estimated by the advanced Vyazovkin method (VA, □) and by the Kissinger–Akahira–Sunose method (KAS, ○). (b) Conversion α in % over time for KLPF resin with accelerator KC at a concentration of 4 wt% at 160°C. Comparison of the isothermal experiment (—) with curves calculated with the VA (□) and with the KAS (○) methods.

method. As with the un-accelerated reaction, this steep increase in activation energy can be attributed to a phase transition that occurs at high levels of conversion.

When using 8% potassium carbonate (Fig. 5a and 5b), the activation energy calculated by both methods showed a steady increase with conversion. This might

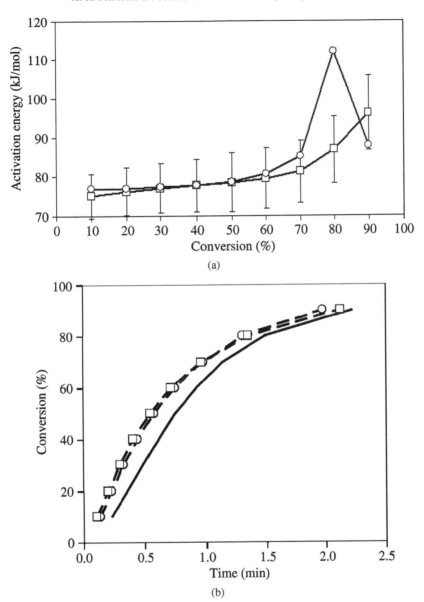

Figure 5. (a) Activation energy E_α for KLPF resin with accelerator KC at a concentration of 8 wt% estimated by the advanced Vyazovkin method (VA, □) and by the Kissinger–Akahira–Sunose method (KAS, ○). (b) Conversion α in % over time for KLPF resin with accelerator KC at a concentration of 8 wt% at 160°C. Comparison of the isothermal experiment (—) with curves calculated with the VA (□) and with the KAS (○) methods.

indicate that the reaction earlier becomes diffusion-controlled as would be expected when the KLPF resin is very rapidly curing and solidifying in the later stage.

Although obviously the VA approach gave better agreement with the experiment in the case without catalyst, both KAS and VA methods gave equally reliable results and showed the same deviations from experimental observations when the resin contained potassium carbonate. Moreover, although the initial stages of the curing were not perfectly predicted, interestingly, at high conversions these deviations were much less pronounced so that both methods seemed suitable to describe the time required to achieve a fully or nearly fully cured, catalyst containing KLPF resin.

4. Conclusion

The thermal properties of freshly prepared pure and additive-modified PF and KLPF resins were studied using dynamic DSC. In an attempt to accelerate the curing of KLPF, the additives zinc borate, potassium carbonate, and propylene carbonate were added to the resin and the curing behaviour was compared to the curing of unmodified KLPF resins. Only potassium carbonate showed a beneficial effect on the curing of the resin. While 4 wt% K_2CO_3 reduced the curing time to about half of the cure time required by pure KLPF resin, doubling the amount of catalyst led to a proportional reduction in curing time. Based on dynamic DSC experiments, isoconversional kinetic analysis of resin curing was performed and it was found that both approaches using the advanced Vyazovkin method or the Kissinger–Akahira–Sunose method lead to similar results with the catalyst containing KLPF resin and both are suitable for predicting the curing kinetics of catalyzed lignin-modified phenol-formaldehyde resin.

References

1. A. Pizzi, in: *Handbook of Adhesive Technology*, A. Pizzi and K. L. Mittal (Eds), p. 394. Marcel Dekker, New York, NY (1994).
2. A. Tejado, G. Kortaberria, C. Peña, M. Blanco, J. Labidi, J. M. Echeverría and I. Mondragon, *J. Appl. Polym. Sci.* **107**, 159 (2008).
3. A. Tejado, G. Kortaberria, C. Peña, M. Blanco, J. Labidi, J. M. Echeverría and I. Mondragon, *J. Appl. Polym. Sci.* **106**, 2313 (2007).
4. T. Yan, Y. Xu and C. Yu, *J. Appl. Polym. Sci.* **114**, 1896 (2009).
5. W. Hoareau, F. B. Oliveira, S. Grelier, B. Siegmund, E. Frollini and A. Castellan, *Macromolecular Mater. Eng.* **291**, 829 (2006).
6. G. Y. Liu, X. Q. Qiu and D. S. Xing, *J. Chem. Eng. of Chinese Universities* **21**, 678 (2007).
7. A. Pizzi, R. Kueny, F. Lecoanet, B. Massetau, D. Carpentier, A. Krebs, F. Loiseau, S. Molina and M. Ragoubi, *Ind. Crops Products* **30**, 235 (2009).
8. A. Effendi, H. Gerhauser and A. V. Bridgewater, *Renewable and Sustainable Energy Reviews* **12**, 2092 (2008).
9. K. G. Forss and A. Fuhrmann, *Forest Products J.* **29**, 39 (1979).
10. I. Enkvist, US Patent 3,864,291 (1975).
11. A. B. Wennerblom and A. H. Karlsson, US Patent 3,949,352 (1976).
12. M. R. Clarke and A. J. Dolenko, US Patent 4,113,675 (1978).

13. M. Olivares, H. Aceituno, G. Neimann, E. Rivera and T. Sellers, *Forest Products J.* **45**, 63 (1995).
14. B. Danielson and R. J. Simonson, *J. Adhesion Sci. Technol.* **12**, 923 (1998).
15. N. E. El Mansouri, A. Pizzi and J. Salvado, *J. Appl. Polym. Sci.* **103**, 1690 (2007).
16. N. E. El Mansouri, A. Pizzi and J. Salvadó, *Holz Roh Werkstoff* **65**, 65 (2007).
17. A. Pizzi, *J. Adhesion Sci. Technol.* **20**, 829 (2006).
18. M. Olivares, J. A. Guzman, A. Natho and A. Saavedra, *Wood Sci. Technol.* **22**, 157 (1988).
19. M. Duval, B. Bloch and S. Kohn, *J. Appl. Polym. Sci.* **16**, 1585 (1972).
20. A. Pizzi, R. Garcia and S. Wang, J. Appl. Polym. Sci. **66**, 255 (1997).
21. C. Zhao, A. Pizzi, A. Kuhn and S. Garnier, *J. Appl. Polym. Sci.* **77**, 249 (2000).
22. B. D. Park, B. Riedl, E. W. Hsu and J. Schields, *Polymer* **40**, 1689 (1999).
23. A. Knop and L. Pilato, *Phenolic Resins*. Springer, Berlin (1985).
24. M. V. Alonso, M. Oliet, F. Rodriguez, G. Astarloa and J. M. Echeverria, *J. Appl. Polym. Sci.* **94**, 643 (2004).
25. E. Kissinger, *Anal. Chem.* **29**, 1702 (1957).
26. T. Akahira and T. Sunose, *Res. Report Chiba Inst. Technol. (Sci. Technol.)* **16**, 22 (1971).
27. S. Vyazovkin, *J. Therm. Anal. Calorimetry* **83**, 45 (2006).
28. A. Kandelbauer, G. Wuzella, A. Mahendran, I. Taudes and P. Widsten, *Chem. Eng. J.* **152**, 556 (2009).
29. G. H. van der Klashorst, in: *Wood Adhesives — Chemistry and Technology*, A. Pizzi (Ed.), vol. 2, p. 155. Marcel Dekker New York, NY (1989).
30. A. Pizzi (Ed.), *Advanced Wood Adhesives Technology*, p. 219. Marcel Dekker, New York, NY (1994).
31. A. W. Christiansen and L. Gollob, *J. Appl. Polym. Sci.* **30**, 2279 (1985).
32. B. D. Park and B. Riedl, *J. Wood Chem. Technol.* **19**, 265 (1999).
33. M. V. Alonso, M. Oliet, J. Garcia, F. Rodriguez and J. Echeverra, *Chem. Eng. J.* **122**, 159 (2006).

Acacia mangium Tannin as Formaldehyde Scavenger for Low Molecular Weight Phenol-Formaldehyde Resin in Bonding Tropical Plywood

Y. B. Hoong [a], **M. T. Paridah** [a,b,*], **Y. F. Loh** [c], **M. P. Koh** [d], **C. A. Luqman** [b]
and A. Zaidon [a]

[a] Faculty of Forestry, Universiti Putra Malaysia, 43400 UPM, Serdang, Selangor, Malaysia
[b] Institute of Tropical Forestry and Forest Products, Universiti Putra Malaysia, 43400 UPM, Serdang, Selangor, Malaysia
[c] Fibre and Biocomposite Development Center, Malaysian Timber Industry Board, Lot 152, Jalan 4, Kompleks Perabot Olak Lempit, 42700 Banting, Selangor, Malaysia
[d] Forest Research Institute Malaysia, 52109 Kepong, Selangor, Malaysia

Abstract

One of the limitations in using low molecular weight phenol-formaldehyde (LmwPF) resin as a binder for wood-based panels is the amount of the free formaldehyde being emitted during soaking, pressing and sometimes during the earlier stage of application. Tannin from bark extracts is rich in phenolic compounds, and thus may be able to absorb this free formaldehyde and at the same time provide strength to the joint. In this study, tannin–phenol-formaldehyde adhesives were prepared by blending *Acacia mangium* bark extracts (40% solids) with low molecular weight phenol-formaldehyde (40% solids at 1:1 ratio). The tannin–LmwPF adhesive produced cured within 4 min at 130°C, reduced the free formaldehyde to level E1 of European norm EN-120. The 3-ply plywood had acceptable shear strength (>1.0 MPa) exceeding the minimum requirements of European norms EN-314-1 and EN-314-2:1993 for interior and exterior applications, respectively. The study has shown that *Acacia mangium* tannin can be used as formaldehyde scavenger in LmwPF resin without compromising the strength of the joints.

Keywords

Formaldehyde scavanger, tannin, *Acacia mangium*, plywood, low molecular weight phenol-formaldehyde

---

* To whom correspondence should be addressed. Tel.: +603 89476977; Fax: +60 389472180; e-mail: parida_introb@yahoo.com

Wood Adhesives
© Koninklijke Brill NV, Leiden, 2010

1. Introduction

1.1. Tannin-Based Adhesives

Phenol-formaldehyde resin is the most common adhesive for exterior applications due to its water resistance, low initial viscosity and its ability to bond various types of wood substrates [1]. Because of its resemblance to phenolic moieties, studies on tannin have been oriented towards an alternative formulation to replace the current synthetic phenol-formaldehyde or phenol-resorcinol-formaldehyde adhesives [2–4]. A few suitable alternative natural resources such as oil palm shell, pecan shell nut, lignin, starch, rice bran and tannin are also available for this purpose. Among these materials, tannins represent the best immediate substitute for phenol in wood adhesive production [1].

Condensed tannins have complex polyphenolic structure with several hydroxylation sites which give higher reactivity toward formaldehyde than phenol. This higher reactivity to formaldehyde makes tannin suitable for application in wood adhesives, producing products with low or nearly zero formaldehyde emission after curing [5, 6]. Studies by Pizzi [1] found that tannin is a suitable replacement for phenol in phenol-formaldehyde adhesive used for particleboard and plywood manufacture. Kreibich and Hemingway [7] also reported that pine tannin can also be used as laminating adhesive to partially replace the resorcinol-formaldehyde through a simple blending prior to adhesive spreading. According to Vazquez *et al.* [8], tannin extracted from *Pinus pinaster* bark, and fortified with phenol-formaldehyde adhesive by copolymerisation of tannin with resols gave satisfactory plywood performance.

Tannin-based adhesives have been studied by many researchers for the past 30 years [3, 7–14]. Nowadays, tannins are commercially produced: e.g., mimosa tannins in Brazil, South Africa, India, Zimbabwe and Tanzania. On the other hand, quabracho tannin is produced in Argentina, mangrove tannin in Indonesia, whilst both Italy and Slovenia produce chestnut tannin [15].

1.2. Low Molecular Weight Phenol-Formaldehyde (LmwPF) Resin

Low molecular weight phenol-formaldehyde (LmwPF) resin has been studied by many researchers as an additional treatment to enhance the properties, particularly the dimensional stability of the material. In such process, the wood raw material (whether in the form of strands, veneers or lumber) would be subjected to soaking, vacuum impregnating or soaking-compressing in order to force the resin into several layers of the wood cells. The treated material would then be dried in an oven until the resin is partially cured, i.e., reaching a gel state. Following this the treated material would undergo a normal manufacturing process, for instance, glue spreading, assembly and cold and hot pressing for plywood, and blending, forming and hot pressing for strandboard. The final products from these processes would have much improved properties, in particular the dimensional stability. This method is most suitable for porous woody materials like oil palm stem, bamboo, kenaf (*Hibiscus cannabinus*) core and alike [16, 17] Paridah *et al.* [17] used LmwPF to enhance the strength and stiffness of oil palm stem veneers, whilst Anwar *et al.* [18] used it for bamboo for producing exterior-grade bamboo flooring. Kajita and Imamura [19] investigated the use of low molecular weight phenol-formaldehyde resin to enhance the properties of particleboards. All these studies found that modulus of rupture (MOR) increased markedly with the amount of resin used. Paridah *et al.* [20] pre-treated strands of *Acacia mangium* with LmwPF resin prior to forming oriented strand board (OSB), and found that the anti-swelling efficiency (ASE) of the resulting OSB was 61% while water absorption gradually reduced by the use of LmwPF resin. The PF-treated OSB also showed better strength properties than those of untreated OSB. Many studies [19–21] suggested that the improvements were due to the smaller-sized PF resin molecules being deposited extensively into the wood cell walls and once the PF resin was cured it was attached chemically to the wall producing a stronger and more intact joint that could resist moisture better. In contrast, the normal PF for plywood is cooked to an advanced stage where relatively larger molecules are formed that cure and fill only the cell lumina, thus providing limited capability for restraining dimensional changes in the boards.

The average molecular weight and molecular weight distribution of the resins significantly influence the viscosity and the ability of the adhesive to melt-flow and penetrate the cell wall, and finally the properties of the bonded products [22]. When cured under heat, LmwPF resins will strengthen the wood and impart excellent dimensional stability [19]. Some data suggest that effective submicroscopic penetration in the wood of the polymers occurs at a molecular weight of 1000 or less [23], which would be about 10 cross-linked phenolic groups of a resol resin.

The presence of high number of methylol groups in the main polymer chains of LmwPF resin is responsible for the longer time required to cure the LmwPF resin. The rate of curing depends on temperature, time and the interaction with wood/fiber surface. Since LmwPF resin normally has an average Mw of <1000, the time to cure the resin is expected to be much longer compared to the normal PF.

One of the drawbacks of using LmwPF is the high amount of formaldehyde emission during soaking and hot pressing. This is understandable since LmwPF resin contains substantial amounts of methylol groups in the oligomeric chains. Some of these methylol groups will be released as free formaldehyde upon being exposed to high temperature and humidity. One of the ways to capture this free formaldehyde is by using tannin as formaldehyde catcher/scavenger. It is anticipated that a reactive tannin, particularly the one containing resorcinolic A-ring, would easily bond with the methylol groups or/and with free formaldehyde and cross-link with the phenolic backbone.

This paper reports the results of shear strength on plywood bonded with a mixture of LmwPF resin and *Acacia mangium* tannin. It is anticipated that the presence of *Acacia mangium* tannin will not only accelerate the curing of LmwPF but also will reduce the amount of free formaldehyde contained in the resin system.

2. Experimental

2.1. Bark Preparation

Barks from 7–8 year old *Acacia mangium* trees were collected from acacia process-ing mills in Telaga, Sabah (Malaysia). The *Acacia mangium* barks were oven dried to 12% moisture content and then chipped into 2–3 cm sized particles. The barks were further ground into fine particles of size 0.5–1 mm using a flaker machine.

2.2. Extraction of Tannin

The tannin extract was prepared by heating a mixture of bark, water, sodium sulfite and sodium carbonate (100:600:2:0.5 w/w) at 75°C in a water bath for 3 h. The solution was initially screened through a fine filter (140 mesh) and further filtered using a sintered glass (40–100 μm). Then, the extracts were concentrated to 40–50% solids under reduced pressure using a Buchi Rotavapor rotary evaporator at 50 to 55°C. The resulting solid was further dried in an oven maintained at 50°C until the weight was constant [14].

2.3. Determination of Tannin Content (Reactive Tannin)

The Stiasny number method was used to determine the polyphenol content of the extracts [24]. Fifty ml of 0.4% w/w tannin solution was pipetted into a 150 ml flask. Aqueous formaldehyde (37%, 5 ml) and hydrochloric acid solution (10 N, 5 ml) were then added and the mixture was heated under reflux for 30 min. The reaction mixture was filtered through a sintered glass filter (40–100 μm) while it was still hot. After that the precipitate was dried in an oven at 105°C to constant weight. The Stiasny number is the ratio of the oven dried weight of the precipitate to the total dissolved solids content of the tannin extract, expressed as a percent-age.

2.4. Preparation of LmwPF Resin

LmwPF resin was prepared by reacting phenol and formaldehyde in an alkaline condition. The resin cooking procedure followed the normal PF resin cooking for plywood but the methylolation period was maintained <4 h.

2.5. Determination of Viscosity

Both tannin solution and LmwPF resin were prepared at 40% solids. The viscosities of LmwPF and the adhesive mixture of tannin and LmwPF were determined using a Brookfield viscometer Model LVT, CP51 at 25°C.

2.6. Determination of Gel Time

Commercial LmwPF (control) and different proportions of tannin and LmwPF mix-ture (both at 40% w/w solid content, 10 g) were placed in a test tube and 0.3 g paraformaldehyde powder (95–100% assay) was added to the test tube. The mix-ture was then heated in a water bath at 90 ± 2°C and stirred with a glass rod until it formed a gel. A soft spiral wire was used to examine the gel state. The time required for formation of gel state is known as the gel time. The test was duplicated for each adhesive formulation:

The proportions of tannin and LmwPF used in this study were:

(1) 100% LmwPF (control);

(2) 10% tannin : 90% LmwPF;

(3) 30% tannin : 70% LmwPF;

(4) 50% tannin : 50% LmwPF;

(5) 70% tannin : 30% LmwPF;

(6) 90% tannin : 10% LmwPF.

2.7. Adhesive Formulation

All the adhesive formulations in the study used tannin solution and resins of 40% solids content. Eight different adhesive formulations were prepared where 7 were tannin-based adhesives. For control (Formulation A), only LmwPF and wheat flour (100 and 15 parts by weight, respectively) were mixed. For formulations B to H, small amounts (10, 30, 50 parts) of LmwPF were first added to the tannin solution (40% solids) followed by wheat flour (15 parts) and paraformaldehyde (3 and 5%). The adhesive formulations used in the study are given in Table 1.

2.8. Fabrication of Plywood

Three-ply plywood samples (size 300 mm × 300 mm) were prepared in this study. The 2-mm thick tropical hardwood veneers were peeled from Mempisang (*Annonaceae spp.*) logs. Prior to gluing, all veneers were dried to 8–10% moisture content. The veneers were assembled by placing tight and loose-side veneers fac-

Table 1.
Adhesive formulations used to bond Mempisang (*Annonaceae spp.*) plywood

Ingredient (parts by weight)	Adhesive							
	A	B	C	D	E	F	G	H
Acacia mangium tannin solution (40% solids)	–	50	50	70	70	90	90	90
LmwPF (40% solids)	100	50	50	30	30	10	10	10
Paraformaldehyde (%)*	–	–	3	–	3	–	3	5
Wheat flour	15	15	15	15	15	15	15	15

Note: *The amount of paraformaldehyde is based on the weight of tannin solids.

ing each other. The gluing conditions were as follows; glue spread amount 300 g/m^2 (double glueline); cold press time: 10 min; open assembly time: 25 min; hot press at: 125°C, 15 kg/cm^2 for 4 min [14].

The adhesive bonding qualities were evaluated according to European norms EN 314-1 [24] and EN 314-2 [25]. All specimens were tested in dry state (interior type bond). The specimens which had already passed the dry test were tested according to exterior-type bond, i.e., cold water test: the specimens soaking in cold water $(20 \pm 3)°C$ for 24 h; and boiling test: soaking specimens in boiling water under normal atmospheric conditions for 72 h, then cooling the specimen in water at $(20 \pm 3)°C$ for 1 h.

2.9. Determination of Free Formaldehyde from Plywood

The formaldehyde emission test on the plywood panels was carried out according to the specification prescribed in European norm EN-210, by the perforator method [26].

2.10. Statistical Analysis

An analysis of variance (ANOVA) was carried out to evaluate the significance of different adhesive formulations on shear strength. The results were further analyzed using the Least Significant Difference (LSD) test at $p \leqslant 0.05$, to further evaluate the effects of adhesive formulations on the physical properties, shear strength and formaldehyde emission of the bonded plywood.

3. Results and Discussion

3.1. Physical Properties of Tannin–LmwPF Adhesives

The results in Table 2 reveal that extraction of *Acacia mangium* bark with aqueous sodium sulfite (2%) and sodium carbonate (0.5%) produced relatively high tannin yields (23.3%) and Stiasny number (90%). The bark extract was slightly acidic with a pH of 5.4. This result clearly indicates that this type of tannin contains higher percentage of reactive tannin, much higher than the minimum requirement (at least 65%) suggested by Yazaki and Collins [27]. The gel time of *Acacia mangium* tannin shown in Table 3 reflects its reactivity towards formaldehyde which is comparable

Table 2.
Properties of tannin extracted from *Acacia mangium* tree bark

Yield	23.3% (0.7)
Stiansy number (SN)	90.0% (3.2)
pH	5.4

Note: Value in parentheses ( ) represents standard deviation.

Table 3.
Comparative gel times (s) of *Acacia mangium* tannin, commercial mimosa tannin and quebracho tannin at different pH levels

pH	*Acacia mangium* tannin	Commercial mimosa tannin[1]	Commercial quebracho tannin[1]
5	245	–	–
5.5	221	543	180
6	161	338	165
6.5	130	214	–
7	110	136	102
7.5	84	86	82
8	60	54	59

Note: Based on 40% tannin solution.
[1] Source [3].

to those of commercial mimosa and quebracho extracts. Hoong and co-workers [14, 28] and Pizzi [3] reported that in this gel time range, the tannin is very suitable for adhesive usage, particularly as plywood adhesive.

Figure 1 shows the comparative gel times of different adhesive formulations used in this study. The control (100% LmwPF resin) had a gel time of 30 min which is much longer than those exhibited by tannin-based adhesives. The addition of *Acacia mangium* tannin into the LmwPF adhesive mix markedly reduced the gel time to less than 26 min even without the addition of paraformaldehyde. Surprisingly, the gel time increased as more paraformaldehyde was added, particularly for mixtures having ⩽30% tannin. According to Pizzi [3], such behaviour can be attributed to the less reactive phenolic compound in the LmwPF chains to form methylene bridges as compared to the more reactive resorcinolic compound in the tannin molecules. Since the amount of tannin is low (⩽30%) the excess paraformaldehyde would have no choice but to react with the LmwPF which occurs at a much slower rate. A drastic decrease in gel time, however, was observed as the amount of tannin was increased, notably for tannin at 50% and above. The gel times for these formulations range from 141 s to 243 s. This is understandable because resorcinolic A-rings of tannin show reactivity towards formaldehyde of 7 or 8 times higher than the reactivity of phenol with the LmwPF [3]. Thus, tannin can easily react with both formaldehyde and methylol groups on the LmwPF in the adhesive system. The *Acacia mangium* tannin is present in the form of oligomer with an average degree of polymerization of around 5 flavonoid units [29]. Thus there are, on average, 6-reactive sites on each tannin molecule. All of these reactive sites react with methylol groups from different LmwPF chains. Thus, once the tannin is added to the LmwPF the methylol group on the LmwPF practically does not react (or to a very small extent) with other phenol molecules, instead it almost exclusively reacts with the tannin. From

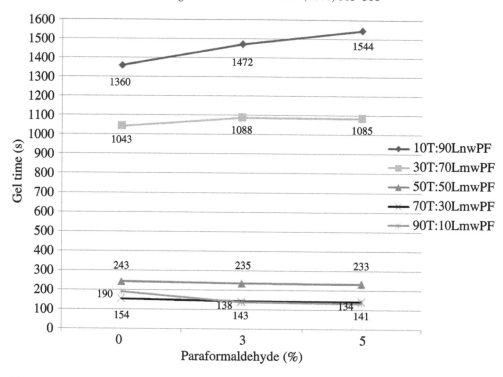

Figure 1. Gel times of different formulations of *Acacia mangium* tannin–LmwPF adhesives at 0%, 3% and 5% paraformaldehyde.

the gel time results it is obvious that tannin is the site of cross-linking and network formation in the system, and the rate of hardening or gelling depends exclusively on its reactivity with the methylol group of the LmwPF, hence resulting in relatively shorter gel time. The shorter gel time would translate into a faster adhesive curing, thus in more economical press time. This behaviour explains why the presence of paraformaldehyde did not reduce the gel time, instead, it increased the gel time. The role of the paraformaldehyde in the adhesive mix is to form bridges between tannin and cellulose in the wood, as well as to accelerate the curing. This function, however, was not seen in this study (as shown by gel time) possibly due to reaction of paraformaldehyde with the phenol molecules in the LmwPF chains which occurred at a much slower rate. In this study, adhesive formulations containing 50% tannin : 50% LmPF, 70% tannin : 30% LmPF and 90% tannin : 10% LmPF are well within acceptable hot press times (i.e., 4–6 min).

Viscosity is an important factor that influences the gluing properties of resin adhesives. The viscosities of 8 different adhesive formulations at 40% solids content are shown in Table 4. The viscosity increases from 45 Poise (in 100% LmwPF) to 80 Poise (in 90% Tannin + 10% LmwPF). Such increment can be assumed due to the increased amount of tannin added. As discussed earlier, *Acacia mangium* tannin belongs to the same group as wattle (mimosa) tannin which is predominantly composed of prorobinetinidin (resorcinol A-ring; pyrogallol B-ring) repeat units

Table 4.
Viscosities of different adhesive formulations

Adhesive formulation	Viscosity (poise)
100% LmwPF	45
10% tannin : 90% LmwPF	50
30% tannin : 70% LmwPF	52
50% tannin : 50% LmwPF	55
70% tannin : 30% LmwPF	65
90% tannin : 10% LmwPF	80

Note: Both tannin solution and LmwPF have 40% solids.

Table 5.
Shear strength and wood failure percentage of Mempisang (*Annonaceae spp.*) plywood after dry test, after 24 h cold water soaking test and 72 h boiling test, and formaldehyde emission

Adhesive type	Dry test (MPa)	Wood failure (%)	Cold water soaking test (MPa)	Wood failure (%)	Boiling test (MPa)	Wood failure (%)	Formaldehyde emission* (mg/100 g)
A (control)	1.71^{ab} (0.39)	80	1.42^a (0.25)	80	1.05^a (0.15)	18	37
B	1.72^{ab} (0.47)	50	0.99^{bcd} (0.18)	45	0.84^{bc} (0.14)	8	17
C	1.52^b (0.19)	68	0.92^{bcd} (0.19)	55	0.70^{cd} (0.06)	22	19
D	1.59^{ab} (0.26)	32	1.11^{bc} (0.29)	0	0.81^{bcd} (0.19)	0	4
E	1.27^b (0.44)	55	0.75^d (0.22)	38	0.69^{cd} (0.06)	17	15
F	1.71^{ab} (0.82)	70	0.74^d (0.28)	18	0.65^d (0.16)	0	0.6
G	1.66^{ab} (0.52)	75	0.85^{cd} (0.35)	50	0.80^{cd} (0.19)	50	7
H	2.12^a (0.62)	82	1.20^{ab} (0.10)	40	1.03^{ab} (0.04)	40	13

Note: Means followed by the same letter a, b, c and d in the same column are not significantly different at $p \leqslant 0.05$.

*Formaldehyde emission of specimens was measured by the perforator method, European norm EN-210 [26].

Values in parentheses ( ) represent standard deviations.

[29]. Hence a fair amount of resorcinol is present which can easily cross-link with methylol groups of the LmwPF chains even at ambient temperature. The presence of non-tannin material, such as sugar and high molecular weight gum as well as other hydroxycompounds, may also contribute to the higher viscosity in tannin-based adhesives [30].

3.2. Plywood Shear Strength

The bond strength of the plywood was determined by carrying out shear tests after 1 week of conditioning at 60% RH and tested according to European norms EN 314-1 [24] and EN 314-2 [25]. Table 5 reveals that plywood bonded with adhesives A to H give dry shear strength values greater than 1.0 MPa. Hence according to EN 314-2 [25], the wood failure percentage becomes irrelevant. Comparing the shear strength and wood failure percentage for different test conditions, it was found that both properties decreased markedly as the severity of test conditions increased from cold water soaking to 72 h boiling tests. The presence of paraformaldehyde apparently has significant influence on the percent wood failure. For instance, in adhesives B, D and F (without paraformaldehyde) the wood failure recorded for plywood bonded with these adhesives ranged from 0–70%. Whilst those with paraformaldehyde (adhesives C, E, G and H) produced wood failures of 17–82%, however it is not significant ($p \leqslant 0.05$). Wood failure percentage measures the strength of a joint in relation to the strength of the wood veneer. High wood failure normally indicates that the bond strength is high and *vice versa*. In this study, three possible situations are occurring when using tannin-based adhesives:

(i) For 50 : 50 tannin–LmwPF (adhesives B and C).

At equal proportions, the cross-linking of tannin–LmwPF is anticipated to be slightly dominated by the tannin since it is more reactive due to the resorcinolic structure of A-ring. Since the LmwPF has more reactive sites (3 as opposed to 1 or 2), more tannin would have reacted, leaving an excess of unreacted LmwPF. However, when paraformaldehyde was added (3%), more linkages were able to be formed with the LmwPF resin. Thus stronger joints were formed as indicated by the increase in wood failure percentage, for example, from 50% to 68% (dry test).

(ii) For 70 : 30 tannin–LmwPF (adhesives D and E).

The high amount of tannin (70 parts) in this formulation accelerates the curing process where all the methylol groups in the LmwPF (30 parts) chains would be used up, leaving a fairly large amount of tannin unreacted due to insufficient amount of LmwPF available. Once the paraformaldehyde was added (3%) to this mixture, further reactions occurred between paraformaldehyde and tannin, resulting in comparable bond quality as indicated by the letter 'b', and higher wood failure (from 32% to 55%) in dry condition.

(iii) For 90 : 10 tannin–LmwPF (adhesives F, G and H).

At this ratio, the adhesive formulation contained a very high amount of reactive tannin (90 parts) as compared to LmwPF (10 parts). Similar situation occurred as in (ii) but the amount of unreacted tannin was much higher. Upon addition of paraformaldehyde (3% and 5%) both the shear strength and wood failure increased tremendously. Clearly more cross-linking had occurred.

All plywood samples bonded with tannin–LmwPF adhesives experienced significant reduction in both shear strength and wood failure percentage after being subjected to cold-water soaking and 72 h boiling tests. Plywood bonded with adhesive H (90 parts tannin, 10 parts LmwPF, 15 parts wheat flour and 5% paraformaldehyde) shows superior performance with shear strength and wood failure percentage meeting the minimum requirements of European norms EN 314-1 and EN 314-2 (Plywood: Bond quality) [24, 25].

3.3. Free Formaldehyde Content

It has been reported by many researchers [17–19] that LmwPF releases high amount of formaldehyde. In this study, *Acacia mangium* tannin was added as formaldehyde catcher to reduce the emission from the bonded products. The results (Table 5) show that the formaldehyde emission from plywood glued using tannin-based adhesive was much lower than from that glued with LmwPF alone. Comparing the values of formaldehyde emissions from the two types of adhesives (LmwPF and tannin–LmwPF), it is obvious that tannin was able to absorb the free formaldehyde: 37 mg/100 g from plywood bonded with 100% LmwPF, 17 mg/100 g with 50 : 50 tannin–LmwPF, 4 mg/100 g with 70 : 30 tannin–LmwPF, and 0.6 mg/100 g with 90 : 10 tannin–LmwPF. This study also indicates that a minimum amount of paraformaldehyde is required when tannin is used with LmwPF to obtain better bonding quality.

4. Conclusion

Tannin from *Acacia mangium* tree bark can be used as formaldehyde scavenger in LmwPF resin for bonding tropical hardwood veneers. The study shows that *Acacia mangium* tannin can be blended with LmwPF to reduce hot press time as well as the formaldehyde emission, and still producing acceptable bond strength and wood failure percentage. The *Acacia mangium* tannin can be used to replace up to 90% (by weight) of the LmwPF resin in plywood adhesive. The plywood panel produced with additional paraformaldehyde (3% or 5%) satisfied the European norms EN 314-1 and EN 314-2 [24–26] requirements and thus can be used as exterior grade plywood adhesive.

References

1. A. Pizzi, in: *Wood Adhesives: Chemistry and Technology*, A. Pizzi (Ed.), Vol. 1, pp. 177–246. Marcel Dekker, New York, NY (1983).
2. A. Pizzi, in: *Advanced Wood Adhesives Technology*, A. Pizzi (Ed.), pp. 149–217. Marcel Dekker, New York, NY (1994).
3. A. Pizzi, in: *Handbook of Adhesive Technology*, A. Pizzi and K. L. Mittal (Eds), 2nd edn, pp. 573–587. Marcel Dekker, New York, NY (2003).
4. R. W. Hemingway and A. H. Conner, *Amer. Chem. Soc. Symp. Ser.* **385**, 510 (1989).
5. E. T. N. Bisanda, W. O. Ogola and J. V. Tesha, *Cement Concrete Composites* **25**, 593–598 (2003).
6. D. Joseph, R. W. Hemingway and V. Tisler, *Holz Roh Werkst.* **58**, 23–30 (2000).
7. R. E. Kreibich and R. W. Hemingway, *Amer. Chem. Soc. Symp. Ser.* **385**, 203–221 (1989).
8. G. Vazquez, G. Antorrena, J. L. Francisco and J. Gonzales, *Holz Roh Werkst.* **50**, 253–256 (1992).
9. W. E. Hillis, in: *Phenolic Resin: Chemistry and Application. Proceedings of Weyerhaeuser Science Symposium*, Tacoma, Washington, USA, pp. 171–188 (1980).
10. Y. Lu and Q. Shi, *Holz Roh Werkst.* **53**, 17–19 (1995).

316   Y. B. Hoong et al. / Wood Adhesives (2010) 305–316

11. L. Calve, G. C. J. Mwalongo, B. A. Mwingira, B. Riedl and J. A. Shields, *Holzforschung* **49**, 259–268 (1995).
12. G. Vazquez, J. Gonzalez-Alvarez, F. Lopez-Suevos and G. Antorrena, *Holz Roh Werkst.* **60**, 88–91 (2002).
13. M. T. Paridah and O. C. Musgrave, *J. Tropical Forest Sci.* **18** (2), 137–143 (2006).
14. Y. B. Hoong, M. T. Paridah, C. A. Luqman, M. P. Koh and Y. F. Loh, *Ind. Crops Prod.* **30**, 416–421 (2009).
15. A. Pizzi, in: *Monomers, Polymers and Composites from Renewable Resources*, M. N. Belgacem and A. Gandini (Eds), pp. 179–199. Elsevier, Amsterdam (2008).
16. A. W. Nor Hafizah, U. M. K. Anwar, Y. F. Loh, Y. B. Hoong and M. T. Paridah, in: *Proceeding of International Conference on Kenaf and Alied Fibres: Viable Biofibres for Future*, Kuala Lumpur, Malaysia, p. 70 (2009).
17. M. T. Paridah, Y. F. Loh, A. Zaidon and E. S. Bakar, in: *Proceeding the Fourth International Symposium on Veneer Processing and Products*, Espoo, Finland, pp. 267–275 (2009).
18. U. M. K. Anwar, M. T. Paridah, H. Hamdan, S. M. Sapuan and E. S. Bakar, *Ind. Crops Prod.* **29**, 214–219 (2009).
19. H. Kajita and Y. Imamura, *Wood Sci. Technol.* **26** (1), 63–70 (1991).
20. M. T. Paridah, L. L. Ong, A. Zaidon, S. Rahim and U. M. K. Anwar, *J. Tropical Forest Sci.* **18** (3), 166–172 (2006).
21. T. Furuno and T. Goto, *Mokuzai Gakkaishi* **25** (7), 488–495 (1979).
22. T. J. Sellers, in: *Handbook of Adhesive Technology*, A. Pizzi and K. L. Mittal (Eds), pp. 599–614. Markel Dekker, New York, NY (1994).
23. A. J. Stamm, *Wood and Cellulose Science*, pp. 175–185. The Ronald Press Company, New York, NY (1964).
24. European Norm EN 314-1. Plywood: bond quality part 1 (1993).
25. European Norm EN 314-2. Plywood: bond quality part 2 (1993).
26. European Norm EN 210. Wood based panel: perforator method (1992).
27. Y. Yazaki and P. J. Collins, *Holz Roh Werkst.* **52**, 307–310 (1994).
28. Y. B. Hoong, M. P. Koh, M. T. Paridah and C. A. Luqman, in: *Proceeding of National Conference on Forest Products*, Kuala Lumpur, Malaysia, pp. 341–352 (2008).
29. Y. B. Hoong, A. Pizzi, M. T. Paridah and H. Pasch, *Eur. Polym. J.* **46**, 1268–1277 (2010).
30. H. Scharfetter, *Forest Prod. J.* **25** (3), 30–32 (1975).

Synthesis of Modified Poly(vinyl acetate) Adhesives

A. Salvini [a,*], L. M. Saija [b], M. Lugli [b], G. Cipriani [a] and C. Giannelli [a,b]

[a] Department of Organic Chemistry, University of Florence, via della Lastruccia 13,
50019 Sesto Fiorentino, Florence, Italy
[b] Cray Valley Italia s.r.l., via Finghè, 42022 Boretto, RE, Italy

Abstract
Modified poly(vinyl acetate) copolymers with drying oils as comonomers have been prepared. The unsaturated triglycerides can produce cross-linking and give a waterproof effect due to their hydrophobicity. The new copolymers synthesized by solution polymerization in an organic medium have been submitted to analytical characterization in order to investigate the role of the drying oils in the polymerization reaction. NMR spectroscopy has been used to obtain information on the structure of the polymerization products and to confirm the eventual C–C bond formation between vinyl acetate and triglycerides. Spectroscopic data are found to be in agreement with copolymer formation.

Copolymers as water dispersion have also been synthesized by emulsion polymerization in presence of poly(vinyl alcohol) as protective colloid and evaluated as wood adhesives. The corresponding polymer films have been subjected to mechanical characterization. An improvement in the water and solvent resistance has been observed due to the presence of drying oils. The cobalt acetate, if present as catalyst in the adhesive formulation, promotes cross-linking and produces a positive effect on the adhesive performance.

Keywords
Adhesives, poly(vinyl acetate) dispersions, drying oils, cross-linking, NMR spectroscopy

1. Introduction

Waterborne dispersions containing poly(vinyl acetate) are widely used as adhesives for wood or wood-based materials. Their success can be attributed to several factors such as easy application, non-flammability, low toxicity and relatively low cost. In fact, high temperatures are generally not required to dry these adhesives, their shelf-life is long and water can be used to remove the residual product from tools. Furthermore, these adhesives show an extremely low toxicity profile due to the presence of water as dispersing medium and to the very low content of volatile organic compounds (VOCs).

* To whom correspondence should be addressed. Tel.: +39-055-4573455; Fax: +39-055-4573531; e-mail: antonella.salvini@unifi.it

Wood Adhesives
© Koninklijke Brill NV, Leiden, 2010

Nevertheless, the adhesives formulated with PVAc-based polymer dispersions suffer from some drawbacks. Due to the high poly(vinyl acetate) thermoplasticity the adhesive joints obtained with PVAc-based formulations are sensitive to high temperatures and have a poor resistance to creep under static load. Moreover, the presence of poly(vinyl alcohol) (PVA), used as protective colloid in the emulsion polymerization phase, makes them also easily affected by both moisture and water.

Various patents and papers report how to improve the performance of poly(vinyl acetate) adhesives. This is generally achieved by using specific functional comonomers in the emulsion polymerization phase, by including polyvalent metal salts in the adhesive formulation, or by post-addition of thermosetting resins such as urea-formaldehyde (UF), melamine-formaldehyde (MF) or polyisocyanates for two-component systems [1–7].

The PVAc formulations currently available on the adhesives market are rather complex systems comprising a poly(vinyl acetate) dispersion, a film forming promoter, a cross-linking agent and/or a hardener. The use of modified poly(vinyl acetate) dispersions generally requires a thermal treatment to obtain the best results and the use of a specific cross-linking comonomer, such as N-methylolacrylamide (NMA), may cause a lower shelf-life of adhesives as well as formaldehyde emissions.

PVAc-based commercial wood adhesives are evaluated using standard tests for non-structural applications, as reported in EN 205 [8], and they are classified in agreement with the standard EN-204 [9]. This standard allows to classify wood adhesives in 4 categories from D1 to D4. D1 adhesives show a good resistance only in dry conditions; D2 adhesives should withstand a rather low water presence, such as in occasional exposure in kitchens and bathrooms; D3 adhesives are suitable to come in contact with cold water, such as for outside windows and doors, kitchen and bathrooms furniture; D4 adhesives are suitable to be used in extreme conditions (resistance to hot water). Vinyl acetate homopolymer can be used to formulate D1 or D2 adhesives. Vinyl acetate based adhesives cross-linked with hardeners and urea-formaldehyde (UF) adhesives belong to class D3. Only the phenol-formaldehyde (PF), resorcinol-formaldehyde (RF) and melamine-formaldehyde (MF) adhesives, some special 2-component polyurethanes (PUs), and cross-linking vinyl adhesives belong to class D4.

In this study the use of drying oils (poppy, linseed, walnut) as comonomers was investigated. These natural compounds have been largely employed as paint binders or as components in alkyd resins.

Several examples of the use of drying oils with PVAc resins have been reported in the literature [10, 11]. Nevertheless, these authors have described the resulting adhesives as '*blends*' in which the oils have the role of a waterproofing agent because of their hydrophobicity.

The copolymerization between vinyl monomers and drying oils was studied in the past and some vinyl acetate/linseed oil copolymers were synthesized by Cristea *et al.* [12]. In their work, the emulsion polymerization was performed using a poly(vinyl alcohol)/water solution as the medium, and the influences of temperature, linseed oil amount and oil preliminary heating on the particle dimension, reaction rate and K-value were studied. The K-value is used to provide information on the molecular weight and it is calculated from dilute solution viscosity measurements. The products were characterized through FT-IR spectroscopy, viscometry, solubility tests and saponification number. The high water resistance of the polymer film was not attributed to the copolymer formation, but to the presence of self cross-linked linseed oil, not copolymerized with VAc.

The affinity for substrates and the water resistance were improved by blending poly(vinyl acetate) with esters of fatty acids such as triglycerides [10]. The water-proofing action and the favourable interaction between these esters and the substrate surface were proposed to explain the performance of these blends.

In the present work, using particular synthesis methodologies, the copolymerization of vinyl acetate and drying oils was studied. The copolymers obtained, functionalized to some extent with unsaturated triglycerides, can produce cross-linking. The presence of the long alkyl chains of fatty acids can also produce an additional waterproofing effect.

2. Experimental

2.1. Materials

Demineralized water, methanol (98%), toluene (99.7%), chloroform (99.4%), acetone (99.5%), ethyl acetate (99.5%), n-hexane (98%), n-pentane (97%), D_2O (99.9%), DMSO-d_6 (99.8%), CD_3OD (99.8%), 2,2'-azobisisobutyronitrile (AIBN), vinyl acetate (>99.0%), methyl linolenate (>99%) were reagent grade and were used without further purification. Poly(vinyl alcohol) (PVA 30-92 grade), t-butyl hydroperoxide (TBHP), sodium formaldehyde sulphoxylate (SFS), potassium persulfate ($K_2S_2O_8$), sodium persulfate ($Na_2S_2O_8$), sodium hydrogen carbonate (NaHCO$_3$) and cobalt acetate were supplied by Cray Valley Italia s.r.l. and were used without further purification. The drying oils (bleached linseed oil, refined walnut oil, poppy oil) were paint oils supplied by Zecchi, Florence, Italy. The ^{1}H, ^{13}C-NMR and FT-IR spectra of drying oils were recorded and are reported in Tables 1–3. The ^{1}H-NMR and FT-IR data were found to be in agreement with those reported in the literature [13–17].

2.2. Instruments and Analytical Techniques

IR spectra were recorded on an FT-IR Perkin–Elmer Spectrum BX spectrometer, using the Spectrum v. 3.02.02 software. The liquid samples were analyzed using KBr disks. Polymers spectra were recorded on films obtained from solutions after solvent evaporation.

^{1}H-NMR, H,H-gCOSY and C,H-gHSQC spectra of solutions were recorded on a Varian Mercury Plus 400 spectrometer and on a Varian VXR 200 spectrometer at

Table 1.
^{1}H-NMR spectral data (δ, ppm; $CDCl_3$ as solvent)

Code[a]	Drying oil	1	2	3	4	5	6	7
CH_3 (saturated, oleic and linoleic chains)	0.88	–	0.85	0.85	0.85	0.85	0.85	0.85
CH_3 linolenic chains	0.97	–	0.97	0.97	0.97	–	–	–
CH_2 (saturated, oleic, linolenic and linoleic chains)	1.30	–	1.30	1.30	1.30	1.30	1.30	1.30
CH_2CH$_2$COOR	1.61	–	1.60	1.60	1.60	1.60	1.60	–
CH$_2$ (polymeric chain)	–	1.76	1.76	1.76	1.76	1.76	1.76	1.76
CH$_2$ (polymeric chain)	–	–	1.85	1.85	1.85	1.85	1.85	1.85
CH_3COO	–	2.02	2.02	2.02	2.02	2.02	2.02	2.02
–CH$_2$CH_2–CH=CH	2.06							
CH$_2$CH_2COOR (saturated, oleic, linolenic and linoleic chains)	2.31	–	2.31	2.31	2.31	2.31	2.31	2.31
CH=CH–CH_2–CH=CH	2.80	–	2.80	2.80	2.80	–	–	2.80
CH_3–OCOR	–	–	–	–	–	–	–	3.66
CH–OCO (polymeric chain)	–	–	4.08	4.08	4.08	4.08	4.08	4.08
CH$_2$–OCO glyceryl groups	4.22	–	4.21	4.21	4.21	–	–	–
CH–OCO (polymeric chain)	–	4.87	4.87	4.87	4.87	4.87	4.87	4.87
CH–OCO glyceryl groups	5.26	–	5.25	5.25	5.25	–	–	–
CH=CH (oleic, linolenic and linoleic chains)	5.36	–	5.34	5.34	5.34	–	–	–

[a] **1** PVAc, **2** VAc/Linseed oil (10%) copolymer, **3** VAc/Linseed oil (20%) copolymer, **4** VAc/Linseed oil (30%) copolymer, **5** VAc/Walnut oil (10%) copolymer, **6** VAc/Poppy oil (10%) copolymer, **7** VAc/Methyl linolenate (10%) copolymer.

399.92 MHz and 199.985 MHz respectively, using deuterated solvents and solvent residual peak as reference.

^{13}C-NMR spectra were recorded at 100.57 MHz on a Varian Mercury Plus 400 spectrometer and at 50.286 MHz on a Varian VXR 200 spectrometer, using deuterated solvents and solvent residual peak as reference. All ^{13}C-NMR spectra were acquired with a broadband decoupler.

The molecular weights (M_n) were evaluated using a modular HPLC-SEC instrument equipped with a Perkin Elmer (LC mod. 250) pump, a Rheodyne 7010 valve with 200 µl injection loop and a Perkin–Elmer LC 30 refractive index detector. A series of three PL gel columns (30 cm long, packed with 5 µm (diameter) particles, Polymer Labs, UK) were used for the separation, using CHCl$_3$ (1 ml/min) as the eluent. The calibration was performed with 10 standard samples of poly(methyl methacrylate) with different molecular weights (Polymer Labs, UK). The products were dissolved in CHCl$_3$ (3–5 mg/ml) and the solutions eluted through a 0.45 µm pore diameter filter.

Glass transition temperatures (T_g) were obtained using a DSC Perkin–Elmer Pyris 1 equipment (temperature range 0–200°C; heating rate 20°C/min). Elemental

Table 2.
^{13}C-NMR spectral data (δ, ppm; $CDCl_3$ as solvent)

[a]Code	Drying oil	1	2	3	4	5	7
CH_3 (saturated, oleic and linoleic chains)	14.0	–	14.0	14.0	14.0	14.0	–
CH_3 linolenic chains	14.2						
CH_3COOR		21.0	21.0	21.0	21.0	21.0	21.0
CH_2 (saturated, oleic, linolenic and linoleic chains)	$22.5 < \delta < 33.9$	–	$22.5 < \delta < 33.9$	$22.5 < \delta < 33.9$	$22.5 < \delta < 33.9$	$22.5 < \delta < 33.9$	$22.5 < \delta < 33.9$
CH_2 (polymeric chain)		39.1	39.1	39.1	39.1	39.1	39.1
CH–OCO (polymeric chain)		66.9	44.0	44.0	44.0	44.0	44.0
			45.0	45.0	45.0	45.0	45.0
			60.5	60.5	60.5	60.5	60.5
			66.6	66.6	66.6	66.6	66.6
CH_2–OCO glyceryl groups	62.0		62.0	62.0	62.0	62.0	–
CH–OCO glyceryl groups	68.9	–	68.8	68.8	68.8	68.8	–
$CH=CH$	127.1	–	124.3	124.3	124.3	124.3	124.4
	127.7		129.9	129.9	129.9	129.9	
	127.8						
	128.0						
	128.2						
	129.6						
	129.9						
	130.1						
	131.8						
CH_3COOR		170.3	170.2	170.2	170.2	170.2	170.2
			170.3	170.3	170.3	170.3	170.3
$COOR$ (saturated, oleic, linolenic and linoleic chains)	172.7	–	172.6	172.6	172.6	–	–
	173.1		173.0	173.0	173.0		

[a]**1** PVAc, **2** VAc/Linseed oil (10%) copolymer, **3** VAc/Linseed oil (20%) copolymer, **4** VAc/Linseed oil (30%) copolymer, **5** VAc/Walnut oil (10%) copolymer, **6** VAc/Poppy oil (10%) copolymer, **7** VAc/Methyl linolenate (10%) copolymer.

Table 3.

FT-IR spectral data (v, cm^{-1})

	OH (stretching)	CH (stretching)	C=O (stretching)	CH (bending)	C–O (stretching), C–C (stretching), C–O–C (stretching)
PVA	3339 (vs)	2941 (m), 2907 (m)			1092 (s)
PVAc		2975 (w), 2967 (vw), 2924 (w)	1737 (vs)	1433 (w)	1373 (s), 1240 (vs), 1122 (w), 1022 (m)
Drying oil (linseed, walnut or poppy oil)		3010 (w), 2925 (vs), 2854 (m)	1745 (vs)	1462 (w)	1163 (w)

analyses were performed with a Perkin–Elmer 2400 Series II CHNS/O analyser. Solids content, pH, viscosity and minimum film forming temperature (MFFT) were determined according to the international standards EN ISO 3251 [18], ISO 976 [19], EN ISO 2555 [20] and ISO 2115 [21].

2.3. Mechanical Tests

The tensile strength and the elongation-at-break of the polymer films were measured according to EN ISO 527 [22]. Mechanical tests were performed on the adhesive joints as described in the standard [8] by adhesively bonding two beech wood specimens using a pneumatic press. In particular, density, moisture content, dimensions, thickness and fabrication conditions are specified in the specific standards.

2.4. Reactivity of Linseed Oil in the Presence of AIBN

Into a double-necked round bottom flask equipped with a magnetic stirrer and a reflux condenser, under nitrogen atmosphere, 1 ml of linseed oil was added to a solution of 12.8 mg of AIBN in 5 ml of ethyl acetate. The solution was stirred and heated at 60°C. Samples of the solution (0.5 ml) were recovered every 30 min, for 6 h. The solvent was evaporated under vacuum at room temperature and the recovered oils were dissolved in CDCl$_3$ and analyzed using ^{1}H-NMR spectroscopy.

2.5. Synthesis of Polymers

2.5.1. Synthesis of Poly(vinyl acetate) (PVAc) (1)

Into a double-necked round bottom flask equipped with a magnetic stirrer and a reflux condenser, under nitrogen atmosphere, 1 ml of vinyl acetate (0.93 g, 10.8 mmol) was added to a solution of 12.1 mg (0.07 mmol) of AIBN in 5 ml of ethyl acetate. The solution was stirred and heated at 60°C for 6 h. 100 ml of *n*-hexane were added at room temperature to the colorless reaction mixture and a white solid residue was obtained.

The solution was decanted and the solid product was recovered and purified by dissolving in chloroform and precipitating by adding *n*-hexane. The white solid obtained was then dried under vacuum and 700 mg of the polymer were recovered (75% yield). The ^{1}H, ^{13}C-NMR and FT-IR spectra of PVAc were recorded and reported in Tables 1–3.

Anal. Calcd for PVAc $[(C_4H_6O_2)_n]$ (*MW* = 829.9 g/mol): C, 55.81; H, 7.02. Found: C, 55.49; H, 7.07%.

2.5.2. Synthesis of VAc/Drying Oils Copolymers in an Organic Solvent (2–7)

The reagent amounts, the reaction conditions and the yields for each copolymer synthesis are reported in Table 4. As an example of the general procedure, the synthesis of VAc/linseed oil (10%) copolymer is reported.

Into a double-necked round bottom flask equipped with a magnetic stirrer and a reflux condenser, 2 ml of vinyl acetate (1.86 g, 21.65 mmol) and 0.2 ml of drying oil (186 mg) were added, under nitrogen atmosphere, to a solution of AIBN (55 mg, 0.33 mmol) in 5 ml of ethyl acetate. The solution was stirred at 60°C. After 6, 24 and 48 h the solution was cooled to room temperature and 55 mg of AIBN were added every time, under nitrogen atmosphere. After 54 h at 60°C the reaction was stopped. The solvent was distilled under vacuum at room temperature and a pale yellow solid was recovered.

The solid was stirred for 12 h with 30 ml of *n*-pentane, then it was filtered and washed with the same solvent. The solid, dried under vacuum, was purified by dissolving in chloroform and precipitating by adding *n*-hexane. The obtained copolymer was dried under vacuum and characterized by NMR and FT-IR spectroscopies (Tables 1–3). T_g (Table 4) and number average molecular weight (Table 4) were determined.

2.5.3. Synthesis of VAc/Linseed Oil (10%) Copolymer as Water Dispersion (8)

The water dispersions of PVAc were prepared under nitrogen atmosphere using a semi-batch emulsion polymerization technique, in a four-neck kettle equipped with a mechanical stirrer, an inlet for feeding streams and a reflux condenser. The reagent amounts, the reaction conditions and the yields for each preparation are reported in Table 5. Solids content, pH and viscosity were measured and reported in Table 5.

The solid materials recovered after water distillation at reduced pressure and room temperature were analyzed by ^{1}H-NMR spectroscopy in order to evaluate the amount of linseed oil with respect to VAc in the copolymer (Table 5).

Mechanical properties, water and solvent resistance of the polymer films obtained from the water dispersions and tensile strengths of the adhesive joints were evaluated (entries **14–18**, Table 6). These results were compared with those obtained for a formulation containing the PVAc homopolymer (entry **19**, Table 6) and for a PVAc homopolymer/linseed oil blend (entry **20**, Table 6).

Table 4.
Synthesis of copolymers in organic solvent. Reagents amounts, reaction conditions, characteristics of polymers

Code[a]	1	2	3	4	5	6	7
VAc	1 ml (0.93 g, 10.8 mmol)	2 ml (1.86 g, 21.65 mmol)	2 ml (1.86 g, 21.65 mmol)	2 ml (1.86 g, 21.65 mmol)	1 ml (0.93 g, 10.8 mmol)	1 ml (0.93 g, 10.8 mmol)	1 ml (0.93 g, 10.8 mmol)
Drying oil	–	0.2 ml (186 mg)	0.4 ml (372 mg)	0.6 ml (558 mg)	0.1 ml (93 mg)	0.1 ml (93 mg)	0.1 ml (90 mg, 0.3 mmol)
AIBN	12.1 mg (0.07 mmol)	(55 mg, 0.33 mmol) × 4	(55 mg, 0.33 mmol) × 4	(55 mg, 0.33 mmol) × 4	28.5 mg (0.17 mmol) × 4	28.5 mg (0.17 mmol) × 4	28.5 mg (0.17 mmol) × 4
Solvent	5 ml	5 ml	10 ml	10 ml	5 ml	5 ml	5 ml
Reaction time (h)	6	54	54	54	54	54	54
T (°C)	60	60	60	60	60	60	60
Yield (%)	75	69.4	52.1	44.2	62.6	58.7	43.6
T_g (°C)	27	17.0	16.7	14.3	31	31	13.7
M_n (g/mol)	38 000	12 300	7650	6400	12 400	11 600	6600

[a] **1** PVAc, **2** VAc/Linseed oil (10%) copolymer, **3** VAc/Linseed oil (20%) copolymer, **4** VAc/Linseed oil (30%) copolymer, **5** VAc/Walnut oil (10%) copolymer, **6** VAc/Poppy oil (10%) copolymer, **7** VAc/Methyl linolenate (10%) copolymer.

Table 5.
Synthesis of copolymers as water dispersion[a] reagents amounts, reaction conditions, characteristics of copolymer water dispersion

Code	Oil/Vac w/w (%)	Radical initiator	Protective colloid	Heating program temperature/time	Solid content (%)	pH	Brookfield viscosity (mPa s)[b]	VAc conversion (%)	Linseed oil mol/VAc mol in the copolymer (%)
8	20	$K_2S_2O_8$	PVA	70°C/5 h 30′	38			92.7	0.3
9	10	$K_2S_2O_8$	PVA	70°C/5 h 30′	36			100.0	1.3
10	10	$Na_2S_2O_8$	PVA	70°C/5 h 30′	36			63.3	0.6
11	10	TBHP/SFS	PVA	70°C/5 h 30′	36			95.2	0.5
12	10	$K_2S_2O_8$	PVA	70°C/3 h 30′, 90°C/2 h	40			100.0	0.3
13	10	TBHP/SFS	PVA	70°C/3 h 30′, 90°C/2 h	40			100.0	0.6
14	10	$Na_2S_2O_8$	PVA	70°C/3 h 30′, 90°C/2 h	39	4.6	4520	100.0	0.2
15	10	$Na_2S_2O_8$	PVA	70°C/3 h 30′, 90°C/2 h	39	3.4	10 200	100.0	0.2
16	10	TBHP/SFS	PVA	70°C/3 h 30′, 90°C/2 h	38	2.6	37 000	100.0	0.2
17	15	TBHP/SFS	PVA	70°C/3 h 30′, 90°C/2 h	39	4.0	3600	100.0	0.2
18	5	TBHP/SFS	PVA	70°C/3 h 30′, 90°C/2 h	40	3.9	5100	100.0	0.2

[a] $NaHCO_3$ was present as buffer. [b] Brookfield viscosity is a simple viscosity value obtained with a Brookfield Viscometer, with a 20 rpm speed of spindle rotation.

Table 6.
Mechanical properties, water and solvent resistance of polymer films obtained from water dispersions

Code	Oil/VAc w/w (%)	D3-1[a] (MPa)	D3-3[b] (MPa)	MFFT[c] (°C)	Tensile strength (MPa, 5 mm/min)	Elongation-at-break (%, 5 mm/min)	Solubility in water after 7 days (%)	Solubility in ethanol after 8 h (%)	Solubility in acetone after 8 h (%)
14	10	9.5	0.6	17	6.8	3.5	4.7	2.4	44
14B	10	9.7	0.8		14.5	4.9	4.1	1.3	21
14 + cobalt acetate (0.1%)									
15	10	10.5	0.3	17	11.3	5.9	4.9	2.3	21
15B	10	10.3	0.5		17.5	7.1	4.9	1.2	18
15 + cobalt acetate (0.1%)									
16	10	11.4	0.6	16	15.8	5.0	2.5	2.3	15.3
16B	10	11.8	0.7		13.7	5.0	2.5	1.4	13.0
16 + cobalt acetate (0.1%)									
17	15	11.5	0.6	–	11.9	4.5	–	–	–
18	5	11.8	0.7	17	17.6	5.0	–	–	–
19	Homopolymer (no oil)	11.4	0.3	17	24.3	5.0	5.9	0.1	48
20	19 + linseed oil (10%) blend	12.7	0.9	17	16.5	5.0	5.0	0.3	50

[a] Adhesive strength in dry conditions, [b] adhesive strength in wet conditions, [c] minimum film forming temperature.

3. Results and Discussion

3.1. Spectroscopic Characterization of the Drying Oils

The reactivity of drying oils depends on the presence of polyunsaturated chains in the triglyceride molecules. In fact, the rate of the cross-linking process is determined by the high reactivity of the bis-allylic CH_2 groups present in the polyunsaturated fatty acid chains. The amount of linoleic and linolenic groups in the different natural oils can be evaluated using GC-MS or ^{1}H-NMR techniques as reported in the literature [14–17, 23]. The number of double bonds in the triglyceride molecules is specific for each kind of oil; however, it is possible to observe small variations on the basis of their geographical origin, the seed quality and the oil extraction method [24, 25].

In order to analyze the products of the copolymerization between VAc and each drying oil, the raw natural materials were characterized using ^{1}H, ^{13}C-NMR and FT-IR spectroscopies and the data obtained are reported in Tables 1–3. The FT-IR, ^{1}H and ^{13}C-NMR data are in agreement with the literature [13–17].

The different acid compositions of each drying oil are in agreement with the ratio between the values obtained from the integration of the ^{1}H-NMR signals of the glyceric protons and those of the fatty acid chains protons [14–17, 23–25].

3.2. Reactivity of Linseed oil in the Copolymerization Conditions

In order to identify the chemical transformations due to the copolymerization, the reactivity of linseed oil with a radical initiator was studied and compared with the known behaviour of this oil in the usual drying process in the presence of air and light [16, 26].

The reaction was performed in ethyl acetate solution, under nitrogen atmosphere, in the presence of the radical initiator (AIBN) and under the same reaction conditions (60°C) as used for the vinyl acetate polymerization. In particular, nitrogen atmosphere was used. Therefore, the reactions involved in the cross-linking process of a drying oil in the oxygen atmosphere were not present in this case.

The reaction was monitored by ^{1}H-NMR spectroscopy. For this purpose, solution samples were recovered after 6, 24, 48 h. After solvent evaporation the residue was dissolved in $CDCl_3$ and analyzed.

Chemical transformations were not observed in these reaction conditions.

3.3. Synthesis of VAc/Drying Oil Copolymers

Several copolymers were synthesized in an organic solvent in order to analyze the final product in the absence of any additive required for a water dispersion. AIBN was used as the radical initiator and the reactions were performed in nitrogen atmosphere. The spectroscopic characteristics of the copolymers were compared with those of an analogous PVAc homopolymer obtained with the same procedure.

The following polymers were synthesized:

- Poly(vinyl acetate) homopolymer (PVAc);

- VAc/linseed oil copolymers with different amounts of linseed oil (10–30%, V/V);

- VAc/walnut oil copolymer (10% of walnut oil, V/V);

- VAc/poppy oil copolymer (10% of poppy oil, V/V);

- VAc/methyl linolenate copolymer (10% of methyl linolenate, V/V).

Finally, some of these copolymers were synthesized as water dispersions in order to obtain water dispersion adhesives.

3.3.1. Synthesis of Poly(vinyl acetate) (PVAc) Homopolymer

PVAc is usually synthesized as water dispersion, for use as an adhesive.

For an easy comparison between the chemical, physical and spectroscopic characteristics of the new copolymers with those of a standard homopolymer, PVAc was synthesized in an organic solvent as reported in the literature [27].

A glass transition temperature (T_g) of 27.0°C was found through DSC analysis (Table 4) and this value is lower than the one of 32°C reported for the PVAc obtained as water dispersion. A number average molecular weight of 38 000 g/mol was found by gel permeation chromatography (GPC), in agreement with the lower molecular weights obtained in solution synthesis, as compared to those obtained in water dispersion. The ^{1}H-NMR (Table 1), ^{13}C-NMR (Table 2) and IR (Table 3) spectroscopic data are in agreement with those reported in the literature for this homopolymer [27].

3.3.2. Synthesis of VAc/Linseed Oil Copolymers

In the first series of syntheses the reaction conditions were modified in order to obtain higher conversion and an effective copolymerization.

- A very low conversion was obtained by heating for 6 or 24 h at 60°C an ethyl acetate solution of vinyl acetate and linseed oil in the 1:1/V:V ratio with AIBN as radical initiator.

- A 16% yield was obtained after 24 h when the amount of linseed oil was reduced (10% linseed oil volume with respect to VAc volume). Therefore, the presence of linseed oil decreases the reactivity of vinyl acetate with AIBN as radical initiator.

- The same reaction was performed by adding AIBN four times during the reaction (6% mol/mol with respect to vinyl acetate). After 54 h a white product was obtained with a 69.4% yield.

At the end of the reaction, after solvent evaporation, the oily residue was washed with *n*-pentane in order to remove the unreacted drying oil and 27% of the initial drying oil (50 mg) was recovered.

The product, insoluble in *n*-pentane, was dissolved in chloroform and then the purified solid was recovered by adding *n*-hexane. The pale yellow solid obtained was dried under vacuum and characterized. A T_g of 17.0°C (Table 4) and a number average molecular weight of 12 300 g/mol (Table 4) were found.

This product was also characterized through ^{1}H-NMR (Table 1, Fig. 1), ^{13}C-NMR (Table 2, Fig. 2), FT-IR (Fig. 3). Two-dimensional NMR spectra gCOSY and gHSQC were used to provide the correlations between ^{1}H-NMR signals, and between ^{1}H and ^{13}C-NMR signals.

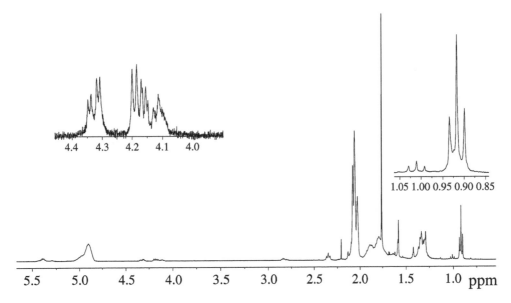

Figure 1. ^{1}H-NMR spectrum (CDCl$_3$ as solvent) of VAc/linseed oil (10%) copolymer.

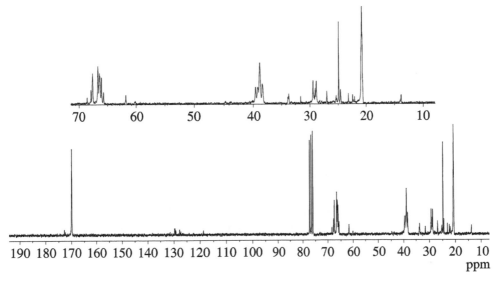

Figure 2. ^{13}C-NMR spectrum (CDCl$_3$ as solvent) of VAc/linseed oil (10%) copolymer.

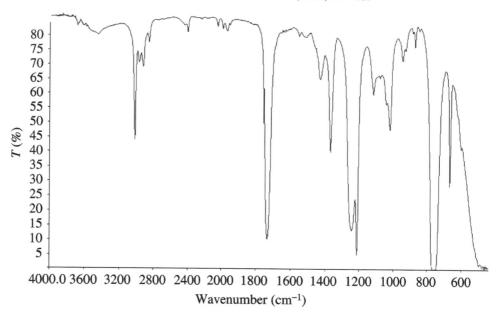

Figure 3. FT-IR spectrum of VAc/linseed oil (10%) copolymer.

The signals at 1.76, 2.02 and 4.87 ppm in the ^{1}H-NMR spectrum are attributable to the VAc fragments of the polymer chain, in agreement with the homopolymer data. Some signals attributable to the linseed oil fragments appeared modified in their intensity with respect to those of the unreacted oil. In particular, the intensity of the signals at 5.35 (CH), 2.80 (–CH=CH–CH_2–CH=CH–), 0.97 (CH$_3$ linolenic chain) ppm decreased with respect to the unreacted oil. Two new signals, which correspond neither to the poly(vinyl acetate) homopolymer nor to drying oil, were present at 4.07 and 1.85 ppm.

In the ^{13}C-NMR spectrum, recorded with CDCl$_3$ as solvent, the new signals at 60.5, 44.0 and 45.0 ppm were present together with the other characteristic signals for VAc and linseed oil fragments. Also in this spectrum the intensities of the signals attributable to the double bond carbons were reduced with respect to the other signals attributable to linseed oil.

At the end, using two-dimensional NMR techniques such as gCOSY and gH-SQC, homo-nuclear and hetero-nuclear correlations were found, respectively, between signals at 4.07 and 1.85 ppm in the gCOSY spectrum and between signals at 60.5 and 4.07 ppm in the gHSQC spectrum.

The spectroscopic data are in agreement with the copolymer formation. In fact, the signal at δ 4.07 ppm (^{1}H) is attributable to the CH near the linolenic chain fragment, in agreement with the H–C correlation observed with the signal at δ 60.5 ppm in the ^{13}C spectrum, and with the H–H correlation with the signal at 1.85 ppm, attributable to the adjacent CH$_2$ (Fig. 4(a) and 4(b)).

The negative effect of the high oil amount on the conversion of vinyl acetate was confirmed using different amounts of linseed oil. In fact, in the presence of

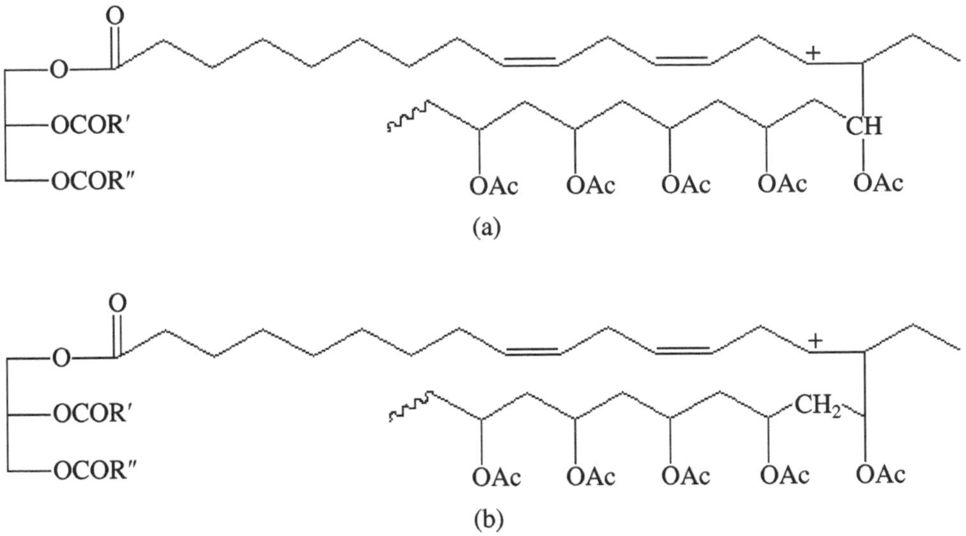

Figure 4. Structure of the radical intermediate on the linolenic chains. In the structure (a), the CH near the linolenic chain fragment is shown. In the structure (b), the CH_2 near this CH is shown.

20 or 30% of linseed oil with respect to VAc (V/V) and using the same reaction conditions, yields of 52.1% or 44.2%, respectively, were obtained. This behaviour is in agreement with a low reactivity of the free triglyceride or the radical intermediate formed after the addition of a triglyceride molecule in the chain. The high steric hindrance and the delocalization of the radical formed on the triglyceride structure can, in fact, reduce the rate of chain growth.

The products of these reactions were characterized through DSC and GPC (Table 4) and by ^{1}H-NMR (Table 1) and ^{13}C-NMR (Table 2). In the ^{1}H-NMR spectra of the products obtained from the VAc/linseed oil copolymerization, the same signals were present but the relative intensity of the linseed oil signals increased with the linseed oil/VAc ratio.

The T_g and the number average molecular weight decreased when the amount of linseed oil increased in the range 10–30% (Table 4). The different molecular weights can be explained by the lower rate of the chain growth in the presence of higher amounts of linseed oil, whereas the decrease observed in the T_g is in agreement with the plasticizing action of triglycerides.

3.3.3. Synthesis of the VAc/Walnut Oil (10%) Copolymer

In order to evaluate the influence of the drying oil composition, the reactions of VAc with other kinds of drying oils were also studied.

The difference between walnut oil and linseed oil compositions is the lower amount of linolenic chains and the higher amount of linoleic ones in the former.

The reaction was performed under the same reaction conditions as used for linseed oil (54 h at 60°C) and with a walnut oil/VAc ratio of 10% (V/V). After the work-up a white solid was obtained with a 62.6% yield and characterized by DSC, GPC (Table 4), ^{1}H-NMR (Table 1), ^{13}C-NMR (Table 2).

In the ^{1}H-NMR spectrum of the product obtained from the VAc/walnut oil copolymerization (Table 1) signals attributable to the VAc and triglyceride fragments were present. Nevertheless, the intensities of the walnut oil signals are lower than those of the linseed oil in its copolymer. In particular, the triplet at δ 0.97 ppm, attributable to the terminal CH$_3$ of linolenic chains, disappears after the reaction, in agreement with the low amount of the linolenic chains in the walnut oil and with the high reactivity of the C=C double bond nearest the methyl groups in the linolenic chains. Moreover, the presence of a signal at δ 4.07 ppm was in agreement with a new C–C bond formation between a triglyceride molecule and the PVAc chain.

In the ^{13}C-NMR spectrum (Table 2), signals attributable to the PVAc chain and triglycerides fragments of the copolymer were present, whereas new signals at δ 60.5, 45.0, 44.0 ppm were attributable to the new C–C bonds formed in the copolymerization.

These data confirmed the formation of the copolymer. Nevertheless, the amount of triglycerides bonded to the polymer chain was lower with respect to that present in the VAc/linseed oil copolymer, in agreement with the lower reactivity of walnut oil because of its lower amount of linolenic chains.

The T_g value of 31°C is similar to the one obtained for analogously synthesized PVAc-homopolymer. This behaviour appeared in agreement with a low amount of triglycerides bonded to the polymer chain. On the other hand, the effect of walnut oil on the chain growth was the same as observed for linseed oil and a number average molecular weight of 12 400 g/mol was obtained.

3.3.4. Synthesis of the VAc/Poppy Oil (10%) Copolymer

The copolymerization reaction between VAc and poppy oil was performed under the same reaction conditions as used for walnut oil. A white solid was obtained with a 58.7% yield. The product was characterized by DSC, GPC (Table 4) and ^{1}H-NMR (Table 1).

A T_g of 31°C and a number average molecular weight of 11 600 g/mol were determined. These data were similar to those obtained for the product of the VAc/walnut oil reaction.

In the ^{1}H-NMR spectrum, signals of PVAc and poppy oil were present, whereas signals attributable to formation of the VAc/poppy oil copolymer (Table 1) were not identified. Therefore, a homopolymer or a copolymer containing a very low amount of triglycerides were formed in this reaction.

Poppy oil, as walnut oil, contains a very low amount of linolenic chains and the C=C double bonds present in linoleic and oleic acyl groups have a lower reactivity than the terminal C=C double bond of linolenic fragment.

3.3.5. Synthesis of the VAc/Methyl Linolenate (10%) Copolymer

In order to evaluate how the structure of triglyceride influenced reaction rate, conversion, T_g and molecular weight, methyl linolenate was used in the copolymerization reaction with VAc. In fact, this molecule has the reactivity of the polyunsaturated linolenate chain but it has a lower steric hindrance than the triglyceride molecule. Under the same reaction conditions as used for walnut oil (10% monoester volume/VAc volume), after work-up, a pale yellow solid was obtained with a 43.6% yield. The product was characterized by DSC, GPC (Table 4) and ^{1}H-NMR (Table 1).

A T_g of 13.7°C and a number average molecular weight of 6600 g/mol were found. Some of these data are similar to those obtained for the product of VAc/linseed oil reaction. In particular, the T_g values and the yellow colour can be connected with the presence of polyunsaturated chains. On the other hand, the low conversion and low molecular weight are in agreement with a negative effect on the reaction rate of the chain growth caused by the presence of these esters.

The resonance effect in the polyunsaturated chain is, therefore, more important than the steric hindrance on the characteristics of the copolymers synthesized and linseed oil or methyl linolenate show similar behaviour.

Signals attributable to the VAc fragment (1.76, 2.02, 4.87 ppm) and methyl linolenate fragment (1.30, 2.30, 3.66) were present in the ^{1}H-NMR spectrum of the VAc/methyl linolenate copolymer (Table 1), whereas the characteristic signal at 0.97 ppm was shifted to 0.85 ppm in agreement with the involvement in the copolymerization of the C=C double bond near the terminal CH$_3$. The other signals of the methyl linolenate fragment were present in the spectrum and the intensities of the signals at 5.35 (*CH*) and 2.80 (–CH=CH–*CH$_2$*–CH=CH–) ppm decreased with respect to the unreacted oil.

Signals at 1.85 (*CH$_2$*) and 4.07 (*CH*) ppm in the ^{1}H-NMR spectrum and signal at 60.5 (*CH*) ppm in the ^{13}C-NMR spectrum were attributable to the copolymer formation.

3.3.6. Synthesis of VAc/Linseed Oil Copolymer as Water Dispersions

In order to see if the copolymerization of VAc and linseed oil could be carried out in water dispersion, several formulations and different reaction conditions were studied (Table 5). PVA was used as protective colloid.

Different radical initiator systems (r.i.s.) and different VAc/r.i.s. ratios were tested. In the presence of K$_2$S$_2$O$_8$ used in the same molecular ratio as reported for the organic solvent synthesis (0.15 mmol r.i.s./10.8 mmol VAc), a yellow copolymer was obtained. The solid, recovered after solvent distillation and washed with n-pentane, contained 60% of the initial linseed oil bonded to polymer chains.

In the following syntheses the radical initiator amount was reduced (37 μmol r.i.s./10.8 mmol VAc) reproducing the molecular ratio similar to that in an industrial process. In these conditions three radical initiator systems were tested: K$_2$S$_2$O$_8$, Na$_2$S$_2$O$_8$ and TBHP/SFS.

As reported above for AIBN, preliminary tests were performed in order to analyze the reactivity of linseed oil in the presence of each selected r.i.s. and no homopolymerization of drying oil was observed in water with PVA as protective colloid and dispersing agent.

The 10 or 20% w/w (1 or 2% mol/mol) of linseed oil with respect to VAc was used and reaction parameters such as temperature (range 60–90°C) and reaction time (range 2–6 h) were varied with the aim to obtain a water dispersion with a solid residue of 35–40%.

By increasing the reaction temperature from 60 to 90°C, the conversion was improved and the formation of grits was reduced. In fact, at high temperatures, the viscosity was lower and the mechanical stirring was more effective.

At the end of the reaction, at room temperature, several formulations required water addition in order to reduce the high viscosity of the reaction mixture. The reaction conditions used in each synthesis and the characteristics obtained for each formulation are reported in Tables 5 and 6.

The VAc conversion was obtained as the difference between the amount of dry solid content and the total weight of solid reagent and linseed oil, whereas the amount of linseed oil in the copolymer was estimated as the ratio between triglyceride and VAc (mol/mol). This ratio was obtained from the integrals of the respective signals in the ^{1}H-NMR spectrum recorded on the solid residue washed with n-pentane to remove the free linseed oil. In particular, signal at 4.87 ppm, attributable to $-C\underline{H}(OAc)CH_2-$ fragment, and multiplet at 4.35–4.10 ppm, attributable to the triglyceride fragment $C\underline{H}_2(OR)-CH(OR')-C\underline{H}_2(OR'')$, were selected as characteristic signals.

Unlike the copolymerization in the organic solvent it is not possible with water as solvent to use the signal at 4.07 ppm (^{1}H-NMR) to evaluate the amount of triglyceride bonded to the chain. In fact, in the copolymer obtained in the organic solvent this signal was attributed to $-C\underline{H}(OAc)-CH_2-$ near the linolenic chain fragment. When the reaction was performed in water at higher temperature (70–90°C) a signal at 4.05 ppm was observed also in the homopolymer spectrum. This result can be explained by the presence of several additives in the mixture. Therefore, the amount of linseed oil bonded in the copolymer was evaluated as the difference between the amount of oil added in the reaction and the amount of the oil removed with the n-pentane extraction.

3.4. Mechanical Tests

Standard tests were used to evaluate the mechanical performances of adhesive joints for non-structural applications. The values of adhesive strength for thin bondlines obtained, as reported in EN 205 [8], permit to classify an adhesive in agreement with the standard EN-204 [9]. The results are reported in Table 6.

The determination of tensile strength and elongation-at-break [21] gives information on the mechanical resistance of non-structural adhesive films. The results are reported in Table 6.

The PVAc homopolymer (**1**) and the VAc/linseed oil (10%) copolymer (**2**) obtained in organic solvent were tested using the EN205-D3 test. In agreement with this test the adhesive strength can be measured in dry conditions (D3-1) or in wet conditions (D3-3). The presence of ethyl acetate as solvent (solid content 50%)

caused several problems in the application on the wood surface. Consequently, the specified conditions were partially modified. For example, the test was performed after a drying period of 24 days, at room temperature, vs 7 days.

The following values of adhesive strength in dry conditions (D3-1) were obtained:

- PVAc (**1**), D3-1, 6.5 ±2.7 N/mm^2;

- VAc/linseed oil (10%) copolymer (**2**), D3-1, 2.7 ± 0.8 N/mm^2.

These data are in agreement with a negative effect of the presence of linseed oil probably due to the lower affinity between wood and the copolymer containing triglycerides fragments. Moreover, neither of the tested products penetrated into wood probably because of the low affinity of the organic solution for the wood surface. The known advantages of the use of water dispersions as wood adhesives are in agreement with this behaviour. The negative effect of the presence of an organic solvent on the wettability and capillary action is increased by the absence of the additives usually employed in the commercial products.

The performance of the copolymers obtained as water dispersions was tested and values of adhesive strength in dry conditions (D3-1) and in wet conditions (D3-3) were obtained (Table 6). The effect of the presence of cobalt acetate as the catalyst for the cross-linking reaction was also studied.

The results obtained with these formulations were in agreement with a positive effect of the presence of linseed oil on the adhesive strength (D3-1 and D3-3) with respect to those of the pure homopolymer, whereas the performance was lower than that obtained with commercial copolymer containing N-methylolacrylamide (NMA) and AlCl$_3$. The presence of cobalt acetate, in order to promote the cross-linking, produces a positive effect.

The tensile strength and the elongation-at-break and minimum film-forming temperature were evaluated. The drying oil present in the adhesive film works as a plasticizer. However, changes in the minimum film-forming temperature were not observed.

The water and organic solvent resistance was tested and the solubility in water, ethanol and acetone was evaluated (Table 6).

4. Summary and Conclusions

PVAc copolymers with drying oils as comonomers were synthesized and characterized. The role of the drying oils was identified with NMR spectroscopy. The presence of several triglyceride molecules chemically bonded to the polymer chain is potentially interesting for the production of self cross-linking wood adhesives with waterproofing properties. The presence of unreacted C=C double bonds in the copolymer products requires a hardener agent or a heat treatment to obtain the cross-linking effect.

The products were initially polymerized in an organic solvent to perform the spectroscopic characterization in the absence of external additives such as those required in an emulsion polymerization process.

The copolymer formation can be confirmed on the basis of the spectroscopic characteristics. In fact, in the ^{1}H and ^{13}C-NMR spectra, signals attributable to copolymers were detected. In the presence of linseed oil, which contains a large amount of linolenic chains, or in the presence of methyl linolenate, new signals attributable to the copolymer were observed in the ^{1}H-NMR (δ 4.07 and 1.85 ppm) and ^{13}C-NMR (δ 60.5, 44.0 and 45.0 ppm) spectra. The residual drying oil not bonded to the chain was separated in the work-up and purification processes.

The reaction mechanism and the structure can be hypothesized. In agreement with the different reactivities between vinyl acetate monomer and triglycerides molecules, in presence of AIBN, the first step would involve the radical activation of vinyl acetate. In the second step, the chain growth was obtained with the radical addition of vinyl acetate or triglyceride. The subsequent addition of the radical intermediate on the triglyceride involved preferentially the terminal double bond of the linolenic chain. This hypothesis is in agreement with the decrease of the signal attributable to the CH_3 of the linolenic chains. This regioselectivity can be explained by the lower steric hindrance of the double bond on the C15 with respect to those on the C9 or C12 in the linolenic chain.

After the addition of other monomers, the final copolymers can be obtained from the standard termination steps as disproportionation or combination of radical intermediates.

The hypothesized mechanism and the consequent copolymer structure are reported in Fig. 5 (in the example here reported the triglyceride contains one linoleic and two linolenic chains).

Also the double bonds present in the oleic and linoleic chains can react with the radical intermediates, however their reactivity is lower than that of the terminal double bond in the linolenic chain.

Signals attributable to the copolymer formation were not identified for the reaction between VAc and poppy oil, whereas in the presence of walnut oil these signals were identified with lower intensities.

In the presence of methyl linolenate a mechanism similar to the one described for linseed oil can be hypothesized. The hypothesized mechanism and the consequent copolymer structure are reported in Fig. 6.

The preparation of modified PVAc adhesives containing drying oils as comonomers showed some advantages, such as a high water and solvent resistance. However, further studies are required in order to obtain the formulations useful for industrial purpose.

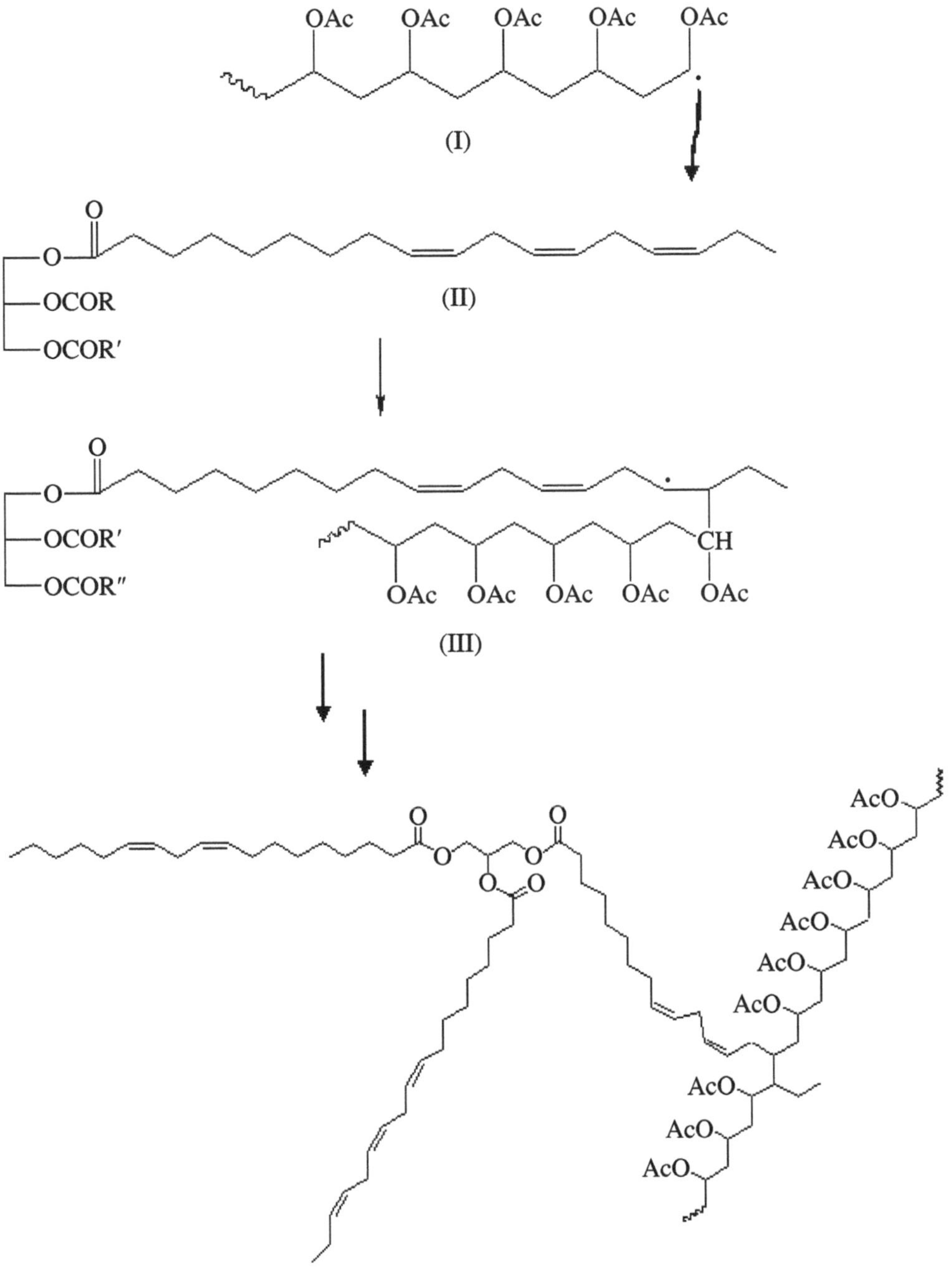

Figure 5. Polymerization mechanism and VAc/linseed oil copolymer structure (in the example, the triglyceride contains one linoleic and two linolenic chains). The poly(vinyl acetate) radical intermediate (I) reacts with a double bond of the triglyceride (II) and produces the radical intermediate (III), which after other reaction steps is transformed into the product (IV).

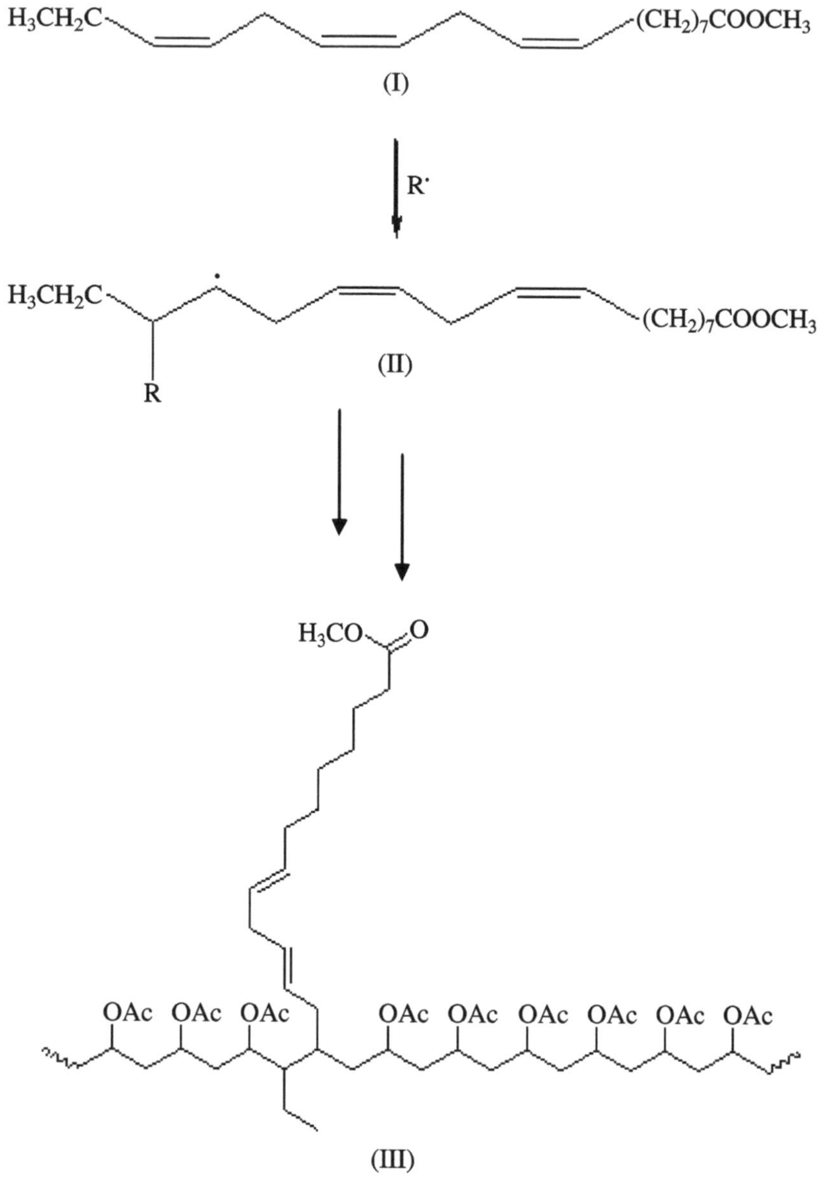

H₃CH₂C$\sim$＝$\sim$＝$\sim$＝$\sim$(CH₂)₇COOCH₃

(I)

R·

H₃CH₂C$\sim$·$\sim$＝$\sim$＝$\sim$(CH₂)₇COOCH₃

(II)

R

$$H_3CO\diagdown C=O$$

OAc OAc OAc   OAc OAc OAc OAc OAc OAc

(III)

Figure 6. Polymerization mechanism and VAc/methyl linolenate copolymer structure. The poly(vinyl acetate) radical intermediate (R·) reacts with a double bond of the methyl linolenate (I) and produces the radical intermediate (II), which after other reaction steps is transformed into the product (III).

Acknowledgements

This research was supported by the University of Florence, Ministero della Istruzione, Università e Ricerca (M.I.U.R.) and Cray Valley Italia. The authors would like to thank the Ente Cassa di Risparmio di Firenze for allowing the use of

400 MHz NMR spectrometer and Mr. Maurizio Passaponti, Department of Organic Chemistry, University of Florence, for elemental analysis.

References

1. R. Walter, *Osterr. Forst- u. Holzwirtsch* **9**, 246 (1954).
2. A. J. Easteal and K. G. K. De Silva, PCT Int. Appl., Patent WO 03046099 (2003).
3. H. Tawada, T. Okamatsu and T. Kamikaseda, *Koen Yoshishu, Nippon Setchaku Gakkai Nenji Taikai* **39**, 67 (2001).
4. G. Feichtmeier and P. S. Willett, PCT Int. Appl., Patent WO 9967343 (1999).
5. H. Gandert, G. Diehm, M. Leubner, C. Fueger and S. Weinkoetz, PCT Int. Appl., Patent WO 2005097931 (2005).
6. L. Qiao, A. J. Eastel, C. J. Bolt, P. K. Coveny and R. A. Franich, *Pigment Resin Technol.* **29**, 152 (2000).
7. P. Ball, W. Bauer, R. Graewe and R. Tangelder, Ger. Offen., Patent DE 102004007028 (2005).
8. EN 205 (2003). Wood adhesives for non-structural applications — determination of tensile strength of lap joints.
9. EN 204 (2002). Classification of thermoplastic wood adhesives for non-structural applications.
10. K. Boege, P. Daute, M. Fies, J. Klauck and J. Klein, Ger. Offen., Patent DE 4430875 (1995).
11. J. E. O. Mayne, H. Warson and R. M. Levine, Patent GB 906117 (1962).
12. I. Cristea, G. Pop and G Ionita, *Buletinul Institutului Politehnic "Gheorghe Gheorghiu-Dej" Bucuresti* **27**, 57 (1965).
13. R. G. Sinclair, A. F. Mckay, G. S. Myers and R. N. Jones, *J. Am. Chem. Soc.* **74**, 2578 (1952).
14. M. D. Guillén and A. Ruiz, *Eur. J. Lipid Sci. Technol.* **105**, 688 (2003).
15. M. D. Guillén and A. Ruiz, *Eur. J. Lipid Sci. Technol.* **107**, 36 (2005).
16. A. Spyros and D. Anglos, *Anal. Chem.* **76**, 4929 (2004).
17. G. Cipriani, A. Salvini, L. Dei, A. Macherelli, F. S. Cecchi and C. Giannelli, *J. Cultural Heritage* **10**, 388 (2009).
18. EN ISO 3251, Determination of non-volatile-matter content (2008).
19. ISO 976, Polymer dispersions and rubber latices — Determination of pH (2008).
20. EN ISO 2555, Resins in the liquid state or as emulsion or dispersions — Determination of apparent viscosity by the Brookfield Test method (1999).
21. ISO 2115, Plastics — Polymer dispersions — Determination of white point temperature and minimum film-forming temperature (1996).
22. EN ISO 527, Plastics — Determination of tensile properties (1993).
23. M. Lazzari and O. Chiantore, *Polymer Degrad. Stabil.* **65**, 303 (1999).
24. J. S. Mills and R. White, *The Organic Chemistry of Museum Objects*. Butterworth and Heinemann, Oxford (1994).
25. A. Wakjira, M. T. Labuschagne and A. Hugo, *J. Sci. Food Agric.* **84**, 601 (2004).
26. H. Wexler, *Chem. Rev.* **64**, 591 (1964).
27. A. Salvini, L. M. Saija, S. Finocchiaro, G. Gianni, C. Giannelli and G. Tondi, *J. Appl. Polym. Sci.* **114**, 3841 (2009).

Acrylated Epoxidized Soy Oil as an Alternative to Urea-Formaldehyde in Making Wheat Straw Particleboards

Mohamad Tasooji [a,*], **Taghi Tabarsa** [a], **Abolghasem Khazaeian** [a] **and Richard P. Wool** [b]

[a] Department of Wood and Paper Engineering, Gorgan University of Agricultural Sciences and Natural Resources, Shahid beheshty St, Gorgan, Iran

[b] Department of Chemical Engineering, University of Delaware, Newark, Delaware 19716, USA

Abstract

Wheat straw particleboards were made using urea-formaldehyde (UF) and acrylated epoxidized soy oil (AESO) resins with two resin content levels: 8% and 13%, and three pressing times: 8, 10 and 12 minutes. Physical and mechanical properties of the boards were investigated. Results showed that AESO bonded particleboards had higher physical and mechanical properties than UF bonded boards, especially in terms of internal bond strength and thickness swelling. All properties of AESO bonded boards increase by increasing both the resin content and pressing time; but for UF bonded boards, there is a reduction in properties at 12 min pressing time due to the UF bonds destruction. AESO bonded boards can compete with wood particleboards according to EN-standards. The good properties of these particleboards are because of high compatibility between wheat straw particles and AESO resin due to its oil based structure.

Keywords

Acrylated epoxidized soy oil, wheat straw particleboards, oil based resin

1. Introduction

As the population grows and demand for wood composite products increases, as well as considering the global concern about nature and environment, developing countries and even the countries with considerable forest resources are focusing on lignocellulosic non-wood sources to produce composite products.

Amongst agricultural productions, wheat is the second most cultivated cereal plant worldwide. Wheat straw is the main by-product of cereal harvest and is primarily used as fodder for animals [1]. It consists of relatively large number of elements, including the actual fibers, parenchymal cells, vessel elements, and epi-

---

[*] To whom correspondence should be addressed. Mohamad Tasooji, 4, Maryam Alley, Haji Pour Amir Avenue, Satarkhan Avenue, Tehran, Iran. Tel.: +989123884075; Fax: +982144217490; e-mail: mtasooji@ut.ac.ir

Wood Adhesives

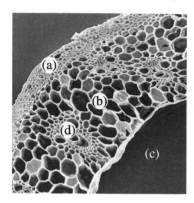

Figure 1. Cross section of wheat straw: (a) epidermis, (b) parenchyma, (c) Lumen and (d) vascular bundles [3].

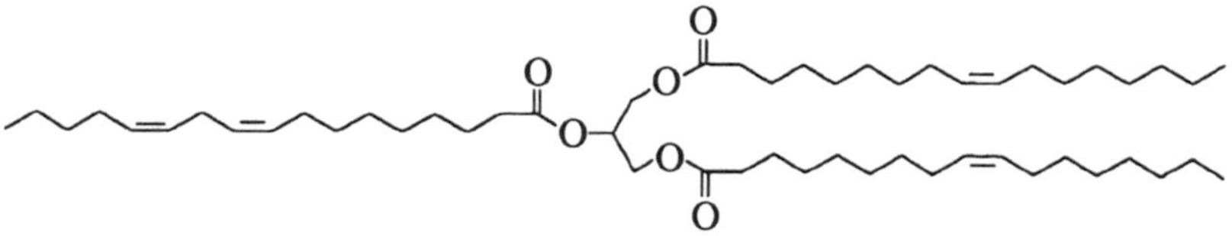

Figure 2. Triglyceride molecule, the major component of plant oils.

dermal cells [2]. As seen in the cross section of straw [3] (Fig. 1), the epidermal cells are the outermost surface cells, covered by a thin wax layer.

Although there are low density acoustical and thermal panels and also structural panels that use pMDI as binder for producing wheat straw particleboards [1], the outer wax layer is a serious problem in making straw particleboards with water based resins. Some investigations have been carried out on destroying or modifying the wax layer to improve the bonding quality between wheat straw particles and water based resins [2, 4–6]. Another alternative is also available for achieving good bonding quality with no changes in the wax layer, i.e., by changing the resin type and utilizing oil based resins, straw particleboards with high quality bonds can be made.

During recent years there have been a large number of studies focusing on the plant oils as a raw material for producing different polymers. The main component of plant oils is triglyceride. A triglyceride consists of three molecules of fatty acids and one molecule of glycerol (Fig. 2) [7].

The fatty acids contribute from 94–96% of the total weight of triglyceride oil [8]. The most common oils contain fatty acids that vary from 14 to 22 carbons in length with 0 to 3 double bonds per fatty acid. Linoleic and oleic are the main fatty acids in soy oil (Table 1) [7].

There are some active sites amenable to chemical reaction in triglyceride structure: the double bonds, the allylic carbons, the ester group, and the carbons alpha to ester group. These active sites can be used to introduce polymerizable groups on triglyceride for producing resins. There are different ways to produce resins

Table 1.
Fatty acid distribution in various plant oils

Fatty acid	# C: # DB[a]	Canola	Corn	Cottonseed	Linseed	Olive	Palm	Rapeseed	Soybean	High oleic[b]
Myristic	14:0	0.1	0.1	0.7	0.0	0.0	1.0	0.1	0.1	0.0
Myristoleic	14:1	0.0	0.0	0.0	0.0	0.0	0.0	0.0	0.0	0.0
Palmitic	16:0	4.1	10.9	21.6	5.5	13.7	44.4	3.0	11.0	6.4
Palmitoleic	16:1	0.3	0.2	0.6	0.0	1.2	0.2	0.2	0.1	0.1
Margaric	17:0	0.1	0.1	0.1	0.0	0.0	0.1	0.0	0.0	0.0
Margaroleic	17:1	0.0	0.0	0.1	0.0	0.0	0.0	0.0	0.0	0.0
Stearic	18:0	1.8	2.0	2.6	3.5	2.5	4.1	1.0	4.0	3.1
Oleic	18:1	60.9	25.4	18.6	19.1	71.1	39.3	13.2	23.4	82.6
Linoleic	18:2	21.0	59.6	51.4	15.3	10.0	10.0	13.2	53.2	2.3
Linolenic	18:3	8.8	1.2	0.7	56.6	0.6	0.4	9.0	7.8	3.7
Arachidic	20:0	0.7	0.4	0.3	0.0	0.9	0.3	0.5	0.3	0.2
Gadoleic	20:1	1.0	0.0	0.0	0.0	0.0	0.0	9.0	0.0	0.4
Eicosadienoic	20:2	0.0	0.0	0.0	0.0	0.0	0.0	0.7	0.0	0.0
Behenic	22:0	0.3	0.1	0.2	0.0	0.0	0.1	0.5	0.1	0.3
Erucic	22:1	0.7	0.0	0.0	0.0	0.0	0.0	49.2	0.0	0.1
Lignoceric	24:0	0.2	0.0	0.0	0.0	0.0	0.0	1.2	0.0	0.0
Average # DB/triglyceride		3.9	4.5	3.9	6.6	2.8	1.8	3.8	4.6	3.0

[a] The number of carbon atoms and C=C double bonds (DB) per fatty acid molecule. For example, 18:2 means that the fatty acid is 18 carbon atoms long and contains 2 C=C double bonds.

[b] Genetically engineered high oleic acid content soybean oil (DuPont).

from triglycerides [7]. Acrylated epoxidized soy oil (AESO) resin can be made by acrylating the epoxidized soy oil. Although acrylated epoxy oils are mostly used in surface coatings, in recent years many researchers have used AESO to produce high modulus polymers and composites [9–15].

In our research we used AESO resin for producing wheat straw particleboards. Since both raw materials are derived from nature, this kind of particleboard is totally a bio-based product, without any use of forestry resources (wood particles) or any formaldehyde emissions (because AESO does not contain any formaldehyde). We made the boards, evaluated their physical and mechanical properties and eventually compared the results to wood particleboard EN-standards.

2. Materials and Methods

2.1. Wheat Straw

Wheat (*Triticum aestivum L.*) straw particles were obtained from local farms around Gorgan, Iran. The length of straw particles was about 2–5 mm. The particles were oven-dried at $103 \pm 2°C$ for 10 h and their moisture content reduced to 2 percent.

Table 2.
Urea-formaldehyde resin properties

Solid content (%)	65
Viscosity (cP)	155
Gel time (s)	52
U:F (molar ratio)	1:1.3

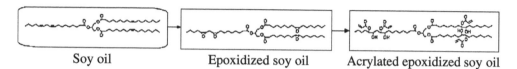

Soy oil   Epoxidized soy oil   Acrylated epoxidized soy oil

Figure 3. Production procedure for AESO.

2.2. Urea-Formaldehyde Resin

For our control samples, we used UF resin to produce straw particleboards. The UF resin was obtained from Chasbsaz company in Sari, Iran. Two percent NH_4Cl based on resin solid content was added as a catalyst. Table 2 shows the UF resin properties.

2.3. AESO Resin

AESO resin was produced in the Wood and Paper Laboratory at Gorgan University of Agricultural Sciences and Natural Resources, Iran.

The production procedure of AESO is shown in Fig. 3. The double bonds are one of the active sites that can react with functional groups. Epoxy groups react with double bonds in an epoxidation reaction to form epoxidized soy oil [16] and the acrylic acid reaction with epoxidized soy oil occurs through a substitution reaction [7].

Here are the details:

Stoichiometric amount of acrylic acid was mixed with hydroquinone, and 1,4-diazobicyclo[2.2.2]octane in a beaker and was stirred to achieve a homogeneous mixture. Hydroquinone and 1,4-diazobicyclo[2.2.2]octane act as free-radical inhibitor and catalyst, respectively. Then the homogeneous mixture was added to epoxidized soy oil (ESO) in a beaker and was stirred by a mechanical mixer for 16 hours in a water bath at 95°C. After that, it was cooled down to room temperature.

ESO was obtained from Sajo O & F Corporation Korea, acrylic acid and catalyst were from Merck, and hydroquinone from Sepidaj, an Iranian pharmacy company. The properties of ESO are shown in Table 3.

2.4. ^1H-NMR Analysis

To assure the acrylation reaction, ^1H-NMR was used for characterizing both ESO and AESO resins. The procedure was carried out according to ASTM-D5292-99

Table 3.
Epoxidized soy oil properties

Appearance	Transparent liquid
Molecular weight (g/mol)	About 1000
Oxirane oxygen content (%)	Minimum 6.6
Iodine value (Wijs method)	Maximum 3.0
Viscosity (30°C, cP)	Maximum 350
Specific gravity (25°C)	0.982–1.002
Volatile matter (%)	Maximum 0.1

Table 4.
Pressing cycle

	Pressure (MPa)	Time (% of total pressing time)
Step 1	3	60
Step 2	1.5	20
Step 3	1	20

standard with Bruker's NMR 400 MHz AVANCE 400 machine at Iran Polymer Institute, Tehran, Iran.

2.5. Curing of AESO Resin

Unlike UF, AESO resin cannot be sprayed on straw particles because of its high viscosity. In order to reduce AESO viscosity, styrene, which also increases the strength, stiffness and glass transition temperature due to an increase in cross-link density of the cured resin, was blended with AESO in the amount of 33 wt% [7, 12]. Since the polymerization process of AESO is addition type, a free radical initiator should be used for the curing. Benzoyl peroxide was used as free radical initiator in the amount of 5 wt% of the AESO + styrene mixture.

2.6. Particleboard Manufacturing

The resins were sprayed onto the straw particles in a laboratory blender. There were two levels of resin content: 8% and 13% based on the dry weight of straw particles, and three levels of pressing times: 8, 10 and 12 minutes. A hot press was used to manufacture the boards. The pressing temperature was 200°C and a three-step pressing cycle was used (Table 4). The manufactured boards were 300 mm side square and 10 mm thick. For AESO bonded particleboards there was a 2-h post-curing process at 150°C. All the boards were conditioned at 20°C and 65% relative humidity for 2 weeks. For each treatment, three replicates were manufactured, and for each physical or mechanical property 10 samples were used. The nominal board density was 0.7 g/cm^3.

2.7. Mechanical Tests

Mechanical properties were measured using a Universal Testing Machine (Shenck 40 kN). Three-point flexural test was carried out according to EN-310 "Wood-based panels — determination of modulus of elasticity in bending and of bending strength" and the modulus of elasticity as well as modulus of rupture were calculated. The dimensions of samples for EN-310 standard were 10 mm × 50 mm × 250 mm. Internal bond (IB) strength test was carried out following EN-319 "Particleboards and fiberboards — determination of tensile strength perpendicular to the plane of the board" and thickness swelling (TS) after 2 and 24 h water immersion was evaluated following EN-317 "Particleboards and fiberboards — determination of swelling in thickness after immersion in water". The dimensions of samples for both IB strength and TS tests were 10 mm × 50 mm × 50 mm according to the standards. Density was also measured following EN-323 "Wood based panels — determination of density".

3. Results and Discussion

3.1. ^{1}H-NMR Results

By comparing Fig. 4 with Fig. 5, a reduction in peaks can be seen at 2.95–3.00 ppm, and new peaks appear at 5.7–6.3 ppm in Fig. 5. This shows that the amount of epoxy groups decreased and acrylic groups appeared after acrylation reaction [17–19].

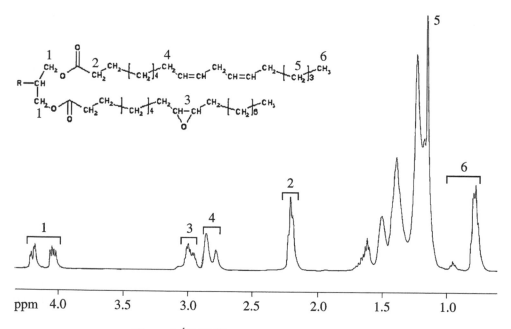

Figure 4. ^{1}H-NMR spectrum of epoxidized soy oil.

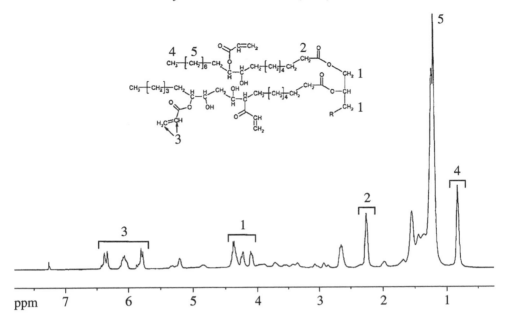

Figure 5. ^{1}H-NMR spectrum of acrylated epoxidized soy oil.

Table 5.
AESO properties

Appearance	Honeylike
Viscosity (cP)	15610 ± 10
Density (g/cm^3)	1.052 ± 0.002
Molecular weight (g/mol)	1300
Gel time (min:s)	2:36

3.2. AESO Properties

The properties of produced AESO resin are shown in Table 5.

3.3. Mechanical and Physical Properties of Boards

There is a tremendous difference between AESO and UF bonded straw particle-boards with respect to internal bond strength and thickness swelling. According to Table 6, AESO bonded boards are better in both 8% and 13% resin contents compared to all UF bonded boards; but for modulus of rupture (MOR) and modulus of elasticity (MOE) the differences between AESO and UF bonded boards are not as much as in IB strength or TS and even in some cases, 13% UF bonded boards are better than 8% AESO. All physical and mechanical properties increase with increasing resin content in both resins. In AESO bonded boards all the properties improve by increasing pressing time; but in UF bonded boards there is an improvement until 10 min of pressing time, and after that, a reduction is seen in mechanical

Table 6.
Mechanical and physical properties of boards

Resin type	Resin content (%)	Pressing time (min)	Internal bond strength (MPa)	Modulus of rupture (MPa)	Modulus of elasticity (MPa)	Thickness swelling (2 h) (%)	Thickness swelling (24 h) (%)
AESO	8	8	0.24	11.8	1681	19.8	35.7
		10	0.32	13.0	1725	12.6	26.2
		12	0.38	14.0	1943	9.6	22.7
AESO	13	8	0.47	14.9	2048	6.5	19.3
		10	0.51	16.1	2111	6.3	19.1
		12	0.55	18.8	2446	6.0	15.5
UF	8	8	0.07	10.4	1321	75.2	85.4
		10	0.08	11.4	1366	65.2	74.4
		12	0.06	9.4	1135	69.8	78.4
UF	13	8	0.12	14.2	1515	39.8	48.4
		10	0.14	15.5	1857	36.2	44.1
		12	0.13	14.7	1626	36.6	46.5

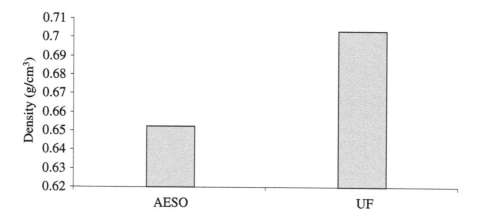

Figure 6. Densities of AESO and UF bonded boards.

properties and an increase in TS at 12 min of pressing. There is also a significant difference between AESO and UF bonded boards densities (Fig. 6).

3.4. Resin Type

The thin nonpolar wax layer covering wheat straw outermost epidermal cells causes low surface free energy of wheat straw particles; therefore, the contact angle between polar UF resin and nonpolar wheat straw surface is very high (Fig. 7) [1]. This leads to low quality bonding between UF resin and straw particles [1, 4]. On the other hand, AESO oil based resin has nonpolar structure and the compatibility between AESO resin and straw particles surface causes stronger bonds. This can be the main reason for the improvement in properties of AESO boards.

Figure 7. High contact angle between polymerized UF resin drops and straw particle.

3.5. Resin Content and Pressing Time

By increasing resin content, surface contact between the resin and straw particles increases [1] and this leads to improved bonding quality in both resins. Since there is a high compatibility between AESO and straw particles, this improvement is greater in AESO particleboards.

All mechanical and physical properties of AESO bonded boards increase with increasing pressing time. As AESO gel time is longer than of UF, and thus a more complete curing process occurs by increasing the pressing time in AESO bonded boards. Due to UF low gel time and low resistance to high temperatures, maybe the bonds between UF and straw particles are disrupted during long pressing time; this can be seen in all UF boards investigated in this research. Also due to low mat permeability, the heat convection from surface to the core takes longer.

Lower permeability of wheat straw particles leads to slower heat convection from surface to core, and less lateral gas/steam escape to the atmosphere and thus less heat loss, which can result in higher end-temperature inside the mat [20].

Since MOR and MOE are related to the resistance of particleboards surface layers and IB strength and TS are more related to the bonds between the resin and particles in core layers, the MOR and MOE reductions at 12 min pressing time are higher than IB strength and TS reductions in UF bonded straw particleboards.

Due to high compression ratio (board density/raw material density) in straw particle mats and also small size of straw particles that were used in producing particleboards, the differences between MOR and MOE in AESO bonded boards and UF bonded boards are not as much as the differences in IB strength and TS.

Table 7.
AESO bonded straw particleboard in comparison with EN standards for wood particleboard

TS (24 h) (%)	MOE (MPa)	MOR (MPa)	IB strength (MPa)	
15.54	2447	18.8	0.55	AESO bonded straw particleboard
16	2300	17	0.4	EN 312-p4
–	–	12.5	0.28	EN 312-p2

The strong bonds between AESO and straw particles due to their compatibility are an important reason for improvements in all AESO bonded straw particleboards.

3.6. Density

Despite 0.7 g/cm^3 nominal density, there is a significant difference in density between AESO and UF bonded boards. Due to the viscoelastic property of AESO resin, there is a springback about 1 mm in manufactured AESO bonded boards; this leads to a reduction in AESO boards density due to increase of volume.

Despite the positive effect of density on physical and mechanical properties, AESO bonded boards with lower density show better characteristics than UF bonded boards. This shows the high strength per unit weight of AESO boards; this is due to strong bonds between AESO and straw particles.

According to Table 6, the best treatment is a straw particleboard made of 13% AESO resin content at 12 min pressing time. In Table 7 there is a comparison between our best treatment and EN standards for wood particleboards (EN 312-p2: "Requirements for general purpose boards for use in dry conditions", EN 312-p4: "Requirements for load-bearing boards for use in dry conditions"). This table shows that wheat straw particleboards with 13% AESO content manufactured at 12 min pressing time can totally compete with wood particleboard and can be a good bio-based substitute for it, and also without any formaldehyde emission.

4. Conclusion

Oil based resins such as AESO are a good choice for making straw particleboards. Wheat straw particleboard bonded with AESO resin has high mechanical and physical properties in comparison to UF bonded straw particleboard. It can be a substitute for wood particleboard; this product is a good solution for countries with little or no forest resources but have considerable agricultural waste. AESO and wheat straw as substitutes for UF and wood, respectively, can decrease the consumption of petroleum and also wood resources.

Acknowledgements

The authors would like to thank the Wood and Paper Laboratory staff, Gorgan Agricultural Sciences and Natural Resources, Iran and also Sarah Nikraftar.

References

1. B. Nicolas, G. Elbez and U. Schonfeld, *J. Wood Sci.* **50**, 230–235 (2004).
2. G. Han, K. Umemura, M. Zhang, T. Honda and S. Kawai, *J. Wood Sci.* **47**, 350–355 (2001).
3. Y. Hui, L. Ruigang, S. Dawa, W. Zhonghua and H. Yong, *Carbohydrate Polymers* **7**, 35 (2007).
4. Y. Zhang, X. Lu, A. Pizzi and L. Delmotte, *Holz als Roh und Werkstoff* **61**, 49–54 (2003).
5. G. Han, K. Umemura, E. D. Wong, M. Zhang and S. Kawai, *J. Wood Sci.* **47**, 18–23 (2001).
6. G. Han, C. Zhang, D. Zhang, K. Umemura and S. Kawai, *J. Wood Sci.* **44**, 282–286 (1998).
7. R. P. Wool and X. S. Sun, *Bio-Based Polymers and Composites*. Elsevier Academic Press 30 Corporate Drive, Suite 400, Burlington, MA 01803, USA (2005).
8. S. F. Guner, Y. Yagci and A. Tuncer, *Prog. Polym. Sci.* **31**, 633–670 (2006).
9. B. Laetitia and R. P. Wool, *J. Appl. Polym. Sci.* **105**, 1042–1052 (2007).
10. M. A. Dweib, B. Shenton, H. W. Hu and R. P. Wool, *Composite Structures* **74**, 379–388 (2006).
11. K. H. Chang and R. P. Wool, *J. Appl. Polym. Sci.* **95**, 1524–1538 (2005).
12. S. Khot, J. Lascala, E. Can, S. S. Morye, G. I. Williams, G. Palmese, S. Kusefoglu and R. P. Wool, *J. Appl. Polym. Sci.* **82**, 703–723 (2001).
13. W. Thielemans, E. Can, S. S. Morye and R. P. Wool, *J. Appl. Polym. Sci.* **83**, 323–331 (2002).
14. W. Thielemans, I. M. McAninch, V. Barron, J. W. Blau and R. P. Wool, *J. Appl. Polym. Sci.* **98**, 1325–1338 (2005).
15. R. P. Wool, S. Kusefoglu, G. Palmese, S. Khot and R. Zhao, US Patent 6,121,398 (2000).
16. K. Eckwert, L. Jeromin and A. Meffert, US Patent 4,647,678 (1987).
17. H. A. J. Aerts and P. A. Jacobs, *J. Am. Oil Chem. Soc.* **81**, 841–846 (2004).
18. J. Lascala and R. P. Wool, *J. Am. Oil Chem. Soc.* **79**, 1–7 (2002).
19. L. Jue, K. Shrikant and R. P. Wool, *Polymer* **46**, 71–80 (2005).
20. C. Dai, W. Wasylciw and J. Jin, *Wood Sci. Technol.* **38**, 529–537 (2004).

Gluten Protein Adhesives for Wood Panels

H. Lei [a,b], **A. Pizzi** [a,*], **P. Navarrete** [a], **S. Rigolet** [c], **A. Redl** [d] **and A. Wagner** [e]

[a] ENSTIB-LERMAB, Nancy University, 27 rue du Merle Blanc, B.P. 1041, 88051 Epinal, France
[b] Southwest Forestry University, Bai Long Si Road, 650224 Kunming, Yunnan, P. R. China
[c] ICSI, University of Haute Alsace, 15 rue Jean Starcky, 68057 Mulhouse, France
[d] Syral, Burchstraat 10, 9300 Aalst, Belgium
[e] Syral, Zone Industrielle Portuaire, 67390 Marckolsheim, France

Abstract
Adhesives for wood particleboards based on hydroxymethylated or glyoxalated hydrolysed gluten protein with addition of an equal proportion of tannin/hexamine resin or without any addition of synthetic resins are shown to yield results satisfying the relevant standard specifications for interior wood boards. Even better are resins based on hydroxymethylated or glyoxalated hydrolysed gluten protein to which has been added a small proportion of an isocyanate (pMDI). Adhesive resins formulations based on the latter in which the total content of natural material was either 70% or 80% of the total resin solids content gave good results. Resins based on the former in which the natural material content was approximately 90–95% also gave good results. The 30/70 pMDI/hydroxymethylated lignin adhesive formulation was tried at progressively shorter pressing times yielding internal bond strength results acceptable for relevant interior grade standards at industrially significant pressing times.

Keywords
Wood adhesives, gluten, NMR, wood, particleboard, panel products

1. Introduction

Recent regulations and environmental legislation have put pressure to stop the use of traditional synthetic adhesives for wood panels. Higher costs of oil from which traditional adhesives are derived have also contributed to this trend. As a consequence, the interest in natural adhesives for wood panels has intensified. This has translated in the revival of the traditional approach of partially substituting in the formulations a small proportion of the synthetic adhesive with a natural material [1]. This approach is rather old and is revived now to conserve traditional synthetic adhesives that might well not be available in the future. Its main disadvantages are

---

* To whom correspondence should be addressed. Tel.: (+33) 329296117, (+33) 329296138; e-mail: antonio.pizzi@enstib.uhp-nancy.fr

(i) still the presence of too high a proportion of synthetic oil-derived polymers in the adhesive formulation, with its consequences on costs, and (ii) the negative environmental impact of the presence of a high proportion of synthetic polymers. This approach, as simple as it may first appear to be, is not a short, or medium or long term solution to the problems being faced now by the wood panels industry.

Wood panel adhesives predominantly, or even exclusively, composed of natural materials already exist. Some of these have now been in commercial/industrial use in a few countries for a rather long time. Among these it is sufficient to briefly mention tannin adhesives [1–3], commercial since the early 1970s, variations on tannin/lignin adhesives that are coming to the market at present [4, 5], some types of lignin adhesives [6–9] and different adhesives based on soy protein flour and hydrolysates [10–13] also on the way to early industrialisation. In all these adhesives, synthetic polymers are present in very small amounts or are totally absent.

Gluten is a protein derived from wheat [14, 15]. The amide groups of the peptide links in gluten proteins are capable to react with aldehydes in the same manner as other amides, e.g., urea [1]. Amine groups in lysine and arginine (aminoacids in which gluten is rich) [14] also present even more marked reactivity with aldehydes, similar to those presented by melamine [1] and phenols [1].

As human allergy to flour products containing gluten is rather widespread, gluten is extracted in great quantities from wheat flour to prepare wheat meal and other gluten-free products. Great quantities of gluten are thus available for other applications. Gluten adhesives for wood panels have already been prepared. However, they need to be applied in powder form to the panel furnish. This is, in general, unsuitable for the industrial application in wood particleboard factories where the equipment can only apply liquid adhesives [16].

This paper thus deals with different approaches to the formulation of liquid wood adhesives predominantly based on gluten for wood particleboards and other particulate wood panels, in an easily applied liquid form.

2. Experimental

The three types of gluten used in this study were donated by Syral (Aalst, Belgium). Two of the gluten types used were gluten protein powder hydrolysates, one produced by chemical and the other by enzymatic hydrolysis, while the third sample was gluten flour. The gluten protein powder hydrolysate obtained by chemical hydrolysis was of the type PX705, which is a partially hydrolyzed functional wheat protein for food application. It appears as a fine, slightly yellowish powder. It contains 86 wt% protein and 4.9% water. Its solubility at pH 7.0 is 96%. This material was used throughout the experiments. The resin preparation procedures described were applied to all three types of gluten.

2.1. Resin Preparation

A gluten-formaldehyde resin, a gluten-glyoxal resin, and gluten that was treated with alkali without forming a resin were prepared.

2.1.1. Gluten-Formaldehyde Resin and NaOH-Treated Gluten

In a three-neck reaction vesssel equipped with a mechanical stirrer, a thermometer and a condenser were charged water (709 g), NaOH (28 g), a phase transfer agent such as ethylene glycol (5.3 g), and silicone oil (10 drops) and the mixture heated to 70°C under continuous mechanical stirring. Gluten protein powder hydrolysate (350 g) was then added to the rapidly stirring solution. The mixture was then heated to 90°C over 15 min, under rapid mechanical stirring, and held between 88°C and 92°C for 1 h. This mix constituted the alkali-treated gluten (GB) used in the formulations as indicated in the tables.

To GB, formaldehyde as a 37% formalin solution (134 g) was added in the hot reaction mixture over a 5 min period. The mixture was maintained under continuous mechanical stirring between 88°C and 92°C for an additional 55 min. The mixture was then cooled to 35°C in an ice bath. This product constituted the gluten-formaldehyde resin GA used in the formulations as indicated in the tables.

2.1.2. Gluten-Glyoxal Resin

In a three-neck reaction vesssel equipped with a mechanical stirrer, a thermometer and a condenser were charged water (709 g), NaOH (28 g), a phase transfer agent such as ethylene glycol (5.3 g), and silicone oil (10 drops) and the mixture heated to 70°C. Gluten protein powder hydrolysate (350 g) was then added to the rapidly stirring solution. The mixture was then heated to 90°C over 15 min, under rapid mechanical stirring, and held between 88°C and 92°C for 1 h. Glyoxal (40% in water) (239 g) was added to the hot mixture over 5 min. The mixture was maintained under continuous mechanical stirring between 88°C and 92°C for an additional 2 h and 55 min. The mixture was then cooled to 35°C in an ice bath. This product constituted the gluten-glyoxal resin GC used in the adhesive formulations as indicated in the tables.

2.1.3. Tannin, pMDI and Final Resins

Mimosa tannin extract (supplied by Silva, S. Michele Mondovi', Italy) of Stiasny value of 92.2 [17–19] was used. The tannin solution in water was prepared at 45% concentration and its pH adjusted to 10.4 with a 33% NaOH aqueous solution. The high pH was chosen as the hardener used performs best at such a pH [3]. The hardener used was 5% hexamethylenetetramine (hexamine) on tannin solids. This was added to the tannin extract solution as a 30% aqueous solution.

To the tannin/hexamine solution were added the modified GA, GB and GC lignin solutions. The proportion of tannin: modified lignin was 50:50 by weight as shown in Table 1. The pMDI (polymeric 4,4′-diphenylmethane diisocyanate) used was Desmodur 44VK from Bayer. The PF resin used in one of the formulations in Table 1 was prepared according to the procedure already presented [20].

2.2. Thermomechanical Analysis

The resins were tested dynamically by thermomechanical analysis (TMA) on a Mettler 40 apparatus. Triplicate samples of beech wood alone, and of two beech wood

Table 1.

Internal bond (IB) strength results of particleboards prepared with different gluten protein adhesive formulations. GA = gluten-formaldehyde. GB = gluten alkali treated. GC = gluten-glyoxal

		Board density (kg/m^3)	IB strength (MPa)
A	Tannin(6% hexamine)/GA-50/50	692	0.28
B	Tannin(6% hexamine)/GB-50/50	689	0.53
C	Tannin(6% hexamine)/GC-50/50	679	0.34
D	pMDI/PF/GA-25/20/55	663	0.42
E	pMDI/GA-30/70	678	0.71
F	Tannin(6% hexamine)/HA-50/50	680	0.28
H	GA:11% resin loading	–	–
EN 312			⩾0.35

plies each 0.6 mm thick bonded with each adhesive system, for sample dimensions of 21 mm × 6 mm × 1.2 mm were tested in non-isothermal mode between 40°C and 220°C at a heating rate of 10°C/min in three-point bending on a span of 18 mm. A force varying continuosly between 0.1 N, 0.5 N and back to 0.1 N was applied on the specimens with each force cycle of 12 s (6 s/6 s). The classical mechanics relation between force and deflection $E = [L^3/(4bh^3)][\Delta F/(\Delta f)]$ (where L is the sample length, b and h are the sample width and thickness, ΔF is the variation of the force applied, and Δf is the deflection obtained) allows the calculation of the modulus of elasticity (MOE) E for each case tested.

2.3. Solid State NMR

GA, GB and GC resins where examined by freeze drying without curing, while the GA + pMDI reaction product was first hardened and then examined. Solid state CP–MAS (cross-polarisation/magic angle spinning) ^{13}C-NMR spectra were recorded on a Bruker Avance II MSL 300 MHz spectrometer at a frequency of 75.47 MHz. Chemical shifts were calculated relative to tetramethylsilane (TMS). The rotor was spun at 12 kHz on a 4 mm Bruker probe. The spectra were acquired with 5 s recycle delays, a 90° pulse of 5 μs and a contact time of 1 ms. The number of transients was 5000.

2.4. Particleboard Preparation and Testing

Duplicate one-layer laboratory particleboards of dimensions 350 mm × 310 mm × 16 mm were prepared by adding 10% total resin solids on dry wood particles. The boards were pressed at a maximum pressure of 28 kg/cm² followed by a decrease in pressure first to 15 kg/cm² and last to 5 kg/cm², at a press temperature of 195°C and for a total press time of 7.5 min. Shorter press times were also used (see Table 3). The target average density for all the panels was 680–700 kg/m³. The panels, after light surface sanding, were tested for dry internal bond (I.B.) strength according to European Norm EN 314 [21].

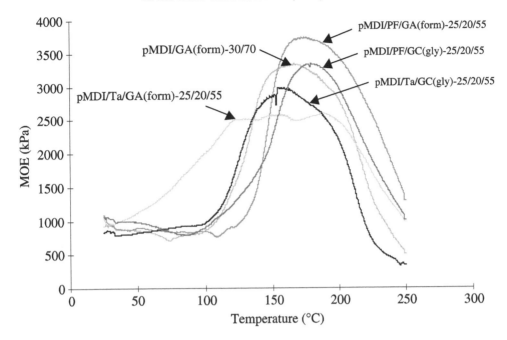

Figure 1. Modulus of elasticity (MOE) variation as a function of temperature obtained by thermomechanical analysis (TMA) of different gluten protein-formaldehyde and gluten protein-glyoxal adhesive formulations. Gluten protein hydrolysate produced chemically (G). GA(form) = gluten reacted with formaldehyde. GC(gly) = gluten reacted with glyoxal. Ta = tannin.

3. Results and Discussion

The thermomechanical analysis results for the PX705 gluten protein powder hydrolysate (prepared by chemical route) are shown in Fig. 1. All the formulations appear to give good results but some of them perform better. Of these the pMDI/PF/GA formulation, pMDI/PF/GC formulation and the pMDI/GA formulation appear to give the highest possible modulus of elasticity (MOE) of the joints between 3300 and 3600 MPa. However, the pMDI/GA 30/70 formulation appears to have two advantages over the other two formulations: (i) the relative proportion of gluten, hence of natural material, is markedly higher, and (ii) the rise of the curve on curing is shifted to lower temperature indicating this to be a faster curing resin, hence more suited for particleboard application. The other two formulations in which the synthetic PF resin is substituted with a tannin(+ hexamine hardener) yield a lower strength of the joint. However, they are as fast curing (the pMDI/Ta/GC formulation) or considerably faster curing (the pMDI/Ta/GA formulation) than the pMDI/GA formulation. Thus, they also appear to be suitable for fast pressing particleboard application, although their maximum possible strength may not be up to standard.

Tannin/gluten resins without any isocyanate added were also tried. The gluten protein powder hydrolysate (prepared by enzymatic route) was used as this did

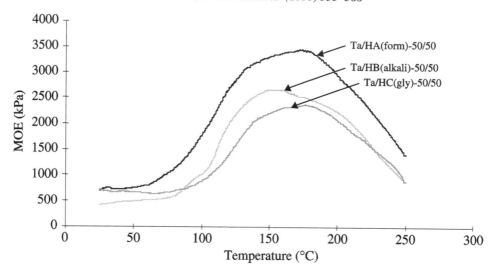

Figure 2. Modulus of elasticity (MOE) variation as a function of temperature obtained by thermome-chanical analysis (TMA) of different gluten protein-formaldehyde and gluten protein-glyoxal adhesive formulations. Gluten protein hydrolysate produced enzimatically (H). HA(form) = gluten reacted with formaldehyde. HB(alkali) = gluten pretreated with alkali. HC(gly) = gluten reacted with glyoxal. Ta = tannin.

not appear to be much different from the chemically hydrolysed one. The ther-momechanical analysis results are shown in Fig. 2. While the maximum possible stength results appear to indicate that the tannin/gluten-formaldehyde resin yields the best results of the three tested; if compared to the results of the other formu-lations in Fig. 1 it is clear that all these formulations are slower-curing than those tested in Fig. 1. For completeness, the more interesting formulations of Figs 1 and 2 were tested with the non-hydrolysed gluten flour. The results obtained are shown in Fig. 3. The faster curing formulations appear again to be the pMDI/GA 30/70, and the pMDI/Ta/GA 25/20/55. However, notwithstanding such promising results from gluten flour, its problem is that it is not soluble hence it is rather difficult to handle and apply with the equipment generally available in particleboard facto-ries.

The results of the TMA tests were confirmed by preparing and testing laboratory particleboard panels. Thus, in Table 1 are shown the results of laboratory particle-board panels prepared using a number of different formulations. Of the different formulations tried, at the very long press time of 7.5 min at 195°C, only three are able to satisfy the dry IB (internal bond) strength requirement of >0.35 MPa of the relevant European Norm EN 312 [21] for interior grade panels. These are:

(i) The pMDI/GA 30/70 which presents an IB strength of 0.71 MPa which is double of that required by EN 312 [21]. This implies that the pMDI/GC 30/70 would be suitable and with no formaldehyde, thus with no emission;

(ii) The tannin + hexamine/GB 50/50;

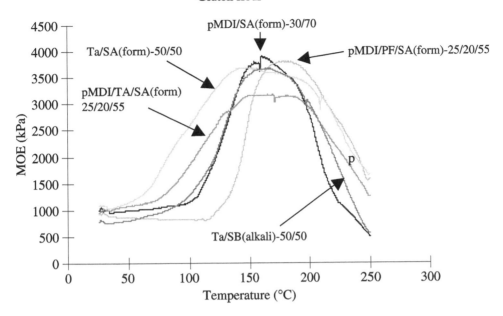

Gluten flour

Figure 3. Modulus of elasticity (MOE) variation as a function of temperature obtained by thermomechanical analysis (TMA) of different gluten protein-formaldehyde and gluten protein-glyoxal adhesive formulations. Gluten flour (S). SA(form) = gluten reacted with formaldehyde. SB(alkali) = gluten pretreated with alkali. Ta = tannin.

(iii) The pMDI/PF/GA 25/20/55, the results of which would be better than the 0.42 MPa in Table 1 if the average board density is close to the 680–700 kg/m^3 considered as industrially acceptable. The results of this last formulation imply from Figs 1 and 3 that the pMDI/Ta/GA 25/20/55 would also satisfy the EN 312 [21] requirements.

As pMDI is relatively expensive and is the only synthetic material present in the best formulation a series of particleboards in which the proportion of pMDI was reduced in relation to the gluten (both GA and GC) were prepared and tested. In Table 2 are shown the results of the pMDI/GA and pMDI/GC formulations when progressively decreasing the proportion of pMDI from 30% to 20% and lower. The results in Table 2 indicate that while IB strength decreases with a decreasing proportion of pMDI it is still possible to decrease the pMDI proportion down to 20% while still comfortably satisfying the dry IB strength requirements. However, at pMDI proportions lower than 20%, starting at 15% pMDI on total adhesive formulation the results no more satisfy the relevant interior grade requirements of the standard. This is true for both the pMDI/GA and pMDI/GC formulations, the latter of these being favorite if a total absence of formaldehyde is required. The pMDI/GB 20/80 formulation gives such poor particleboards that they have no strength (Table 2). This is expected as the reaction of pMDI with any aldehyde-based resin has been shown to be due to the reaction of the isocyanate group with the methylol group introduced in the resin by the previous reaction with an aldehyde [6]. In GB this group is not present and thus the presence of pMDI is of no benefit at all.

As the press time used is one of the most important parameters in particleboard manufacture, the shorter it is the more acceptable is the adhesive, so a series of particleboards prepared by using progressively shorter press times were tested. Thus, in Table 3 the press time of the pMDI/GA 30/70 formulation was progressively decreased from the long 7.5 min down to 5.5, 4.5, 3.5 and 2.5 min at 195°C. The results in Table 3 indicate that even at 2.5 min press time (10.7 s/mm board thickness) and at a board density on the low side, the results are well above the requirements

Table 2.

Internal bond (IB) strength results of particleboards prepared with different gluten protein adhesive formulations. Effect of the decrease in pMDI proportion

		Board density (kg/m^3)	IB strength (MPa)
A	pMDI/GA-30/70	661	0.71
B	pMDI/GA-25/75	659	0.69
C	pMDI/GA-20/80	670	0.45
D	pMDI/GC-30/70	664	0.61
E	pMDI/GC-25/75	659	0.51
F	pMDI/GC-20/80	659	0.45
	pMDI/GB-20/80	–	–
G	pMDI/GC-15/85	669	0.32
H	pMDI/GC-10/90	666	0.15
EN 312			⩾0.35

Table 3.

Internal bond (IB) strength results of particleboards prepared with a gluten protein/formaldehyde adhesive formulation. Effect of the decrease of press time

	Press time (min)	Board density (kg/m^3)	IB strength (MPa)
A	7.5 min, pMDI/GA-30/70	661	0.60
B	5.5 min, pMDI/GA-30/70	659	0.62
C	4.5 min, pMDI/GA-30/70	670	0.58
	4.5 min, pMDI/GA-20/80	672	0.30
	4.5 min, pMDI/GA-15/85	685	0.16
D	3.5 min, pMDI/GA-30/70	684	0.58
E	2.5 min, pMDI/GA-30/70	662	0.51
F	7.5 min, pMDI/GA-30/70, +10%(NH$_4$)2SO$_4$	667	0.43
EN 312			⩾0.35

of the relevant interior grade boards standard [21]. A press time of 10.7 s/mm is still long, and just on the border of the industrially useful range for interior panels, where press times between 3 and 6 s/mm are the norm today. However, this was achieved at 195°C while modern presses function at the much higher temperature of 220°–230°C. Thus, the effective press time of such a formulation will be much shorter than that achieved in the laboratory and will be in the industrially useful press time range. However, lowering the proportion of pMDI to 20% decreases already at 4.5 min press time the IB strength to values lower than that required (Table 3). Thus, the conclusion is that if industrially viable press times on modern presses are required, the pMDI/GA and pMDI/GC 30/70 formulations are the most convenient to use. Only when longer press times are allowed then one can decrease the proportion of pMDI in the adhesive.

To better understand what occurs in the reaction of gluten with an aldehyde, the resins were examined by solid state CP–MAS ^{13}C-NMR analysis. GA, GB and GC were freeze dried without curing, while the GA + pMDI reaction product was first hardened and then examined (Figs 4–7). The spectrum of GB is not reported as it

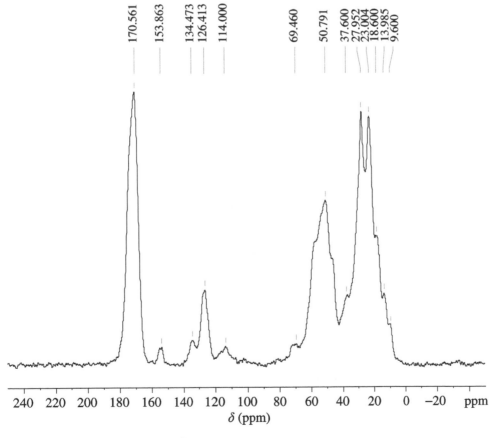

Figure 4. CP–MAS ^{13}C-NMR spectrum of glutein protein hydrolysate.

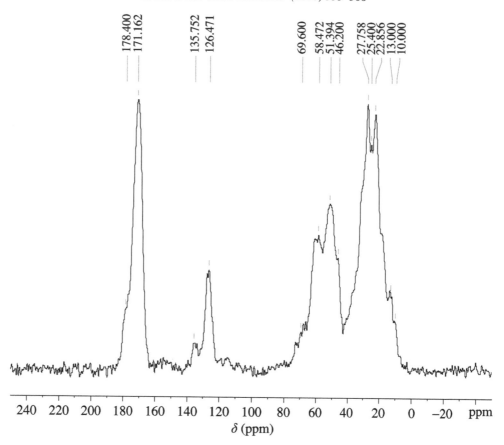

Figure 5. CP–MAS ^{13}C-NMR spectrum of glutein protein hydrolysate-formaldehyde reaction product.

is practically identical to that of the unmodified gluten hydrolysate. Comparing the unmodified gluten hydrolysate (Fig. 4) with GA (Fig. 5) the most evident features differentiating the two are the appearance of a definite peak at 58.5–59 ppm which is due to the formation of the methylol group as a consequence of the reaction of formaldehyde with the protein [1, 22–24]. As the reaction was carried out under alkaline conditions there is no evidence of methylene groups in the 35 to 45 ppm range [23, 24], as it would be expected. However, their presence could be masked by the dominant peaks belonging to the protein in this same region of the spectrum. Some condensation has, however, occurred as can be noticed by the increase in the 69.5–70 ppm band corresponding to the appearance of methylene ether bridges ($-CH_2OCH_2-$) [23, 24] and is even more evident in the spectrum of GC (Fig. 6) as the 57.5–58.5 peak, although appearing quite clearly, is less pronounced than for GA (Fig. 5). This is also expected as the reactivity of formaldehyde is about five time higher than that of glyoxal [24]. A third feature of interest is the disappearance of the unmodified gluten band at 153.8 ppm (GA and GC) confirmed by the disappearance (GA) or marked decrease (GC) of the 113–114 ppm band. These small

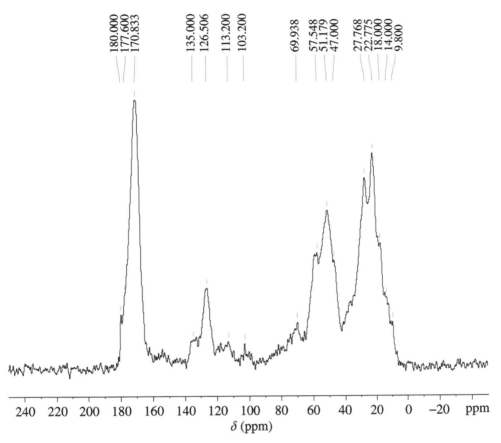

Figure 6. CP–MAS ^{13}C-NMR spectrum of glutein protein hydrolysate-glyoxal reaction product.

bands indicate reaction with the aldehyde of the few aromatic rings present in the protein.

The most interesting NMR spectrum is that of the pMDI/GA resin after hardening (Fig. 7). The features of this are the dominant peaks at 117, 126.8 and 134 ppm [23, 24]. These are all due to the pMDI used, but the most important of these is the 126–127 ppm peak that corresponds to the unreacted isocyanate group (–N=C=O) [23, 24]. The 134 ppm peak belongs to the aromatic carbon of pMDI reacted with the secondary amide group of the ureas obtained by the reaction of pMDI with water. The peak at 117 ppm indicates the presence of the relatively unstable carbamic acid [23, 24], its presence indicating the reaction between isocyanate and methylolated protein is diffusion-controlled, hence limited, possibly due to the too high viscosity of the system. Lower viscosity would favour cross-linking. Thus, even after curing there appears to be a marked proportion of unreacted isocyanate (126 ppm). This might be due to a too fast immobilisation of the network during hardening. This confirms the TMA analysis showing that low proportions of pMDI can be used in the adhesive. The peak at 152 ppm in Fig. 7 is due to the carbonyl

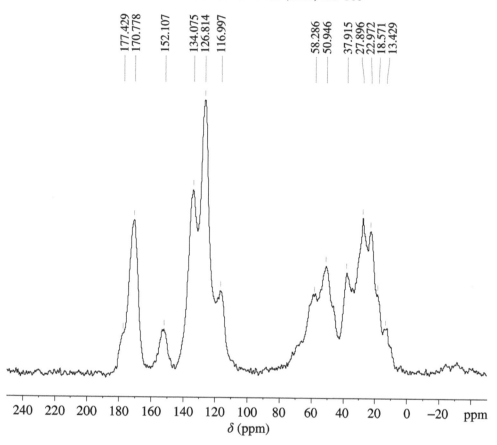

Figure 7. CP–MAS ^{13}C-NMR spectrum of the reaction product of pMDI with glutein protein hydrolysate-formaldehyde intermediate.

groups from the polyureas, biuret, urethane and intermediate carbamic acid produced by the reaction of the isocyanate with both methylolated protein hydrolysate as well as with water. Two further peaks in Fig. 7 are of interest: The higher peak at 51 ppm and the appearance of the peak at 37.9 ppm. The latter is due to the –CH$_2$-bridge between the aromatic rings of the pMDI [23, 24]. The former is the –CH$_2$-that belonged to the methylol group on the gluten that has now reacted with the isocyanate to give a urethane bridge (R–NHCOO–CH$_2$–R′). The increase of this peak shows that the urethane cross-links have effectively formed and that the modified gluten has indeed, at least partially, reacted and cross-linked with the isocyanate.

4. Conclusion

Adhesives for wood particleboards were obtained giving results satisfying the relevant standard specifications for interior grade wood boards and the following were found:

(1) Hydrolysed gluten protein reacted with formaldehyde or with glyoxal to which was added an equal proportion of tannin/hexamine resin and without any addition of synthetic resins. The natural material content was approximately 90–95%.

(2) Hydrolysed gluten protein reacted with formaldehyde or with glyoxal to which was added a smaller proportion (20–30%) of an isocyanate (pMDI). The natural material content was approximately 70–80%.

(3) Shorter, industrially significant press times were obtained for the 30/70 pMDI/hydroxymethylated gluten formulation.

Compared to previous gluten based adhesives the present resins are applicable in liquid form, thus without any need for modifications of the application systems in particleboard factories. Relative to other protein adhesives such as soy-based adhesives based on reaction with formaldehyde some of the resins presented here have several advantages: (i) they cannot and do not produce any aldehyde emission as neither formaldehyde nor any other volatile aldehyde was used in some of the formulations; (ii) the percentage of natural materials was increased up to 70% for one type of formulation and up to 95% for others. Furthermore, in relation to resin formulations based on different cross-linking reactions other than those with formaldehyde the resins presented here have other advantages: they are competitive with alternate natural resin systems such as those based exclusively on tannins and/or lignins.

Acknowledgements

The Chinese author thanks the Natural Science Foundation of China 30800870 for the financial support and the Embassy of France for a bursary.

The French authors gratefully acknowledge the financial support of the CPER 2007–2013 "Structuration du Pôle de Compétitivité Fibres Grand'Est" (Competitiveness Fibre Cluster), through local (Conseil Général des Vosges), regional (Région Lorraine), national (DRRT and FNADT) and European (FEDER) funds.

References

1. A. Pizzi (Ed.), *Wood Adhesives: Chemistry and Technology*. Marcel Dekker, New York, NY (1983).
2. A. Pizzi, *Forest Prod. J.* **28**, 42–48 (1978).
3. F. Pichelin, M. Nakatani, A. Pizzi, S. Wieland, A. Despres and S. Rigolet, *Forest Prod. J.* **56**, 31–36 (2006).
4. H. R. Mansouri, P. Navarrete, A. Pizzi, S. Tapin-Lingua, B. Benjelloun-Mlayah and S. Rigolet, *European J. Wood Technol.*, in press.
5. A. Pizzi, R. Kueny, F. Lecoanet, B. Massetau, D. Carpentier, A. Krebs, F. Loiseau, S. Molina and M. Ragoubi, *Ind. Crops & Prod.* **30**, 235–240 (2009).
6. A. Pizzi and T. Walton, *Holzforschung* **46**, 541–547 (1992).

7. A. Pizzi, J. Valenzuela and C. Westermeyer, *Holzforschung* **47**, 69–72 (1993).

8. N.-E. El-Mansouri, A. Pizzi and J. Salvado', *J. Appl. Polym. Sci.* **103**, 1690–1699 (2007).

9. N.-E. El-Mansouri, A. Pizzi and J. Salvado', *Holz Roh Werkstoff* **65**, 65–70 (2007).

10. Y. Liu and K. Li, *Int. J. Adhesion Adhesives* **27**, 59–67 (2007).

11. X. Geng and K. Li, *J. Adhesion Sci. Technol.* **20**, 847–858 (2006).

12. G. A. Amaral-Labat, A. Pizzi, A. R. Goncalves, A. Celzard and S. Rigolet, *J. Appl. Polym. Sci.* **108**, 624–632 (2008).

13. J. M. Wescott, C. R. Frihart and L. Lorenz, *Proceedings Wood Adhesives 2005*. Forest Products Society, Madison, Wisconsin (2006).

14. P. R. Shewry, N. G. Halford, P. S. Belton and A. Thatham, *Phil. Trans. R. Soc. Lond. B* **357**, 133–142 (2002).

15. M. P. Lindsay and J. H. Skerritt, *Trends Food Sci. Technol.* **10**, 247–253 (1999).

16. A. Redl, European Patent EP 1925641 (A2) assigned to Syral Belgium NV (2008).

17. E. Stiasny, *Der Gerber*, Issue 740, 186; Issue 775, 347 (1905).

18. W. E. Hillis and G. Urbach, *J. Appl. Chem.* **9**, 665–673 (1959).

19. L. Suomi-Lindberg, *Paperi Ja Puu* **67**, 65–69 (1985).

20. C. Zhao, A. Pizzi and S. Garnier, *J. Appl. Polym. Sci.* **74**, 359–378 (1999).

21. European Norm EN 312, Wood particleboard — specifications (1995).

22. A. Pizzi, *J. Adhesion Sci. Technol.* **20**, 829–846 (2006).

23. A. Despres, A. Pizzi and L. Delmotte, *J. Appl. Polym. Sci.* **99**, 589–596 (2006).

24. S. Wieland, A. Pizzi, S. Hill, W. Grigsby and F. Pichelin, *J. Appl. Polym. Sci.* **100**, 1624–1632 (2006).

25. A. Ballerini, A. Despres and A. Pizzi, *Holz Roh Werkstoff* **6**, 477–478 (2005).

Wood Panel Adhesives from Low Molecular Mass Lignin and Tannin without Synthetic Resins

P. Navarrete [a], **H. R. Mansouri** [b], **A. Pizzi** [a,*], **S. Tapin-Lingua** [c],
B. Benjelloun-Mlayah [d], **H. Pasch** [e] **and S. Rigolet** [f]

[a] ENSTIB-LERMAB, Nancy Université, 27 rue du Merle Blanc, BP 1041, 88051 Epinal, France
[b] Department of Wood Science and Technology, Faculty of Natural Resources,
University of Zabol, P. O. Box 98615-538 Zabol, Iran
[c] FCBA, Centre Technologique Forêt, Cellulose, Bois, Ameublement,
Domaine Universitaire, BP251, 38004 Grenoble, France
[d] CIMV, Compagnie Industrielle de la Matiere Vegetale, 134 rue Danton,
92300 Levallois-Perret, France
[e] Department of Chemistry and Polymer Science, University of Stellenbosch,
Private Bag X1, 7602 Matieland, Stellenbosch, South Africa
[f] Matériaux à Porosité Contrôlée, Institut de Science des Matériaux de Mulhouse (IS2M),
LRC CNRS 7228, University of Haute Alsace, rue Jean Starcky, 68057 Mulhouse, France

Abstract

Mixed interior wood panel tannin adhesive formulations were developed in which lignin was in considerable proportion, 50%, of the wood panel binder, and in which no 'fortification' with synthetic resins, such as isocyanates or phenol-formaldehyde resins as used in the past, was necessary to obtain results satisfying relevant standards. A low molecular mass lignin obtained industrially by formic acid/acetic acid pulping of wheat straw was used. Environment-friendly, non-toxic polymeric materials of natural origin constitute up to 94% of the total panel binder. The wood panel itself is constituted of 99.5% natural materials, the balance 0.5% being composed of glyoxal, a non-toxic and non-volatile aldehyde, and of hexamine already accepted as a non-formaldehyde-yielding compound when in presence of very reactive chemical species such as a flavonoid tannin. Particleboards and two types of plywoods were shown to pass the relevant interior standards with such adhesive formulations. Moreover, the much cheaper non-purified organosolv lignin showed the same level of results as the more expensive purified type.

Keywords

Wood adhesives, tannin, lignin, particleboard, plywood, wood panels

---

* To whom correspondence should be addressed. Tel.: +33-329296117; Fax: +33-329296138; e-mail: antonio.pizzi@enstib.uhp-nancy.fr

Wood Adhesives
© Koninklijke Brill NV, Leiden, 2010

1. Introduction

Biocomposites have been an area of growing interest and a subject of active research for quite some time. This is due to both environmental concerns as well as anticipated future scarcity of oil and oil-derived products. A class of composites in which resins of natural origin have already had a commercial/industrial impact is in the field of rigid wood panels, such as particleboards. Natural-origin resins have already been used commercially for the last 30 years for these wood panels, and their use is still growing, although still relatively slowly [1]. In such an application, the binder is never more than 10% by weight of the whole composite panel. This is sufficient to conform to the performance and costs required by the wood panels industry and their respective product standards.

The main natural resins used as wood panel binders are vegetal tannin adhesives, lignin adhesives and more recently also soy protein adhesives [1]. Of these, tannin-based adhesives have been used commercially the longest, since 1971. They offer the advantage over the other two types of not needing any reinforcement with an oil-derived synthetic resin [1]. Lignin [2–5] and soy binders [1, 6–8], however, still require between 20% and 40% of the total resin to be either phenol-formaldehyde or most often PMDI (polymeric isocyanate) to satisfy the requirements of relevant board standards.

The use of synthetic resins limits somewhat the environmental attractiveness of such adhesives based on purely natural materials, while the use of tannins alone is limited at present by the relatively limited supply of these materials [1]. Thus, the aim is to prepare an adhesive based on materials of natural origin, satisfying international standards for both performance and emission, which does not emit or even better does not contain any formaldehyde, the composition of which does not include any synthetic resins, and that is less costly and uses widely available materials. This will render wood panel adhesives based on natural materials much more acceptable both economically and environmentally.

This paper, thus, deals with the preparation of lower cost wood panel adhesives with good performance based on natural materials and needing no fortification by any synthetic resins.

2. Experimental

2.1. Preparation of Glyoxalated Lignin Resin

The lignin used was a low molecular mass lignin obtained industrially as a byproduct of pulp and paper production from wheat straw by an acetic acid-based organosolv process [9]. Two types of lignins were used, one washed and purified and the other without washing. In this article we call this second type 'impure' lignin. These two lignins were supplied by their industrial manufacturer, CIMV (Compagnie Industrielle de la Matiere Vegetale, Reims Factory, France). The characteristics of these two lignins were determined by the FCBA (Table 1) and by CIMV (Table 2).

Table 1.
Characteristics of wheat straw organosolv lignin

	Formyl (mmol/g)	Acetyl (mmol/g)	Free hydroxyls (mmol/g)	Total (mmol/g)
Total –OH groups	0.6	0.5	2.90	4.00
Phenolic –OH groups	0.20	0.05	0.85	1.10
Aliphatic –OH groups	0.40	0.45	2.05	2.90

Table 2.
Additional characteristics of wheat straw organosolv lignin

	Klason lignin Insoluble (%)	Soluble (%)	Total lignin (%)	Ashes (%)	Glucan (%)	Xylan (%)	Mannan (%)	Galactan (%)	Arabinan (%)	Total without glucan (%)
Non-purified	79.3	3.1	82.4	3.49	5.57	5.80	0.15	1.01	1.58	8.54
Purified	82.7	2.22	84.9	1.83	4.49	5.98	0.38	0.97	1.45	8.78

295 parts by mass of this lignin powder (96% solid) were slowly added to 477 parts of water while sodium hydroxide solution (30%) was added from time to time, thus, keeping the pH of the solution between 12 and 12.5 for better dissolution of the lignin powder which was also facilitated by vigorous stirring with an overhead stirrer. A total of 141 parts by mass of 30% sodium hydroxide aqueous solution were added which resulted in a final pH close to 12.5.

A 2-liter flat-bottom flask equipped with a condenser, a thermometer and a magnetic stirrer bar was charged with the above solution and heated to 58°C. 87.5 parts by mass glyoxal (40% in water) were added and the lignin solution was then continuously stirred with a magnetic stirrer/hot plate for 8 h. The solids content for all glyoxalated lignins was around 31%. Glyoxal is a non-volatile non-toxic aldehyde that has been tested in lignin [5, 6], tannin [10] and other adhesives [8] for application to wood panels such as particleboards. Glyoxal is a non-toxic (LD50 rat ≥ 2960 mg/kg; LD50 mouse ≥ 1280 mg/kg) [11], non-volatile aldehyde but less reactive than formaldehyde which is toxic (LD50 rat ≥ 100 mg/kg; LD50 mouse ≥ 42 mg/kg) [12].

2.2. Tannin, Final Resin and Glue-Mix

Mimosa tannin extract (origin: Tanzania. Supplied by: Silva, S. Michele Mondovi', Italy) of Stiasny value of 92.2 [13–16] was used. The tannin solution in water was prepared at 45% concentration and its pH adjusted to 10.4 with 33% water solution of NaOH. The high pH was chosen as the hardener used performs best at such a pH [15, 17–19]. The glue-mix was composed of (i) the tannin extract solution to which was added as hardener 5% hexamethylenetetramine (hexamine) (hexamine solids on tannin solids) as a 30% hexamine solution in water and (ii) the glyoxalated lignin solution. The proportion of tannin solids: glyoxalated lignin solids was 50:50 by weight. Hexamine hardener has now been accepted by JIS A5908 as not being a formaldehyde source in presence of a polyflavonoid condensed tannin [17, 21–23]. The glue mix consisted of the solution of the 50:50 tannin/hexamine: glyoxalated lignin resin, adjusted to pH = 10.0, to which was added 200 mesh olive stone flour (38% on total resin solids) as filler.

   The total amount of natural material on dry resin solids was almost 94% and the adhesive so prepared did not contain any synthetic resins.

2.3. Wood Panels (Particleboards and Plywoods) Manufacture and Testing

Duplicate one-layer laboratory particleboards of dimensions $350 \times 300 \times 14$ mm were prepared using a mixture of core particles of beech (*Fagus sylvatica*) and Norway spruce (*Picea abies*) woods at 28 kg/cm^2 maximum pressure and 190°–195°C press temperature. The wood particles before resin addition had a moisture content of 2%. The resin solids loading on dry wood was maintained at 10% of the total mix of tannin + glyoxalated lignin. The total press time was maintained at 7.5 min. All particleboards were tested for dry internal bond (IB) strength. The IB strength test is a relevant international standard test [24] done on 5 board specimens and is a tension test perpendicular to the plane of the board. Thus, each IB and panel density result in the tables is the average of 10 specimens.

   Three-ply triplicate wood panels of dimensions $500 \times 300 \times 14$ mm and of $400 \times 400 \times 6$ mm were prepared. The former were composed of pine wood (*Pinus sylvestris*) veneers of, respectively, 3.5 mm and 7 mm thicknesses for the two lower plys and of an oak (*Quercus* spp.) veneer of 3.5 mm thickness for the upper veneer giving 14 mm total plywood thickness. These panels were suitable for floor covering. The second type of plywood panels were composed of three 2 mm thick beech (*Fagus sylvatica*) veneers.

   The liquid glue-mix spread used was 380 g/m^2 d.g.l. (double glueline), press temperature was 130°C, and pressure was 15 kg/m^2 for both types of panels. Press time was 5 min for the thicker panel type, this being rather short, and 4 min for the thinner one. These press times were chosen because industrial plywood panels are manufactured with press times calculated as 1 min/mm panel thickness to the furthest glueline from the press hot platen + 2 min. The panels were tested for dry tensile strength according to European Norm EN 314 [25].

2.4. CP-MAS ^{13}C NMR Spectroscopy

The glyoxalated organosolv lignin resins were hardened at 105°C for 2 h in an oven before being ground finely for NMR analysis. The hardened lignin resins were analyzed by solid state CP MAS ^{13}C NMR. Spectra were obtained on a Bruker AVANCE II 400 MHz spectrometer at a frequency of 100.6 MHz and a sample spin of 12 kHz, using a recycling delay time of 1 s and a contact time of 1 ms. Number of transients was about 15 000, and the decoupling field was set at 78 kHz. Chemical shifts were determined relative to tetramethylsilane (TMS) used as control. The spectra were accurate to 1 ppm. The spectra were recorded with suppression of spinning side bands.

2.5. Thermomechanical Analysis (TMA)

The hardening reaction of a single resin system or mixtures can be evaluated by TMA, by monitoring the rigidity of a bonded wood joint as a function of temperature. Thus, glue-mixes of glyoxalated purified and non-purified lignins with tannin/hexamine were analysed by TMA. The compositions of different glue-mixes are given in the tables. All experiments were conducted under the same conditions: heating rate = 10°C/min, 30 mg resin, in the temperature range 25–250°C. The thermomechanical analyzer used was a Mettler Toledo TMA40. The software used for data treatment was STARe. Deflection curves that allow modulus of elasticity (MOE) determination were obtained in the three-point bending TMA mode. The MOE of the wood joints bonded with different resin systems gave a good indication of the final strength of the adhesive system tested.

2.6. Molecular Weight and Molecular Weight Distribution

The number average molecular masses of the two lignins studied were determined by GPC. These were obtained as 1034 and 1028 for the non-purified and purified lignins, respectively. The effluent of the industrial CIMV pilot plant in Reims, France, after washing in water was monitored at 280 nm wavelength with a Beckman UV detector. The column (TSKgel G3000 PWXL) was calibrated with poly(ethylene glycol) in the 138–40 000 g/mol range. The flow rate of the $NaNO_3$ solution (0.1 M at pH 7) was 0.6 ml/min, and the samples were dissolved in $NaNO_3$ at a concentration of 3 mg/ml. The detected signals were digitalized at a frequency of 2 Hz, and the molecular weight distribution was calculated from the recorded signals with normal GPC calculation procedures.

Examination by MALDI-TOF (Matrix assisted laser desorption/ionisation time-of-flight) mass spectrometry showed, however, that GPC determination of the lignin molecular mass produces much higher values than the real values, thus, the values of Mn from GPC result also from molecular associations as already shown for other water soluble polymers [26]. MALDI-TOF results are shown in Fig. 1, indicating that molecular masses only as high as that of tetramers are present and, thus, are lower than those obtained by GPC. The oligomers composing the non-purified and purified lignins are the same and in the same proportion. Their types are listed in Table 3.

The MALDI-TOF spectra were recorded on a KRATOS Kompact MALDI AXIMA TOF 2 instrument. The radiation source was a pulsed nitrogen laser with a wavelength of 337 nm. The time period of one laser pulse was 3 ns. The measurements were carried out using the following conditions: polarity: positive; flight path:

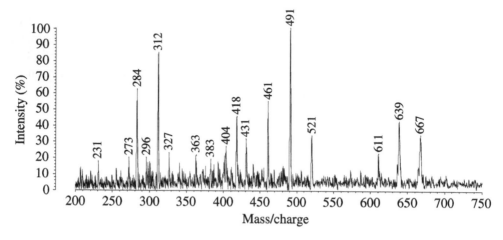

Figure 1. MALDI-TOF mass spectrum in the 200–700 Da range showing distribution of main oligomers in the low mlecular weight organosolv lignin used.

linear; mass range: high (this meant using a high acceleration voltage = 20 kV), and 100–150 pulses per spectrum. The delayed extraction technique was used by applying delay times of 200–800 ns.

For sample preparation, the lignin water solution samples were dissolved in acetone (4 mg/ml, 50/50 vol%) and mixed with an acetone solution (10 mg/ml in acetone) of the matrix. As the matrix 2,5-dihydroxy benzoic acid was used. For the enhancement of ion formation, NaCl was added to the matrix (10 mg/ml in water). The solutions of the sample and the matrix were mixed in the proportion 3 parts matrix solution + 3 parts sample solution + 1 part NaCl solution, and 0.5 to 1 µl of the resulting solution mix were placed on the MALDI target. After evaporation of the solvent the MALDI target was introduced into the spectrometer. The dry droplet sample preparation method was used.

2.7. *Phenolic Hydroxyl Groups Determination by UV Spectroscopy (Δε Method)*

The concentrations of the various phenolic units in the lignin samples were determined by UV spectroscopy, as described by Zakis [27]. This method is based on the difference in the absorptions at 300 and 360 nm between phenolic units in neutral and alkaline solutions. The concentration of ionizing phenolic hydroxyl groups can be quantitatively evaluated by comparing with the respective model compounds with a free ortho position in their aromatic ring.

3. Results and Discussion

Organosolv lignin obtained with a mixed formic and acetic acids pulping technique [9] is bound to present a number of formylated and acetylated phenolic and aliphatic hydroxyl groups. The results in Table 1 confirm this by the evidence of esterification by formyl and acetyl groups mainly of the aliphatic but also of the phenolic

Table 3.
MALDI-TOF derived distribution of oligomers in the organosolv lignin used for the adhesive

	Structure	Calculated MW	Calculated MW + Na	Experimental MW + Na	Peak intensty (%)
S	monomer	210	233	231	18
H-β1-H-1x-OH	dimer	261	284	284	63
H-β1-G G-β1-H	dimers	273	296	296	18
326-1x-CH$_2$– (demethylation)	dimers	289	312	312	85
H-β1-S G-β1-G S-β1-H	dimers	303	326	327	22
H-βO$_4$-S G-βO$_4$-G S-βO$_4$-H	dimers	360	360	383	18
G-$\beta\beta$-G G-β5-G G-55$'$-G	dimers	394.4	417.4	418	43
G-βO$_4$-S G-αO$_4$-S	dimers	407	430	431	32
S-βO$_4$-S S-αO$_4$-S	dimers	437	460	461	54
S-βO$_4$-S + 2x-OH S-αO$_4$-S + 2x-OH	dimers	468	491	491	100
				521	
S-βO$_4$-S-$\beta\beta$-H G-βO$_4$-G-$\beta\beta$-S G-βO$_4$-G-β5-S G-βO$_4$-S-55$'$-G	trimers	585	608	611	21
G-βO$_4$-S-$\beta\beta$-S G-βO$_4$-S-β5-S G-βO$_4$-S-55$'$-S	trimers	615	638	639	40
S-βO$_4$-S-$\beta\beta$-S S-βO$_4$-S-β5-S S-βO$_4$-S-55$'$-S	trimers	644.7	667.7	667	34

G = lignin guaiacyl units; S = lignin siringyl units; H = lignin phenolic units. Linkage between units indicated as βO$_4$, αO$_4$, $\beta\beta$, β5, 55$'$ [30].

hydroxyl groups of lignin. These aliphatic and aromatic formic and acetic acid ester groups are of interest as the presence of esters, even in small amounts, has been
shown to accelerate the reaction and curing of phenolic/aldehyde resins [3, 4, 28]
and this was found also in the case of slow curing lignin-based resins [3, 4].

Their presence in this type of organosolv lignin is, therefore, of some benefit. The proportions of acid soluble and Klason lignin shown in Table 2 indicate that in the case of both the purified and non-purified wheat straw organosolv lignins the level of total lignin is high. Not much differences are seen between the two lignins, and even the values of their number average molecular masses are very similar at 1028 and 1034 g/mol. The percentage distribution of residual sugar oligomers is also similar (Table 2) with glucans and xylans from hemicelluloses and also from cellulose predominating. The major difference being observed is in the percentage of inorganic ashes. In these cases the ashes are the cations of residual acetic and formic acid salts which derive from the precipitation and washing of the residual lignin with an aqueous NaOH solution [9].

A word of caution must be given on the GPC results regarding molecular masses. MALDI-TOF examination of the same two lignins (Fig. 1) showed that the masses of the highest oligomers are lower than those observed by GPC. Associations of molecules by secondary forces often cause molecular aggregates to form. This leads to finding higher mass values, which are incorrect, when using GPC for mass determination [26, 29]. It is nonetheless of interest to observe the type and relative proportion of the oligomers existing in this organosolv lignin as listed in Table 3. From this one can see that dimers and trimers of different types predominate, but that no oligomers of higher masses are present, confirming that the Mn results obtained by GPC are too high and are due to association. As lignin is composed of guaiacyl (G), siringyl (S) and phenolic (H) repeat units, so in Table 3 the interpretation of the compounds detected is indicated as G for guaiacyl units, S for siringyl units, and H for phenolic units not containing methoxy groups. The linkages between them are indicated with the 'classical' nomenclature [30] as βO_4, αO_4, $\beta\beta$, $\beta 5$, $55'$. Thus, as an example, the G-βO_4-S and G-αO_4-S peak at 431 Da in Fig. 1 corresponds to the structure and the H-β1$-$G peak at 296 Da corresponds to the structure

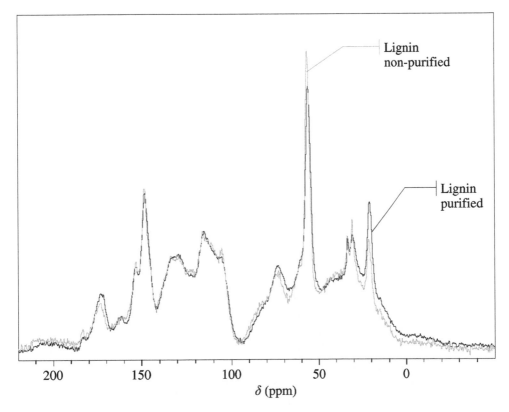

The ^{13}C NMR analysis confirms the similarity of the two lignins. In Fig. 2 the NMR spectra show that the differences between the two lignins exist but they are rather small. Thus, in Fig. 2 the band at 55 ppm characteristic of the methoxy groups ($-OCH_3$) of lignin bound to the aromatic ring is slightly less intense for the purified lignin. As this was purified at 50°–60°C some demethoxylation occurred, a very common reaction in lignin [30]. Equally, the higher relative proportion of

Figure 2. Comparative CP-MAS ^{13}C NMR spectra of purified and non-purified organosolv lignins derived from formic acid/acetic acid wheat straw pulping.

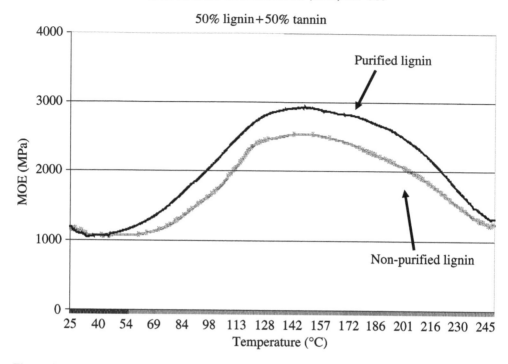

Figure 3. Comparison of thermomechanical analysis curves of beech wood joints bonded with tannin/lignin 50/50 wood adhesive. Variation of the joint MOE as a function of temperature.

acetic/formic acid esters in purified lignin are different as shown by the peaks at 21 and 30 ppm of the carbon of the methyl group $-CH_3$ of, respectively, the acetic acid esters and free acetic acid carboxyl ion, and the less noticeable increase of the C=O peak of the formic acid esters at 161 ppm and of the free formic acid carboxyl ion at 172 ppm.

The results of the thermomechanical analysis indicate that the purified organosolv lignin gives higher Modulus of Elasticity (MOE) values than the non-purified one (Fig. 3). However, the result for non-purified lignin is only lower in appearance. In reality, the impurities present in the non-purified lignin decrease the amount of actual binding material on the beech laminae in the TMA tests. Such a slightly lower amount of actual binding resin yields a lower TMA maximum MOE value for the non-purified lignin, due to the well-known effect of TMA to exaggerate the differences in results [31, 32]. That this is indeed the case is shown by the results of actual wood panels bonded with these adhesives (Tables 4 and 5). The MOE starts to decrease after 170°C in the TMA curves (Fig. 3) due to the start of the degradation of wood constituents, as shown previously [23]. This does not affect board strength as at the very short press times used for panel manufacture the maximum temperature reached in the board core, which determines IB strength, is no higher than 120°C.

The results for wood particleboard prepared with the mixed formulations of tannin/hexamine + glyoxalated lignin are shown in Table 4. The internal bond (IB)

Table 4.
Results on particleboards bonded with tannin/lignin adhesives

	Proportions	Density (kg/m^3)	Dry IB strength (MPa)
Tannin/hexamine + glyoxalated lignin (non-purified)	50/50	702	0.52 ± 0.05
Tannin/hexamine + glyoxalated lignin (purified)	50/50	700	0.45 ± 0.04
Tannin/hexamine + glyoxalated lignin (non-purified)	40/60	710	0.35 ± 0.04
Tannin/hexamine + glyoxalated lignin (non-purified)	30/70	700	0.18 ± 0.03
European Norm EN 314-2 requirement			⩾0.35

IB: Internal bond.

Table 5.
Results on plywoods bonded with tannin/lignin adhesives

	Proportions	Dry tensile strength (MPa)	Wood failure (%)
Beech Plywood			
Control tannin/hexamine	100	1.4 + 0.10	0
Tannin/hexamine + glyoxalated lignin (non-purified)	50/50	1.2 + 0.20	10
Thick floor plywood			
Tannin/hexamine + glyoxalated lignin (non-purified)	50/50	1.0 ± 0.19	46
EN 314-2 requirements		⩾1.0	⩾0
		0.6⩽ and <1.0	⩾40

strength of a panel is a direct measure of the performance of the adhesive. The results in Table 4 indicate that no significant difference exists as regards internal bond (IB) strengths of the panels prepared using the two lignins according to relevant interior panel standards [24]. When the two resins are mixed 50/50 by weight the results are adequate to satisfy international standard requirements for interior grade boards (Table 4). However, a decrease in the relative proportion of tannin/hexamine results in a rapid decrease in the IB strength of the panels (Table 4). As the panel at tannin: lignin = 40:60 has an IB strength equal to the minimum value required by the standard, the indications are that to use formulations in which the amount of tannin is less than about 45% of the total would not be feasible at the industrial level. One consideration that is of particular interest is that the results obtained with the non-purified lignin are comparable and equally good as those of the formulations using the purified one. This is a considerable advantage as the unpurified, unwashed lignin is considerably cheaper to produce.

The results for two types of plywoods, 6 mm thick standard beech plywood, and 14 mm thick flooring type mixed veneers plywood are reported in Table 5. The European Norm EN 314-2 [25] for interior grade plywood specifies that the percentage wood failure and dry tensile strength obtained on testing of the panel be evaluated on a scale. Thus, the percentage wood failure for a dry tensile strength can be 0% for any strength results $\geqslant$ 1.0 MPa. Equally, if the dry tensile strength is higher than 0.6 MPa but lower than 1.0 MPa the percentage wood failure required is $\geqslant$40%. Table 5 shows that both plywoods pass well such standard requirements, and that the tensile strength value of tannin/lignin bonded panel is only slightly lower than that obtained with the tannin alone.

It is of interest to understand the main mechanisms according to which the formulations work. As lignin is of much lower reactivity than a flavonoid tannin, a series of parallel reactions occur to yield coreaction of tannin with lignin. Thus, tannin reacts very rapidly in the hot press with hexamine to form a network based on the known reactions of tannin with hexamine [20–23]. Lignin, much slower, is prereacted in a reactor with glyoxal [1, 2]. As some condensation of glyoxalated lignin occurs during the reaction, so if the reaction is prolonged it will increase unduly the viscosity of the glyoxalated lignin, so condensation needs to be minimized and addition of glyoxal in the lignin maximized. The glyoxalated lignin is, thus, rich in methylol-type groups obtained by glyoxal addition, but the aromatic nuclei of lignin are still too slow for glyoxalated lignin alone to condense to a sufficiently hardened network in just the very brief period the board remains in the hot press. The methylol-type groups obtained by glyoxal addition to lignin are, thus, forced to react with the much more reactive flavonoid tannin, thus, introducing the lignin in the final copolymer network. A simplified scheme explaining in brief the reactions involved is shown in Fig. 4. It must also be pointed out that at pH 10 and higher both A- and B-rings of a flavonoid unit are able to react and cross-link [15, 18], as

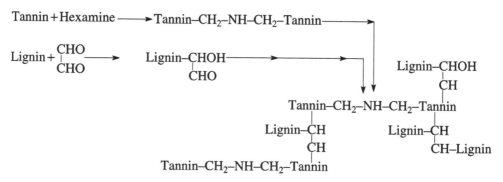

Figure 4. Schematic representation of the series of reactions occurring in the formation of the tannin/lignin hardened network.

shown in the following formula,

not only the A-ring, thus, giving better board strength. Hardening is then controlled by the decomposition of the hexamine into reactive species [17, 20–22]. Such a system is usable with prorobinetinidin/profisetinidin flavonoid tannins such as mimosa and quebracho tannins [17], but yields poor results for the more reactive procyanidin tannins such as pine tannin [34], where the A-ring of the flavonoid unit is so reactive as to prevent reaction of the less reactive B-ring with hexamine.

In Table 4 the decrease of board IB strength when decreasing the relative proportion of tannin in the formulation is due to the decrease of the reactive sites at which the glyoxalated lignin can react in the brief period the panel is in the hot press. In the hot press, the maximum temperature reached in the board core is no higher than 120°C, rendering the reaction even more sensitive to reagents proportions.

4. Conclusions

1. For the first time mixed wood panel adhesive formulations for interior-grade applications were developed in which lignin is in considerable proportion, 50%, of the wood panel binder.

2. For the first time wood adhesives in which no 'fortification' with synthetic resins, such as the isocyanates or phenol-formaldehyde resins used in the past, is necessary to obtain results satisfying relevant standards were developed.

3. In the resins developed environment-friendly, non-toxic polymeric materials of natural-origin constitute as much as 94% of the total panel binder formulation. The wood panel itself is, thus, constituted of 99.5% natural materials.

4. The 0.5% non-natural material is composed of glyoxal, a non-toxic and non-volatile aldehyde, and of hexamine that has already been accepted as a non-formaldehyde-yielding compound when in presence of a flavonoid tannin [17].

The tannin/hexamine binders alone produced panels with zero formaldehyde emission when tested by the desiccator method [33].

References

1. A. Pizzi, *J. Adhesion Sci. Technol.* **20**, 829–846 (2006).
2. H. Lei, A. Pizzi and G. Du, *J. Appl. Polym. Sci.* **107**, 203–209 (2008).
3. A. Pizzi and A. Stephanou, *Holzforschung* **47**, 439–445 (1993).
4. A. Pizzi and A. Stephanou, *Holzforschung* **47**, 501–506 (1993).
5. N.-E. El Mansouri, A. Pizzi and J. Salvado, *Holz Roh Werkst.* **65**, 65–70 (2007).
6. L. Lorenz, C. R. Frihart and J. M. Wescott, in: *Proceedings of Wood Adhesives 2005, Forest Products Society*, Madison, Wisconsin, pp. 501–506 (2006).
7. J. M. Wescott, C. R. Frihart and L. Lorenz, in: *Proceedings of Wood Adhesives 2005, Forest Products Society*, Madison, Wisconsin, pp. 263–270 (2006).
8. G. A. Amaral-Labat, A. Pizzi, A. R. Goncalves, A. Celzard and S. J. Rigolet, *J. Appl. Polym. Sci.* **108**, 624–632 (2008).
9. G. Avignon and M. Delmas, US patent 7,402,224 (2000).
10. A. Ballerini, A. Despres and A. Pizzi, *Holz Roh Werkst.* **63**, 477–478 (2005).
11. NIOSH, National Institute for Occupational Safety and Health, The Registry of Toxic Effects of Chemical Substances, (December 2000).
12. NTIS** — National Technical Information Service. (Springfield, VA 22161) Formerly US Clearing House for Scientific & Technical Information. AD-A125–539.
13. E. Stiasny, Der Gerber Issue 740, 186 (1905).
14. E. Stiasny, Der Gerber Issue 775, 347 (1905).
15. W. E. Hillis and G. Urbach, *J. Appl. Chem.* **9**, 665–673 (1959).
16. L. Suomi-Lindberg, *Paperi Ja Puu* **67**, 65–69 (1985).
17. F. Pichelin, M. Nakatani, A. Pizzi, S. Wieland, A. Despres and S. Rigolet, *Forest Prod. J.* **56**, 31–36 (2006).
18. A. Pizzi, *J. Appl. Polym. Sci.* **22**, 2397–2399 (1978).
19. M. Theis and B. Grohe, *Holz Roh Werkst.* **60**, 291–296 (2002).
20. C. Kamoun, A. Pizzi and M. Zanetti, *J. Appl. Polym. Sci.* **90**, 203–214 (2003).
21. C. Kamoun and A. Pizzi, *Holzforsch. Holzverw.* **52**, 16–19 (2000).
22. C. Kamoun and A. Pizzi, *Holzforsch. Holzverw.* **52**, 66–67 (2000).
23. C. Kamoun, A. Pizzi and R. Garcia, *Holz Roh Werkst.* **56**, 235–243 (1998).
24. European Norm EN 312 (1995), Wood particleboard — specifications.
25. European Norm EN 314-2 (1993), Plywood — specifications.
26. A. Despres and A. Pizzi, *J. Appl. Polym. Sci.* **100**, 1406–1412 (2006).
27. G. F. Zakis, *Functional Analysis of Lignins and Their Derivatives*, p. 65. Tappi, Atlanta (1994).
28. C. Zhao, A. Pizzi, A. Kühn and S. J. Garnier, *J. Appl. Polym. Sci.* **74**, 359–378 (1999).
29. D. G. Roux, *Wattle Bark and Mimosa Extract*. Leather Industries Research Institute, Grahamstown, South Africa (1976).
30. D. Fengel and G. Wegener, *Wood: Chemistry, Ultrastructure, Reactions*. Walther de Gruyter, Berlin (1983).
31. A. Pizzi, M. Beaujean, C. Zhao, M. Properzi and Z. Huang, *Holz Roh Werkst.* **61**, 419–422 (2003).
32. M. Zanetti, A. Pizzi and P. J. Faucher, *J. Appl. Polym. Sci.* **92**, 672–675 (2004).
33. Japanese Standard JIS A5908 (1994).
34. P. Navarrete and A. Pizzi, unpublished results (2009).

Part 4

Wood Welding and General Paper

Wood-Dowel Bonding by High-Speed Rotation Welding — Application to Two Canadian Hardwood Species

G. Rodriguez [a], **P. Diouf** [a], **P. Blanchet** [a,b] **and T. Stevanovic** [a,*]

[a] Centre de Recherche sur le Bois, Département des Sciences du Bois et de la Forêt, Faculté de Foresterie et Géomatique, Université Laval, QC, G1K 7P4, Canada

[b] FPInnovations, 319 rue Franquet, Québec, QC, G1P 4R4, Canada

Abstract

The aim of this work was to investigate the possibility to apply high-speed rotation-induced wood-dowel welding technique to two Canadian hardwood species commonly used for furniture and structural applications, sugar maple (*Acer saccharum*) and yellow birch (*Betula alleghaniensis*). Different factors have been evaluated such as the wood species, the grain orientation, the rotation rate, as well as the receiver hole diameter. The results indicate that high-speed rotation-induced wood-dowel welding can be suitable for these two wood species with average tensile strength values comparable to their respective PVAc-glued joints. Additionally, wood-welded joints presented higher water resistance than their glued-joint counterparts. The results of the temperature measurements confirm that the softening and degradation temperatures of wood components have been reached during the welding process. The X-ray microdensitometry analyses show an increase of density for the interfacial material between the wood substrate and the dowel, the profile of which is more uniform for sugar maple than for birch. The scanning electron micrographs of the interfacial contact zone confirm the different extents of wood-to-wood welding.

Keywords

Wood welding, parameter interaction, maple, birch, wood joints, wood polymers

1. Introduction

Assembly techniques by dowel insertion, with or without the use of adhesives, are common in joining solid wood in furniture and wood joinery industries. The traditional way to insert dowels in good quality solid timber furniture is by percussion, pneumatic or manual means.

Recently, high-speed rotation-induced wood-dowel welding, which is an adhesive-free wood joining technique, has been developed [1–3] and represents one of the innovative aspects of wood bonding. The process consists in insertion

---

[*] To whom correspondence should be addressed. Tel.: (+1) 418 656 2131 # 7337; Fax: (+1) 418 656 2091; e-mail: tatjana.stevanovic@sbf.ulaval.ca

of dowels in smaller diameter pre-drilled holes at fast rotational speed. The friction induced causes rise in temperature and, consequently, the softening and flowing of some amorphous polymer material, such as lignin and hemicelluloses which bond together the wood cells in the structure of the wood. High-quality welded joints, suitable exclusively for interior grade applications such as furniture [4, 5] and structural elements [6, 7], are achieved. So far, most of the studies on high-speed rotation-induced wood-dowel welding have been done with beech (*Fagus sylvatica*) and Norway spruce (*Picea abies*) woods which are European wood species most commonly used for furniture and structural applications. Less attention has been paid to other species including the North American species.

The aim of this study was the investigation of the wood welding performance of two Canadian hardwood species commonly used by the furniture industry. These species are sugar maple (*Acer saccharum*) and yellow birch (*Betula alleghaniensis*). Wood welding technique allows the possibility of developing eco-friendly wood dowel assembly without use of any adhesive for high value wood product. The influence of parameters such as rotation rate, diameter of receiving hole, grain orientation and wood species and their interactions on the tensile strength was evaluated.

2. Materials and Methods

2.1. Preparation of Dowel Joints by High-Speed Rotation Mechanically-Induced Wood Fusion Welding

Commercial smooth wood dowels of birch and maple woods (A. Lapointe & Fils Ltée, Saint-Romain, Quebec, Canada), dried at 12% moisture content (MC), 9.68 mm in diameter and 80 mm in length, were inserted using a high-speed fixed-base, manually-operated drill (model IMA A60 type 90/2-4-75, Berg Man Borr-AB, Sweden) with pre-drilled holes of various diameters (7.14, 7.37 or 7.67 mm), in pieces of birch and maple woods (40 × 20 × 20 mm), respectively. The holes were drilled precisely at a rapid feed speed of 2500 revolutions per minute (rpm), in order to avoid charring the surfaces of the holes. The dowel was inserted perpendicularly to the transverse direction (Fig. 1) in three different directions (radial, tangential and diagonal), and welded to the substrate by a fast rotational movement at three rates (1000, 1500 or 2500 rpm). Once the fusion and bonding were achieved, approximatively between 2 and 4 s, the rotation of the dowel was stopped. Poly (vinyl acetate) (PVAc) adhesive (Mastercraft, Toronto, Canada) was used to bond dowels to birch and maple woods as control. According to the producer's recommendations, double gluing technique was used in which the surfaces of both receiving holes (9.68 mm diameter) and dowels were coated with glue prior to insertion of the dowels (15 mm in depth). The welded samples as well as the controls were conditioned for at least one week in an environmental chamber at 20°C and 65% relative humidity before testing.

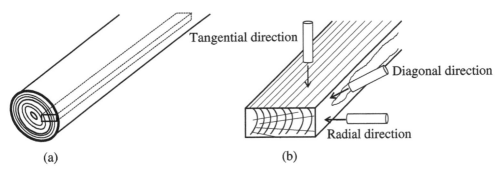

Figure 1. (a) Schematic representation of a wood log. (b) Insertion directions of the dowel.

2.2. Evaluation of Welding Performance

By using two wood species, three rotation rates, three grain orientations and three hole diameters as parameters, a total of 540 welded specimens ($2 \times 3 \times 3 \times 3 \times 10$) were prepared, with 10 replicates for each parameter. The tests were carried out according to the ASTM-D 1037 standard with an MTS universal testing machine QT-5 (Intertechnology Inc., Don Mills, ON, Canada). Dowels were pulled out of the substrate at a rate of 2 mm/min. The specimens were compared to PVAc-glued controls. The loading was continued until the separation occurred on the surface of the test samples. From the maximal observed load (F_{max}), bonding surface of sample (S), the depth (h) and the hole radius (r), the withdrawal strength (σ_k) was calculated from the following equation:

$$\sigma_k = \frac{F_{max}}{S} = \frac{F_{max}}{\pi r (2h + r)}. \tag{1}$$

2.3. Water Resistance

In addition to dry condition, the tensile strength of the welded joint was assessed under the following wet conditions:

(i) Cold water soaking test: samples were immersed completely in cold water ($22 \pm 2°C$) for a period of 24 h.

(ii) Accelerated condition (boiling water test): samples were immersed in boiling water for 2 h, and then cooled in the cold water ($22 \pm 2°C$) for 1 h.

For the two conditions described above, samples were taken out of the cold water and subjected to the tension test as described above. Ten replicates were made for each condition and compared to PVAc controls (total 80 specimens).

2.4. Thermal Behaviour

During the welding process, temperature was monitored with fast response thermocouples placed in thin predrilled holes placed at 1 mm and 2 mm distance from the inserted dowel as presented in Fig. 2. A VISHAY data acquisition station (model

5100, Intertechnology Inc., Don Mills, ON, Canada) with 100 Hz in frequency was used to measure the rise in temperature of the system (Fig. 3).

2.5. X-ray Microdensitometry

Before analysis, the samples were sliced into 1.5–1.8 mm-thick slices parallel to the plane of the substrate as illustrated in Fig. 3. The samples were conditioned at 20°C and 65% RH in order to obtain an equilibrium moisture content of 12% before scanning. Approximately 2-mm thick samples were analyzed using a QTRS-01X QMS Tree Ring Analyzer (Quintek Measurement Systems, Knoxville, TN, USA). Density was measured with a finely collimated (0.04 mm wide) X-ray beam operating at 17.5 keV and 0.75 mA. The equipment was calibrated in order to estimate the density of the wood substrate sample (sugar maple and yellow birch wood species).

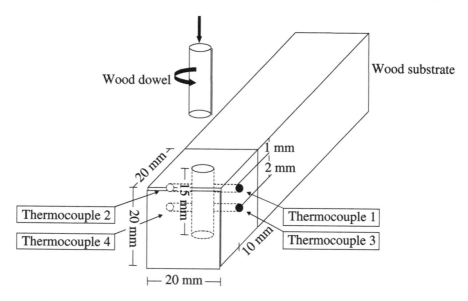

Figure 2. Schematic representation of the thermocouples in the heat-affected zone.

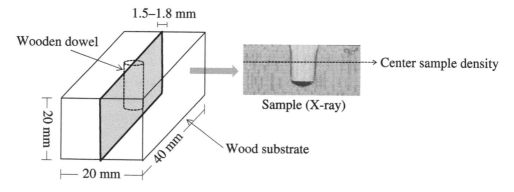

Figure 3. Schematic representation of the sample preparation for X-microdensitometry study.

A relative density profile was obtained in order to compare the welded zones of the two wood species.

2.6. Scanning Electron Microscopy

Before scanning electron microscopy (SEM) investigation, samples were prepared as described before for X-ray microdensitometry sample preparation (Fig. 3). The investigation was carried out after metallizing with gold–palladium. The SEM equipment used was a JEOL 6360LV microscope (Tokyo, Japan).

2.7. Statistical Analysis

The mechanical test results were analysed using the analysis of variance technique (ANOVA) with the SAS software.

3. Results and Discussion

3.1. Welding Performance

It has been demonstrated in previous studies performed on European species [1–3] that the dowel/receiving hole diameter ratio is the most important parameter influencing wood joint strength. The interactions between other parameters such as wood species, insertion direction, rotation rate, or wood dowel moisture content were shown to influence the welded joint quality. The results on North American species seem to be different and they are presented in Table 1. It can be observed from this table that there is a significant effect of the quadruple interactions among the wood species, the insertion direction, the rotation rate and the hole diameter on tensile strength. This means that the effect of wood species on welding performance depends on the other three factors. However, the F-values obtained for the factor related to the wood species were 20, 10 and 4 times higher than the values for rotation rate, insertion direction and hole diameter, respectively. Thus, the variation observed in the tensile strength could be explained by the difference in wood species followed by rotation rate rather than insertion direction and hole diameter. This may be due to the variability in chemical composition and the structure of the wood polymers of the two studied species, particularly the lignin and the hemicelluloses which ultimately impact the quality of welded joints, as well as the anatomical structure. Contrary to previous study [1], the rotation rate factor seems to have a high influence on the tensile strength. The F-values for rotation rate were 9 and 3 times higher than for grain orientation and hole diameter, respectively. This can be due to the effect of the rotation rate on the temperature rise rate to reach the softening temperature of the amorphous wood polymers, mainly lignin. A recent study on linearly welded wood joints [8] reported that the shorter is the welding time the lower is the degree of deterioration of the weldline of the wood joint. Similar effect can be expected for rotation-welded wood joints. Previous studies on beech wood [1, 3] reported that the higher the dowel/hole diameter ratio, the greater is the friction and thus a better welded joint is achieved. As in previous

Table 1.
Results of the analysis of variance (F-values) for physical and mechanical properties of wood-dowel bonding by high-speed rotation welding of birch and maple woods

Source of variation	F-value	Pr > F[1]
Species	878	<0.0001
Insertion direction (ID)	32.8	<0.0001
Rotation rate (RR)	284	<0.0001
Hole diameter (HD)	81.9	<0.0001
Species × ID	13.3	<0.0001
Species × RR	7.40	0.0007
Species × HD	10.6	<0.0001
ID × RR	18.4	<0.0001
ID × HD	49.9	<0.0001
RR × HD	32.3	<0.0001
Species × ID × RR	3.78	0.0048
Species × ID × HD	3.88	0.0041
Species × RR × HD	7.38	<0.0001
IP × RR × HD	10.6	<0.0001
Species × ID × RR × HD	3.02	0.0025

[1] Indicates that the factors (source of variation) have a significant impact on the dependent variable, when $p < 0.001$ is indeed highly significant.

studies, the present results confirm the near 5/4 dowel/hole diameter ratio (1.26 in our case) as the optimal ratio but a dowel/hole diameter ratio higher than 5/4 did not yield better results under the conditions used. The average values of the tensile strength obtained for birch and maple wood species are given in Tables 2, 3 and 4. A stepdown Bonferroni test was performed and the Homogeneity Groups (HGs) were found with the objective of determining whether or not there was a significant difference between the tensile strength values obtained on the basis of the variables. The highest tensile strength was obtained for sugar maple wood sample (9.2 MPa) with the following parameter combination: dowel inserted in radial direction in a 7.67 mm diameter hole, at 1000 rpm. The lowest value was obtained for yellow birch wood sample, inserted in diagonal direction and in a 7.37 mm diameter hole at 1000 rpm. For birch samples, the best welding performance with an average tensile strength value of 7.1 MPa was obtained with the following combination: radial direction, 1500 rpm and 7.67 mm diameter hole. Our results for tensile strength are similar to those obtained for European species [1–3] (values of tensile strength between 7–10 MPa for beech and between 2–5 MPa for spruce). Comparing welded joints with glued joints, no significant difference in the average values of tensile strength was found between the two best welded dowel joints (whose parameters are presented in Table 3) and the best glued joint in the case of birch. Moreover, in the case of maple (Table 4), the tensile strength of the best welded joint is significantly

Table 2.
Average values of tensile strength (MPa) of dowel welded wood samples

Welding parameter		Tensile strength
Wood species	Birch	3.4
	Maple	6.2
Insertion direction	Tangential	4.5
	Diagonal	4.4
	Radial	5.4
Rotation rate (rpm)	1000	3.7
	1500	4.2
	2500	6.4
Hole diameter (mm)	7.14	4.3
	7.37	4.3
	7.67	5.7

higher than that of the best glued joint. We can also observe that 15 other welded joints presented comparable average values of tensile strength at 0.05 probability level than the best PVAc-glued joint.

3.2. Water Resistance

Table 5 presents the average values of tensile strength after wet condition treatments. After 24 h cold water soaking, the tensile strengths of the best welded joints of both wood species are comparable at 0.05 probability level, but better than their respective PVAc-glued joints. The loss of tensile strength was lower for welded dowels than glued dowels (Table 5). A 2 h boiling test is often used as an indication of the exterior-grade potential of these joints [9]. Maple wood welded joint exhibits a higher resistance than birch wood welded joint after 2 h boiling water treatment. Birch had similar loss of tensile strength as dowel glued in maple. The dowel glued in birch presented the most significant loss of tensile strength after 2 h boiling (Table 5). This indicates that the maple-dowel welded joint is more suitable to weather exposure than the birch joint. However, both welded wood species showed a better tensile strength retention after water exposure than glued dowel. In cold water soaking condition, the percentages strength losses for both birch and maple samples were both 66%. In accelerated condition, % losses of strengths for birch and maple samples were 72% and 68%, respectively. Thus, welded wood joints can only be destined for interior applications such as for furniture and for interior-grade wood joints as was indicated also in previous European studies.

3.3. Thermal Behaviour

Friction between the dowel and the substrate during insertion by high-speed rotation causes temperature-induced lignin softening and deformation and wood weld-

Table 3.

Average values of the tensile strength (average of 10 replicates for each measurement) in dry condition obtained for birch wood

Species	Type of joint	Insertion direction	Rotation rate (rpm)	Hole diameter (mm)	Tensile strength Mean ± SD[2] (MPa)	Homogeneity group
Birch	*Glued*	*Radial*	–	*9.8*	*7.8 ± 0.8*	*A*
Birch	Welded	Radial	1500	7.67	7.1 ± 1.1	AB
Birch	Welded	Radial	1000	7.67	6.9 ± 0.9	AB
Birch	Welded	Diagonal	2500	7.67	5.9 ± 0.9	BCD
Birch	Welded	Diagonal	2500	7.37	5.9 ± 0.5	BC
Birch	Welded	Diagonal	2500	7.14	5.8 ± 0.8	BC
Birch	Welded	Radial	2500	7.37	5.7 ± 1.7	ABCDE
Birch	*Glued*	*Diagonal*	–	*9.8*	*5.5 ± 0.6*	*CD*
Birch	Welded	Tangential	2500	7.14	5.2 ± 1.1	CDE
Birch	Welded	Radial	2500	7.67	5.1 ± 1.9	ABCDEFG
Birch	Welded	Radial	2500	7.14	5.1 ± 1.6	BCDEF
Birch	Welded	Tangential	2500	7.37	4.8 ± 0.6	DE
Birch	*Glued*	*Tangential*	–	*9.8*	*4.5 ± 0.4*	*EF*
Birch	Welded	Tangential	1500	7.14	3.3 ± 1.2	FGH
Birch	Welded	Diagonal	1500	7.67	3.0 ± 0.8	GH
Birch	Welded	Tangential	2500	7.67	3.0 ± 0.9	GH
Birch	Welded	Radial	1500	7.14	2.6 ± 1.8	FGHIJ
Birch	Welded	Tangential	1500	7.37	2.2 ± 0.2	HI
Birch	Welded	Tangential	1500	7.67	1.9 ± 0.5	HIJ
Birch	Welded	Tangential	1000	7.37	1.9 ± 0.8	HIJ
Birch	Welded	Diagonal	1500	7.14	1.8 ± 1.0	HIJ
Birch	Welded	Diagonal	1500	7.37	1.8 ± 0.9	HIJ
Birch	Welded	Radial	1000	7.14	1.5 ± 0.8	IJ
Birch	Welded	Diagonal	1000	7.67	1.5 ± 0.7	IJ
Birch	Welded	Diagonal	1000	7.14	1.4 ± 0.6	J
Birch	Welded	Tangential	1000	7.14	1.4 ± 0.4	J
Birch	Welded	Tangential	1000	7.67	1.3 ± 0.6	J
Birch	Welded	Radial	1500	7.37	1.3 ± 0.2	J
Birch	Welded	Radial	1000	7.37	1.2 ± 0.4	J
Birch	Welded	Diagonal	1000	7.37	1.2 ± 0.3	J

[2] Standard deviation.

The different letters show whether or not there is a difference between the tensile strength values for the sub-variables related to the main variables. The variables without any differences among them are shown with the same letters. At the same time, these letters are given in an order from the best value (A) to the worst value (J).

ing [1]. Based on previous work [3], the thermocouple system selected and used in this study was a fast response system to ensure adequate measure of the welding temperature. Figure 4 illustrates the temperature curves of the best welded joints obtained for the two wood species, as a function of welding time in the contact

Table 4.
Average values of the tensile strength (average of 10 replicates for each measurement) in dry condition obtained for maple wood

Species	Type of joint	Insertion direction	Rotation rate (rpm)	Hole diameter (mm)	Tensile strength Mean ± SD[1] (MPa)	Homogeneity group
Maple	Welded	Radial	1000	7.67	9.2 ± 1.0	A
Maple	Welded	Radial	1500	7.67	9.0 ± 1.2	ABC
Maple	Welded	Radial	2500	7.67	8.9 ± 0.9	AB
Maple	Welded	Radial	2500	7.37	8.0 ± 0.8	ABCD
Maple	Welded	Diagonal	2500	7.37	7.7 ± 1.2	BCD
Maple	Welded	Tangential	1500	7.67	7.7 ± 0.2	ABCDE
Maple	Welded	Diagonal	2500	7.67	7.6 ± 1.6	BCDE
Maple	Welded	Tangential	2500	7.37	7.6 ± 1.0	ABCDEF
Maple	Welded	Tangential	2500	7.14	7.5 ± 1.2	ABCDE
Maple	Welded	Diagonal	2500	7.14	7.4 ± 1.1	BCDE
Maple	*Glued*	*Radial*	–	*9.8*	*7.3 ± 1.1*	*CDEF*
Maple	Welded	Radial	2500	7.14	7.1 ± 0.9	DEF
Maple	Welded	Diagonal	1500	7.67	7.0 ± 1.2	DEFG
Maple	*Glued*	*Tangential*	–	*9.8*	*6.5 ± 1.1*	*EFG*
Maple	Welded	Tangential	2500	7.67	6.5 ± 1.0	DEFGH
Maple	Welded	Tangential	1000	7.67	6.4 ± 2.2	BCDEFGHIJKL
Maple	Welded	Tangential	1000	7.37	6.0 ± 1.2	EFGHIJ
Maple	*Glued*	*Diagonal*	–	*9.8*	*5.9 ± 1.0*	*FGHI*
Maple	Welded	Tangential	1000	7.14	5.9 ± 0.8	EFGHIJM
Maple	Welded	Radial	1500	7.37	5.4 ± 1.0	GHIJK
Maple	Welded	Tangential	1500	7.37	5.2 ± 0.6	HIJMN
Maple	Welded	Diagonal	1000	7.14	5.0 ± 0.3	IJK
Maple	Welded	Diagonal	1500	7.37	4.5 ± 09	JKL
Maple	Welded	Diagonal	1000	7.37	4.2 ± 07	KL
Maple	Welded	Tangential	1500	7.14	4.1 ± 1.6	IJKL
Maple	Welded	Radial	1000	7.14	4.0 ± 0.8	KL
Maple	Welded	Radial	1000	7.37	3.7 ± 1.1	L
Maple	Welded	Diagonal	1000	7.67	3.7 ± 1.0	KL
Maple	Welded	Diagonal	1500	7.14	3.7 ± 1.0	KLM
Maple	Welded	Radial	1500	7.14	3.5 ± 1.5	KLN

[1] Standard deviation.
   The different letters show whether or not there is a difference between the tensile strength values for the sub-variables related to the main variables. The variables without any differences among them are shown with the same letters. At the same time, these letters are given in an order from the best value (A) to the worst value (N).

zone. The welding temperatures are reached around 263°C and 274°C in less than 1 s, for birch and maple, respectively. After the very rapid increase, the temperature appears to stabilize at the maximum between 0.7 s and 2 s. This is associated to the lower friction due to the molten material at the dowel/hole interface [10]. Once fast

Table 5.
Average values of the tensile strength (average of 10 replicates for each measurement) in wet conditions

Species	Type of joint	Condition	Mean ± SD (MPa)	% loss[1]	Homogeneity group
Maple	Welded	Cold water soaking	3.1 ± 0.5	66	A
Birch	Welded	Cold water soaking	2.4 ± 0.8	66	A
Maple	Glued	Cold water soaking	1.1 ± 0.2	85	B
Birch	Glued	Cold water soaking	0.8 ± 0.3	90	B
Maple	Welded	Accelerated	2.9 ± 0.3	68	A
Birch	Welded	Accelerated	2.0 ± 0.4	72	B
Maple	Glued	Accelerated	1.7 ± 0.8	77	B
Birch	Glued	Accelerated	0.7 ± 0.1	91	C

[1] Loss of the tensile strengths in wet conditions compared to that in dry condition.

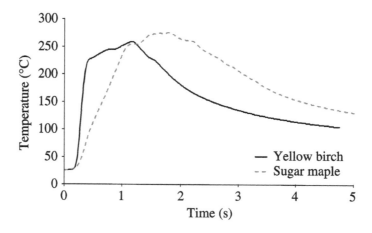

Figure 4. Temperature profile during welding.

rotation of the dowel is stopped, the temperature then decreases as shown in Fig. 4. However, the temperature curves as well as the maximum temperatures reached, in this study, approach more the temperature profiles obtained for linear welding than those obtained for fast rotational dowel welding reported in previous study [11]. No temperature decrease was observed after reaching the maximum temperature when fast rotation of the dowel was maintained. In contrast, a stabilization of the temperature as in linear welding can be observed. In addition, the temperatures reached in the present study are higher than those found for beech (183°C) [3], which confirms that the temperature measured by the thermal camera system is only indicative and lower than the real temperature reached by the dowel during the welding process. For the ranges of temperatures observed in this study (Table 6), thermal softening of wood occurs as well as thermal decomposition [12–14]. It has been reported that the

Table 6.
Experimental values of temperature T obtained during rotational welding

Rotation rate (rpm)	Insertion direction	Maple		Birch	
		Mean ± SD (°C)	Max T (°C)	Mean ± SD (°C)	Max T (°C)
1000	Tangential	269.3 ± 20.1	301	247.9 ± 43.1	309
	Diagonal	266.7 ± 29.8	314	243.6 ± 33.7	286
	Radial	273.8 ± 28.3	324	252.6 ± 32.6	286
1500	Tangential	279.9 ± 21.6	301	277.0 ± 28.1	333
	Diagonal	281.2 ± 30.7	311	268.9 ± 33.6	318
	Radial	276.9 ± 26.1	342	262.9 ± 44.9	326
2500	Tangential	323.0 ± 19.4	368	306.1 ± 39.2	363
	Diagonal	312.5 ± 15.7	333	304.6 ± 38.4	365
	Radial	311.0 ± 53.3	388	308.0 ± 40.9	380

temperature progression at the interface is of major importance [1, 3]. Higher average values of temperature have been reached with maple (266–323°C) during the welding process than with birch (244–308°C). Previous studies [1, 3] reported that the higher the temperature is reached, or the temperature of softening is reached at a higher rate, or both, the higher is the weld strength which is in accord with the results of this study. As presented in Table 1, the differences between average values of tensile strengths of sugar maple and yellow birch species are significant. According to this, the tensile strength value of maple is higher compared to yellow birch. Birch wood (628.6 ± 10.3 kg/m^3) is similar to sugar maple wood (655.4 ± 23.1 kg/m^3) in density, thus the effect of wood species on the welding temperature seems to be related to its anatomical structure. Even if both species are classified as diffuse-porous woods, the main difference related to cell structure are the rays. In birch the rays appear as fine lines similar to softwoods whereas in sugar maple the rays are wider and more distinct, and appear as noticeable flecks on the tangential surface [15]. This could explain the higher tensile strength of maple than birch. In fact, the friction between the dowel and the substrate during insertion results in a better welded-joint than when the contact zone is more uniform.

3.4. X-ray Microdensitometry

Previous studies [1, 16] reported that the densification and the thickness of the interfacial zone play an important role in the quality of the wood-welded joint. Thus, X-ray microdensitometry proved to be a valuable technique to determine the extent of wood welding [16]. As wood species was found to be the most significant parameter in our study, the relative density (obtained by dividing density values by the mean density of the specimen) was used because it is more appropriate for interspecies comparison. In the case of best welded joints, the density increases in the welded interphase from 628.6 ± 10.3 to 1185.6 ± 58.9 kg/m^3, and from 655.4 ±

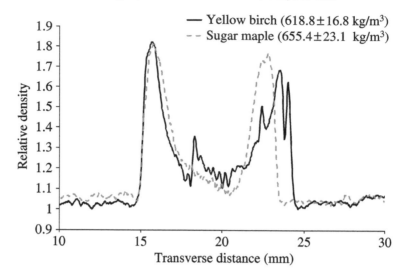

Figure 5. X-ray microdensitometry, average density profiles for sugar maple and yellow birch, allowing comparison of their performances.

23.1 to $1176.6 \pm 24.0 \, kg/m^3$, for birch and maple, respectively. The density profiles obtained by X-ray microdensitometry are presented in Fig. 5. The two curves show clearly that wood-to-wood welding assembly is accompanied by a densification in the interfacial contact zone due to: (1) the loss of the intercellular structure of the wood at the interface; and (2) the decrease of empty spaces in its cellular structure [16]. Even if the two species present similar increase in terms of relative density, Fig. 6 shows clearly why sugar maple welded-joint performs better than yellow birch wood. Maple welded-joint is uniform, while birch welded-joint presents an irregular interface that can be ascribed to the difference in wood anatomical structure mentioned before. If one compares best welded-joint with poorly welded joint relative to wood species, their density profiles show similar pattern but slight density increase at the interface can be noticed for well welded-joint (figures not shown). The higher is the increase of density, the higher is the quality of the welded-joint. This is in accord with previous studies [1, 16] which indicate better mechanical performance of a welded-joint when the density increase at the interface is more regular and higher.

3.5. Scanning Electron Microscopy

Previous studies [1, 2] examined the welded dowel surface after pull-out from the substrate. In this work, the SEM study focused on the welded-joints (Fig. 6(a–d)). Figure 6(a) shows a poorly welded joint between the dowel inserted in the hole (left side) and the maple substrate (right side), whereas Fig. 6(b) presents a well-welded bond between the dowel (right side) and the maple substrate (left side). When examining in more details the micrograph for maple, one can see the absence of visible cracks at the interface between the dowel and the substrate, as

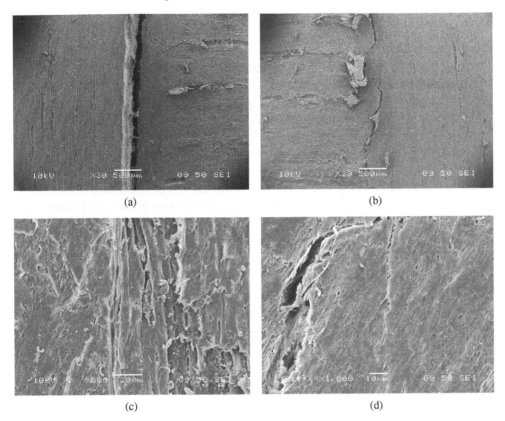

(a)   (b)

(c)   (d)

Figure 6. SEM micrographs of the welded joints: (a) poorly-welded joint (30× magnification); (b) well-welded joint (30× magnification); (c) magnified view of (b) (800× magnification); (d) well-welded joint with low tensile strength (1000× magnification).

illustrated by Fig. 6(c). The extent of flow observed in this micrograph is a clear evidence of the temperature-induced softening or melting of the intercellular material as mentioned in the previous studies. However, even if the interface is well welded, there are cases, particularly for birch, where tensile strength values are low. This seems to be due to the presence of fractures occurred on wood substrate, illustrated in Fig. 6(d) (birch substrate on the left side, and dowel on the right side). It is thus possible that the differences observed in mechanical performance (tensile strength) between the two wood species are due to the differences in their anatomical structure [15]. In addition, Sun *et al.* [17]. have shown differences in chemical composition of birch and maple woods and have pointed out less condensed lignin structure of birch wood. This suggests that welding parameters are specific to a given species. In addition, Fig. 6(b) can also explain why wood welded joints are not water resistant: the weldline is not continuous around the dowel, leaving space for water to deeply penetrate in the weldline, swelling the reorganized fibres and decreasing the bonding effect of the molten material.

4. Conclusion

It has been demonstrated in the present research that the Canadian wood species, sugar maple and yellow birch, can be successfully welded by high-speed rotation of wood dowels, without any adhesives. The strength of the joints obtained for the birch is comparable to that obtained by gluing with PVAc adhesive, whereas for the maple it is superior to that obtained with PVAc adhesive. Wood species has been shown to be the most significant parameter determining the tensile strength of the welded samples. For both wood species, the water resistance performance of the welded wood samples was found to be superior to that of the PVAc adhesive glued samples. This study opens the possibility of using Canadian wood species providing high-quality and eco-friendly joints, for interior-grade applications such as furniture.

Acknowledgements

The authors are very grateful to the Fond québécois de recherche sur la nature et les technologies (FQRNT) for the financial support to this project and to Pr Alain Cloutier for scientific discussions. The technical support of Luc Germain, Sylvain Auger and Éric Rousseau is also gratefully acknowledged.

References

1. A. Pizzi, J.-M. Leban, F. Kanazawa, M. Properzi and F. Pichelin, *J. Adhesion Sci. Technol.* **18**, 1263–1278 (2004).
2. C. Ganne-Chedeville, A. Pizzi, A. Thomas, J.-M. Leban, J.-F. Bocquet, A. Despres and H. R. Mansouri, *J. Adhesion Sci. Technol.* **19**, 1157–1174 (2005).
3. F. Kanazawa, A. Pizzi, M. Properzi, L. Delmotte and F. Pichelin, *J. Adhesion Sci. Technol.* **19**, 1025–1038 (2005).
4. J.-F. Bocquet, A. Pizzi and L. Resch, *Holz Roh Werkst.* **65**, 149–155 (2007).
5. L. Resch, A. Despres, A. Pizzi, J.-F. Bocquet and J.-M. Leban, *Holz Roh Werkstoff* **64**, 423–425 (2006).
6. C. Segovia and A. Pizzi, *J. Adhesion Sci. Technol.* **23**, 1293–1301 (2009).
7. A. Renaud, *Eur. J. Wood Prod.* **67**, 111–112 (2009).
8. P. Omrani, A. Pizzi, H. R. Mansouri, J.-M. Leban and L. Delmotte, *J. Adhesion Sci. Technol.* **23**, 827–837 (2009).
9. P. Omrani, H. R. Mansouri and A. Pizzi, *Holz Roh Werkst.* **66**, 161–162 (2008).
10. A. Zoulalian and A. Pizzi, *J. Adhesion Sci. Technol.* **21**, 97–108 (2007).
11. H. R. Mansouri, P. Omrani and A. Pizzi, *J. Adhesion Sci. Technol.* **23**, 63–70 (2009).
12. S. Z. Chow and K. J. Pickles, *Wood Fiber Sci.* **3**, 166–178 (1972).
13. E. L. Schaffer, *J. Testing Evaluation* **1**, 319–329 (1973).
14. E. L. Back and N. L. Salmen, *Tappi J.* **65**, 107–110 (1982).
15. A. J. Panshin and C. De Zeeuw, *Textbook of Wood Technology*, 4th edn. McGraw-Hill Book Co., New York, NY (1980).
16. J.-M. Leban, A. Pizzi, S. Wieland, M. Zanetti, M. Properzi and F. Pichelin, *J. Adhesion Sci. Technol.* **18**, 673–685 (2004).
17. Y. Sun, M. Royer, P. N. Diouf and T. Stevanovic, *J. Adhesion Sci. Technol.*, in press (2010).

Chemical Changes Induced by High-Speed Rotation Welding of Wood — Application to Two Canadian Hardwood Species

Y. Sun, M. Royer, P. N. Diouf and T. Stevanovic [*]

Centre de Recherche sur le Bois, Département des Sciences du Bois et de la Forêt,
Faculté de Foresterie et Géomatique, Université Laval, Quebec City, PQ G1V 0A6, Canada

Abstract

The chemical changes occurring upon rotational welding with dowels of Canadian wood species sugar maple (*Acer saccharum*) and yellow birch (*Betula alleghaniensis*) were examined by pyrolysis-GC/MS, DSC and XPS anlayses. The analyses performed separately on wood substrate (reference wood) and welded material by pyrolysis-GC/MS, DSC and XPS indicate that the differences in mechanical performances of the two welded woods are due mainly to the differences in original lignin structures as well as in the welding temperatures determined for the two wood species. The more pronounced guaiacyl character of the maple wood lignin seems to explain the preferential condensation reactions of the guaiacyl moieties in maple lignin with formaldehyde and furanic compounds released from lignin and carbohydrates during the fast pyrolysis associated with the welding process. The higher temperature determined for maple wood welding than for birch could be responsible for enhanced miscibility of wood polymers in the welding zone, explaining, therefore, the more significant presence of xylan polymer together with newly formed lignin carbohydrate complex (LCC) in the welded material. The detailed analysis of the compounds identified by pyrolysis-GC/MS, together with the results of the other two methods applied in this study, has confirmed that the S/G ratio cannot be taken as the sole criterion for the discussion of the chemical changes in lignins during welding of wood.

Keywords

Wood welding, sugar maple, yellow birch, wood joints, pyrolysis-GC/MS, DSC, XPS, lignin carbohydrate complex (LCC)

1. Introduction

Assembly techniques by dowel insertion, with or without the use of adhesives, are commonly applied for joining solid wood in furniture and wood joinery industries. High-speed rotation-induced wood dowel welding has been shown to rapidly yield wood joints of considerable strength for European wood species [1–3].

The chemical reactions involved in the formation of the welded joint are believed to occur in a similar way to those involved in pyrolytic processes. In effect, pyrol-

---

[*] To whom correspondence should be addressed. Tel.: (+1) 418 656 2131, ext. 7337;
Fax: (+1) 418 656 2091; e-mail: tatjana.stevanovic@sbf.ulaval.ca

Wood Adhesives

ysis is the transformation of a nonvolatile compound into a volatile degradation mixture by heat in the absence of oxygen. The wood components which are first affected by the rise of temperature are the hemicelluloses (predominantly C5 sugars (pentosans) in case of hardwoods), but also the C6 (hexosans), the degradation of which starts from around 100°C. The weight loss of hemicelluloses takes place mainly in the temperature range of 220–315°C. It is clear that O-acetyl chemical groups play a significant role in hemicelluloses thermal stability, as they yield acetic acid which can then catalyze the depolymerisation. Hemicelluloses depolymerization results in formation of oligosaccharides and monosaccharides, the dehydration of the latter yielding furfural and 5-hydroxymethylfurfural [4]. Cellulose is the most thermally stable wood component due to its high crystallinity. Although its degradation occurs at higher temperature, the thermal changes in the amorphous zone of the cellulose are similar to those of hemicelluloses. Crystalline cellulose is really degraded at 300–340°C [5]. The products of cellulose thermal degradation are levoglucosan and furan and its derived compounds. The effects on lignin generally start at lower temperatures than on cellulose and hemicelluloses and are confined within a rather wide temperature range. Previous studies have demonstrated that endothermic reactions occur from 50°C to 200°C [6] and that lignin rearrangement during the temperature-induced softening involves a small exothermic reaction at 200°C.

However, only a few chemical investigations on the process itself have been carried out so far, and most of the chemical reactions and rearrangements which occur on natural polymers which constitute wood are not well understood. A recent study on two Canadian hardwood species (yellow birch and sugar maple wood) [7] reports that the differences in the measured tensile strengths could be explained more by the difference in species than by other studied parameters (rotation rate, angle of dowel insertion, hole diameter). This result has been attributed to the differences in anatomical and chemical characteristics between the two species. Different analytical tools have been used to evaluate thermal modification of wood and its polymer components: cellulose, hemicelluloses and lignins. This is, however, for the first time that the chemical changes occurring during wood-to-wood welding process are monitored by pyrolysis-GC/MS, differential scanning calorimetry (DSC) and X-ray photoelectron spectroscopy (XPS). These analytical techniques have been shown to be well-suited to characterize changes in functional groups and structure of wood components. The results of the present study complement therefore the chemical analysis of wood welded material previously investigated by a host of analytical techniques such as CP/MAS ^{13}C NMR, FTIR and GC/MS.

2. Materials and Methods

2.1. Sample Preparation

The chemical analyses were carried out with samples of yellow birch (*Betula alleghaniensis*) and sugar maple (*Acer saccharum*). For both species, the reference material (Ref) represents the wood which was not subjected to the welding process

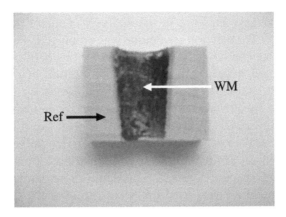

Figure 1. Sampling for XPS analysis: Ref and WM denote wood reference and welded material, respectively.

while WM represents the welded material collected from the sample which showed the highest tensile strength [7]. The samples for pyrolysis-GC/MS and DSC analyses were collected by scraping the wood from the unaltered wood substrate (Ref) and from the welded zone (WM: Welded Material). For XPS analysis, the samples were prepared by transversal cutting of the wood substrate as illustrated in Fig. 1.

2.2. Pyrolysis-GC/MS

Samples were pyrolyzed at 550°C in helium atmosphere using a Pyroprobe 2000 pyrolyzer (CDS Analytical, Inc., Oxford, PA) which was interfaced to a gas chromatograph (Varian CP 3800) and a mass spectrometer (Varian Saturn 2200 MS/MS, 30–650 a.m.u.). The flow rate of the carrier gas was 1.0 ml/min. Pyrolysis interface and GC injector were kept at 250°C. Pyrolysis products were separated on a Varian FactorFour Capillary Column (VF-5 ms, 30 m × 0.25 mm × 0.25 μm). The pyrolysis was carried out from 250°C (holding for 10 s) to pyrolysis temperature (holding for 30 s) with a heating rate of 6°C/ms. The GC oven was kept at 45°C for 4 min and then heated to 280°C at 4°C/min. The final temperature was held for 15 min. Mass spectrometer was operated in the electron impact mode using 70 eV energy and the mass range m/z 40–400 was scanned in 36 s. The compounds were identified by comparing the obtained mass spectra with those from Wiley and NIST computer libraries as well as those reported in the literature [8, 9]. Relative compound distributions were calculated for each carbohydrate and lignin derived pyrolysis product from the relevant peak areas. The summed areas of the relevant peaks were normalized to 100% (all products), and the data from 12 pyrolysis experiments were used to determine the average values and standard deviations.

2.3. XPS

X-ray photoelectron spectroscopy (XPS, Kratos Axis-Ultra) was used to identify the elements present at the sample surfaces, i.e., welded material and reference wood. The take-off angle of the emitted photoelectrons was adjusted to 30° to the surface normal. The base pressure in the analysis chamber during XPS analysis was below 5×10^{-10} Torr. Hybrid lens mode was employed. The XPS spectra were recorded using a monochromatic Al K_α source operating at 300 W. High-resolution spectra were recorded with a nominal energy resolution of 0.5 eV (10 eV pass energy and 0.025 eV steps). These high-resolution spectra were used for chemical analysis. The survey spectra used to perform quantitative elemental analyses of non-welded and welded materials were recorded in 1 eV steps and 160 eV pass energy. The binding energy scale was corrected by referring to the polyaromatic peak in the C_{1s} spectrum as being at 284.6 eV. Shirley and linear backgrounds were subtracted from the C_{1s} and O_{1s} spectra. A mixed Gaussian–Lorentzian product function and a single Gaussian function were used for curve fitting of the C_{1s} and O_{1s} spectra, respectively.

2.4. DSC Analysis

A Mettler Toledo DSC 822e equipment was used to carry out the tests. 3.5 ± 0.5 mg of powdered specimens, in standard 40 µl aluminum crucibles, were heated under air flow (50 ml/min) at a rate of 10°C/min.

3. Results and Discussion

3.1. Pyrolysis-GC/MS

A typical pyrogram obtained by pyrolysis (Py)-GC/MS of wood samples at 550°C is presented in Fig. 2. Peak assignments are presented in Table 1. The compounds identified by Py-GC/MS have been classified as carbohydrate-derived (C)

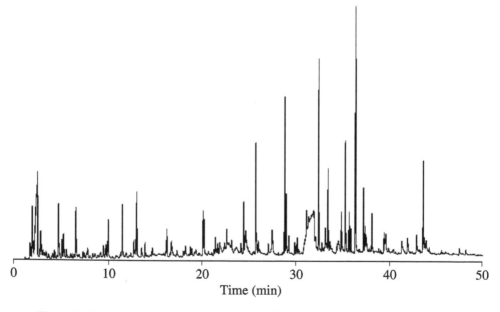

Figure 2. Pyrolysis chromatogram obtained at 550°C for maple wood reference sample.

Table 1.
Identification of main pyrolysis products in maple and birch wood species

Compound name	Retention time (min)	Molecular formula	Molecular weight (g/mol)
Lignin-derived compounds			
Guaiacyl-type unit			
Guaiacol	16.24	G	124
4-Methyguaiacol	20.13	$G-CH_3$	138
4-Ethylguaiacol	23.15	$G-CH_2-CH_3$	152
4-Vinylguaiacol	24.46	$G-CH=CH_2$	150
Eugenol	25.82	$G-CH_2-CH=CH_2$	164
4-Propylguaiacol	26.13	$G-CH_2-CH_2-CH_3$	166
Vanillin	27.43	$G-CHO$	152
Cis-isoeugenol	27.52	$G-CH=CH-CH_3$	164
Trans-isoeugenol	28.94	$G-CH=CH-CH_3$	164
Homovanillin	29.22	$G-CH_2-CHO$	166
1-(4-hydroxy-3-methoxyphenyl) propyne	29.86	$G-CC-CH_3$	162
1-(4-hydroxy-3-methoxy phenyl) allene	30.03	$G-CH=C=CH_2$	162
Acetovanillone	30.13	$G-CO-CH_3$	166
Guaiacylacetone	31.35	$G-CH_2-CO-CH_3$	180
Trans-coniferylalcohol	37.48	$G-CH=CH-CH_2OH$	180
Syringyl-type unit			
Syringol	25.73	S	154
4-Methylsyringol	28.78	$S-CH_3$	168
4-Ethylsyringol	31.12	$S-CH_2-CH_3$	182
4-Vinylsyringol	32.42	$S-CH=CH_2$	180
4-Allyl syringol	33.40	$S-CH_2-CH=CH_2$	194
4-Propylsyringol	33.60	$S-CH_2-CH_2-CH_3$	196
Cis-4-propenylsyringol	34.83	$S-CH=CH-CH_3$	194
Syringaldehyde	35.25	$S-CHO$	182
4-propynylsyringol	35.66	$S-CC-CH_3$	192
4-propynylsyringol isomer	35.84	$S-CC-CH_3$	192
Trans-4-propenylsyringol	36.33	$S-CH=CH-CH_3$	194
Acetosyringone	37.19	$S-CO-CH_3$	196
Syringylacetone	38.07	$S-CH_2-CO-CH_3$	210
Syringic acid methyl ester	38.19	$S-COOCH_3$	212
Propiosyringone	39.58	$S-CO-CH_2-CH_3$	210
Dihydrosinapyl alcohol	41.31	$S-CH_2-CH_2-CH_2OH$	212
Trans-sinapaldehyde	43.62	$S-CH=CH-CHO$	208
Trans-sinapylalcohol	43.89	$S-CH=CH-CH_2OH$	210
Carbohydrate-derived compounds			
Acetic acid	2.52	$C_2H_4O_2$	60
Furfural	6.56	$C_5H_4O_2$	96
Furfuryl alcohol	7.30	$C_5H_6O_2$	98
2-Cyclopent-1,4-dione	8.33	$C_5H_4O_2$	96
5H-Furan-2-one	9.44	$C_4H_4O_2$	84

Table 1.
(Continued.)

Compound name	Retention time (min)	Molecular formula	Molecular weight (g/mol)
2,3-Dihydro-5-methylfuran-2-one	9.96	$C_5H_6O_2$	98
4-Hydroxy-5,6-dihydro-2-pyran-2-one	13.00	$C_5H_6O_3$	114
3-Hydroxy-2-methyl-2-cyclopente-1-one	13.50	$C_6H_8O_2$	112
2-Hydroxy-3-methyl-2-cyclopenten-1-one	13.90	$C_6H_8O_2$	112
3-Hydroxy-2-methyl-4H-pyran-4-one	17.29	$C_6H_6O_3$	126
5-Hydroxymethyl-2-furfuraldehyde	21.86	$C_6H_6O_3$	126
Levoglucosan	31.62	$C_6H_{10}O_5$	162

Table 2.
Percentages of lignin and carbohydrates-related products released from reference woods and welded materials of sugar maple and yellow birch

	Sugar maple				Yellow birch			
	Ref	SD	WM	SD	Ref	SD	WM	SD
G-unit	22.0	1.2	20.3	3.0	14.9	1.0	16.3	1.6
S-unit	50.2	1.2	47.6	2.1	46.6	2.4	45.6	2.7
C	27.8	1.8	32.1	2.7	38.6	3.2	38.1	2.2
S/G	2.3	0.2	2.2	0.1	3.1	0.1	2.8	0.4
L/C	2.6	0.2	2.1	0.3	1.6	0.2	1.6	0.2

Ref: wood reference; SD: standard deviation; WM: welded material; G-unit: guaiacyl-unit lignin; S-unit: syringyl-unit lignin; L: lignin; C: carbohydrates.

and lignin-derived (L) products. For lignin-derived products, the distinction was made between the guaiacyl (G) and the syringyl (S) types of compounds. A total of 45 compounds were identified, 12 corresponding to carbohydrate degradation products, and 33 to lignin, of which 15 were guaiacyl-type and 18 syringyl-type compounds. The percentages of carbohydrates derived, and lignin G- and S-type derived products determined for reference wood (Ref) and welded material (WM) are presented in Table 2. The syringyl/guaiacyl (S/G) and lignin/carbohydrate (L/C) ratios were determined for each sample and the averages from 12 measurements were calculated. The pyrolysis results obtained in this study seem to correspond to the results obtained by wet chemical methods reported in the literature [10] which are presented in Table 3. The L/C ratio was determined to be higher for maple wood than for birch wood (reference wood) in this study as can be seen from Table 2, indicating somewhat higher lignin and lower carbohydrate content in maple wood. For lignin of both species, the S-type units seem to be the major source of lignin pyrolysis-derived compounds, which is in agreement with generally easier

Table 3.
Chemical composition (in %) of the two studied Canadian hardwoods [10]

	α-cellulose	Glucans	Xylans	Galactans	Arabinans	Mannans	Acetyl	Lignin
Maple	45	52	15	<0.1	0.8	2.3	2.9	22
Birch	47	47	20	0.9	0.6	3.6	3.3	21

release of the syringyl type of moieties from hardwood lignins. The slight difference in lignin contents between the two studied wood species seems to be reflected by the pyrolysis products patterns derived from the lignins from the two species: the G-type compounds are significantly more released by pyrolysis of maple wood (Table 2). Consequently, the S/G ratio is lower for maple reference wood than for birch. This finding indicates that maple wood native lignin probably had more condensed structure than the native lignin of birch wood. The same observation has been made for the welded material: the S/G ratio for the lignin-derived products released by pyrolysis from the welded material of maple wood remains lower than that of birch wood. When comparing only between the reference wood sample and the welded material of the same wood, S/G ratio seems to remain unchanged, which is true for both species studied in this work. The G–S (hardwood) lignins with higher percentage of G-units have generally more condensed structures because of the availability of the aromatic C5 position for coupling, which is at the origin of the formation of very resistant types of C–C bonds (β-5 and 5–5) and diaryl ether (4-O-5) bonds. The S-units are mainly linked by more labile ether bonds (mainly at the C4 position of the aromatic rings) as described in various lignin structural models [11]. Therefore, the hardwood lignins containing higher proportion of S-units should be more prone to structure degradation or modification. One of the reactions described for the G–S lignins during European beech wood welding is the demethylation of syringyl units [3]. The higher thermal sensitivity of hardwood lignin S-moieties than G-moieties has been reported in yet another study of hardwood pyrolysis [12]. One of the explanations for the higher susceptibility to thermal degradation of S-units is related to their lower redox potential than G-units due to the presence of the second methoxyl group in the C5, as well as to the generally lower probability for condensation processes to occur with syringyl units. Consequently, fewer inter-units linkages, involving the syringyl units, facilitate their release [13]. In order to gain a better insight into the lignin structure changes occurring during welding process, we investigated in detail the changes in contents of S- and G-type pyrolysis products. The sum of relative peak areas of S- and G-type lignin pyrolysis products was normalized to 100% lignin, and the separate summations were performed for lignin-derived compounds containing oxygenated functions (Oxy) in their lateral chains (aldehydes, ketones, alcohols and esters) and compounds with only alkyl (Alk) side chains. Oxy/Alk ratios for both G and S types of compounds were, thus, calculated for reference and welded wood samples. The results obtained for both reference wood species demonstrate similar Oxy/Alk ratios for both S- and G-types lignin products. However, the Oxy/Alk ratio is significantly reduced in G-type lignin-derived products identified in welded material of maple wood, while it remained unchanged in birch. For S-type units, the Oxy/Alk ratio remained unchanged after welding, for both maple and birch wood species. The decrease of Oxy/Alk ratio in G-type lignin-derived products from sugar maple welded material could be specifically related to the reduction (of 2.2% on average) of *trans*-coniferylalcohol (Table 4). This result indicates a temperature-induced loss of the terminal hydroxymethyl groups from the G-type lignin side chains in the form of formaldehyde, as already observed in previous studies [12], which could then efficiently contribute to condensation reactions of lignin fragments in the welding zone of maple wood. Since our Py-GC/MS results on reference wood samples indicate that G-units are more important in native lignin of maple than of birch, this could serve as an explanation for the more efficient welding of maple wood as determined by tensile strength measurements [7]. The release of formaldehyde from the terminal –CH$_2$OH groups from syringyl type lignin seems to be less important, even though a slight decrease of syringyl-derived alcohols from welded material as compared to reference wood has been determined for both maple and birch wood species (Table 4). The rearrangements of S-type lignin structures have been detected by following the individual lignin-derived products originating from reference and welded maple wood samples. *Cis*-4-propenylsyringol content increased by 0.7% on average in the welded maple wood while *trans*-4-propenylsyringol content decreased by 2.7%, the latter constituting the major change related to an individual compound used to quantify lignin changes caused by welding of sugar maple. The variations of contents of *cis* (Z) and *trans* (E) isomers of propenyl phenols during thermal treatment of wood have already been reported from a study on lignin structure modifications during wood combustion [15]. It is interesting to note that the *trans*-configuration of the propenyl phenol is confirmed to be the major configuration in S-type lignin-derived products determined by Py-GC/MS of reference and welded wood samples of both studied species (Table 4). For birch wood, the changes in contents of these compounds between reference and welded wood samples were determined to be insignificant. However, a significant decrease of syringaldehyde and *trans*-sinapaldehyde contents between the reference and welded wood samples was determined for maple, thus, contributing to the overall decrease of S-type compounds from the welded material of maple wood (reduction of 1% on average). Taken together with the decrease of the G-units compounds from the welded material, this could explain the almost unchanged S/G ratios between the wood reference and its welded counterpart determined for both species (Table 2). This is an important result of the present study which indicates that the S/G ratio, often taken as a sole parameter, is insufficient to provide the explanation for the studied phenomena. The better welding performance of sugar maple wood may be related to the more significant release of formaldehyde from the guaiacyl moieties of lignin which is documented by the most significant decrease of

Table 4.
Lignin-related pyrolysis products (in %) from sugar maple and yellow birch reference woods (Ref) and welded materials (WM)

	Maple		Birch	
	Ref	WM	Ref	WM
G-type derivatives				
Guaiacol	1.8*	2.0*	1.5	1.6
4-Methylguaiacol	2.0	2.4	2.1	2.6
4-Ethylguaiacol	0.8	1.0	1.1	1.5
4-Vinylguaiacol	3.6	4.6	2.6	3.0
Eugenol	1.2	1.2	1.1	1.1
4-Propylguaiacol	0.3	0.3	0.3	0.4
Vanillin	1.6	1.1	1.4	1.3
Cis-isoeugenol	1.0	0.9	0.8	1.0
Trans-isoeugenol	3.5	3.5	3.0	3.2
Homovanillin	1.2	0.9	1.1	1.2
1-(4-hydroxy-3-methoxyphenyl) propyne	0.8	0.6	0.7	0.7
1-(4-hydroxy-3-methoxyphenyl) allene	0.6	0.4	0.5	0.8
Acetovanillone	1.8*	1.4*	1.4	1.3
Guaiacylacetone	5.9	3.7	4.7	5.4
Trans-coniferylalcohol	4.4*	2.2*	2.0*	1.1*
G-aldehyde	2.9	2.0	2.5	2.5
G-ketone	7.7	5.1	6.0	6.8
G-alcohol	4.4*	2.2*	2.0*	1.1*
Oxy/Alk	1.1*	0.6*	0.9	0.9
S-type derivatives				
Syringol	5.2	5.7	5.0	5.0
4-Methylsyringol	5.7	6.7	6.9	7.4
4-Ethylsyringol	8.1	9.8	7.4	7.3
4-Vinylsyringol	10.5	11.2	9.7	10.7
4-Allyl syringol	2.8	3.1	3.4	3.4
4-Propylsyringol	0.7	0.8	0.9	0.7
Cis-4-propenylsyringol	1.6	2.3	1.9	2.0
Syringaldehyde	6.1	5.2	6.9	6.1
4-Propynylsyringol	1.9	1.7	2.1	1.7
4-Propynylsyringol isomer	1.0	1.0	1.2	1.1
Trans-4-propenylsyringol	15.9	13.2	13.9	14.6
Acetosyringone	2.9	2.6	3.4	2.8
Syringylacetone	1.8	2.1	2.1	2.2
Syringic acid methyl ester	0.5	0.5	0.3	0.2
Propiosyringone	1.3	1.3	1.6	1.4
Dihydrosinapyl alcohol	0.4	0.4	1.2	0.8
Trans-sinapaldehyde	5.6	5.3	6.7	5.4
Trans-sinapylalcohol	1.8	1.1	1.4	0.8
S-aldehyde	11.7	10.5	13.6	11.5
S-ketone	4.3	3.9	5.0	4.2
S-alcohol	2.2	1.5	2.5	1.6
S-ester	0.5	0.5	0.3	0.2
Oxy/Alk	0.4	0.3	0.5	0.4

Ref: wood reference; WM: welded material; G-unit: guaiacyl-unit lignin; S-unit: syringyl-unit lignin; Oxy/Alk: ratio between lignin-derived compounds containing oxygenated functions (Oxy) in their lateral chains (aldehydes, ketones, alcohols and esters) and compounds with only alkyl (Alk) side chains; * significantly different ($p < 0.05$).

guaiacyl–propane–lignin-derived products (G-C3: guaiacyl with three carbon side chain) from welded material *versus* reference wood. Thus, released formaldehyde can then participate in the condensation reactions with the phenolic structures of lignin released from thermally modified lignins in the welding zone. The release of formaldehyde is accompanied by the formation of the substructures in lignin containing strong vinyl ether linkages, their presence could also be of importance for the better mechanical performance of the welded products from maple.

As for carbohydrates, the composition and relative distribution of polysaccharide-derived products for the two studied wood species are different, as can be seen from Tables 2 and 5, which can be related to the different chemical compositions of reference woods already shown by literature data (Table 3). The results (calculated on the total carbohydrates basis of 100%) for the percent of carbohydrate-derived products: acetic acid, furfural, 5-hydroxy-methyl-furfuraldehyde and levoglucosan are presented in Table 5. The major carbohydrate product determined from both reference wood species is anhydroglucose (levoglucosan), which is related to cellulose (Table 2). Levoglucosan content was determined to remain stable after welding for both species (Table 5) which could mean that cellulose is only weakly altered during the welding process, especially keeping in mind that we are dealing with fast rather than slow pyrolysis at lower pyrolysis temperature. At the temperatures reached in the welding process applied in the present study (266–323°C for maple and 244–308°C for birch) [7], the chemical changes of cellulose could be related mainly to the amorphous regions of microfibrils. In effect, we have determined a slightly higher release of levoglucosan from maple wood welded material, which could perhaps be related to the higher temperature range reached in maple welding, resulting in a slightly increased alteration of cellulose. Release of acetic acid during wood thermal treatments has been shown to depend on wood species, wood acidity, temperature, and wood moisture content [16, 17]. Acetic acid contents in the pyrolysis products were found to be comparable between the reference wood and welded material for both wood species studied in this work (Table 5). However, the release of furfural increased after welding, which would indicate increase of xylan concen-

Table 5.
Carbohydrates-related pyrolysis products (in %) from reference wood and welded material

	Maple				Birch			
	Ref	SD	WM	SD	Ref	SD	WM	SD
Acetic acid	21.7	2.5	21.8	5.4	20.1	1.7	21.5	2.4
Furfural	6.5	0.4	8.8	0.8	4.3	0.5	6.5	0.7
Hydroxymethylfurfural	2.5	1.1	1.2	0.5	1.4	1.3	2.8	1.6
Levoglucosan	62.9	2.8	65.3	3.9	49.9	4.0	47.9	5.8
Others	6.4	2.4	2.9	6.6	24.3	2.5	21.3	1.9

Ref: reference wood; SD: standard deviation; WM: welded material.

tration in the welding zone for both wood species. One could speculate that some of this xylan has been introduced into the welding zone along with the LCC formed during the welding process by the reactions between lignin intermediates and xylan fragments. Indeed, the covalent bonds existing between lignins and xylans in the natural state in hardwoods are well documented and they are at the origin of the lignin–carbohydrate complexes (LCC) in hardwoods and they are determinant for wood reactivity [18]. The significant decrease of the release of HMF from the welded material *versus* reference maple wood could also be related to the higher temperatures reached in maple wood welding, causing some loss of hexosan sugars through thermal reactions. Therefore, the most significant changes which can be deduced from the results on the carbohydrate-derived pyrolysis products concern xylan introduction into the welding zone. It has been demonstrated in the study on the miscibility of the isolated wood polymers (beech xylan and dioxane lignin from beech) that it increased in the presence of LCC (isolated from the same beech wood as the previous two polymers used for the study) and with an increase of temperature [19]. As the conditions under which the welding process is performed do favor the formation of LCC (which is in accord with our DSC results discussed below) one can speculate that the xylan fragments become more miscible with the rest of polymer materials found in the welding zone and this is how the xylan-derived furfural increase, determined in both birch and maple welded materials, can be explained.

3.2. XPS

Changes in oxygen/carbon (O/C) ratio between the reference wood and welded material should allow identification of the main reactions involved in wood welding. Friction between the dowel and the substrate during insertion by high-speed rotation causes temperature-induced lignin softening and degradation which leads to wood welding [1]. Depending on the heating rate and temperature, the onset of wood pyrolysis occurs approximately at 190°C and the most significant thermal modification occurs between 300 and 400°C [20]. A recent study reported that temperature values of 266–323°C and 244–308°C have been reached during the wood-to-wood dowel welding process of maple and birch, respectively [7]. It is well known that in these temperature ranges, only the amorphous components of wood are affected, particularly lignin and hemicelluloses, and to a lower extent the amorphous cellulose. Results obtained for the surface composition are presented in Table 6, along with the corresponding O/C atomic ratios. The increase of O/C ratios of welded materials, more pronounced in the case of maple wood, indicates that chemical substances which contain oxygenated functionality are formed and are moved to the welding zone. The increase of O/C can be related to the increase of free phenolic groups and the decrease of the typical (ether) bonds between the phenolpropane units in lignin, as found in previous study [21] in the case of linear mechanical friction. The wood-to-wood welding process phenomena could be related rather to those involved in fast pyrolysis than to slow or conventional py-

Table 6.
Surface composition (in %) of the studied samples determined by XPS

Sample	Elements (%)				Carbon C1 components (%)				Oxygen O1 components (%)		Ratio
	C	O	N	Others	C1	C2	C3	C4	O1	O2	O/C
Birch Ref	72.2	27.4	0.3	0.1	47.8	41.2	6.9	4.1	1.8	98.2	0.38
Birch WM	69.1	28.3	–	2.6	46.8	42.9	7.1	3.1	6.8	93.2	0.41
Maple Ref	69.0	27.2	0.1	3.6	48.4	42.3	6.7	2.6	3.3	96.7	0.39
Maple WM	65.2	30.4	–	4.5	35.8	50.7	10.9	2.5	1.3	98.7	0.47

Ref: wood reference; WM: welded material.

rolysis. After the slow pyrolysis, such as involved in wood heat treatment, the O/C ratio tends to decrease due to more significant cross-linking *via* methylenic bridges leading to carbon-enriched material. During fast pyrolysis, the pyrolysis liquids, rich in oxygenated compounds, are formed by rapidly and simultaneously depolymerizing and fragmenting cellulose, hemicelluloses, and lignin during the rapid increase in temperature [22], which also favours the miscibility of these polymers as discussed previously. Rapid quenching, as is the case when welding process is stopped, 'freezes in' the intermediate products of the fast pyrolysis degradation of hemicelluloses, cellulose and lignin.

The C_{1s} and O_{1s} peaks for the reference and welded materials are presented in Fig. 3, while the results of peak fitting for the C_{1s} and O_{1s} peaks are presented in Table 6. The changes in the intensities of components of the C_{1s} peak induced by wood welding confirm the formation of oxygenated compounds discussed above. In particular, the proportion of the C1 component (284.6 eV, corresponding to carbon in C–C and C–H bonds) decreased, while those of C2 (286.5 eV, carbon in C–OH and C–O–C) and C3 (287.9 eV, carbon in C=O and O–C–O) increased. This could be due to the formation of furanic compounds from amorphous polysaccharides degradation as well as to the release of phenolic/carbonyl compounds from lignin depolymerization. Additionally, it is known that the depolymerization of the lignins produces phenolic aldehydes [9]. During the first step of pyrolysis, OH– functions are thermo-oxidized into corresponding aldehydes which can explain the increase of C3 components. Concerning C4 component (289.2 eV, carbon in O–C=O), this study shows that the welded materials contained less carboxyl and/or acetyl groups than the reference which could mean that the thermo-degradation of hemicelluloses might have involved acid catalyzed depolymerization and dehydration which lead to increased furfural formation which could explain C3 component increase in welded material. The ratio of the O1 component (531.8 eV, mainly oxygen in C=O type bonds) to the O2 component (532.9 eV, mainly oxygen in C–O bonds) of maple and birch wood samples show different patterns. During welding, maple wood shows an increase of O1/O2 ratio while this ratio decreases in the case of

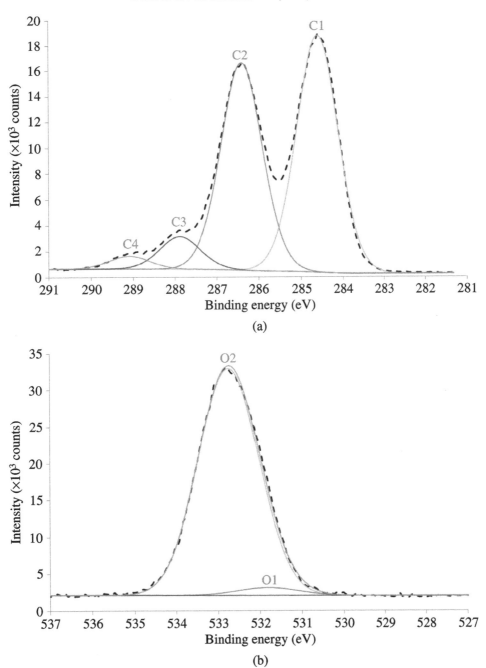

Figure 3. XPS C_{1s} (a) and O_{1s} (b) spectra of birch wood reference sample.

welded birch wood. Thus, XPS study can partly explain the differences observed between birch and maple wood welding mechanical performances, since the chemical reactions involved in wood welding could be related to the differences in chemical composition of the reference materials (Table 3). During the pyrolysis process, the main effects of thermal modification on the chemical structure of wood are degradation of hemicelluloses and amorphous cellulose, and demethylation and depolymerisation of lignin with the formation of low molecular weight fragments (monomers and formation of oligomers with high reactivity). The degradation of hemicelluloses starts by deacetylation, and thereby acetic acid is released from the acetylated xylans of hardwoods, which acts as a depolymerization catalyst which further enhances the polysaccharide decomposition yielding furanic material. In addition, lignin is subjected to the cleavage of the ether linkages, especially β-O-4, leading to the formation of highly reactive free phenolic hydroxyl groups and α- and β-carbonyl groups. These structures are important for the cross-linking *via* methylenic bridges between the lignin fragments and formaldehyde in the presence of organic acids released from hemicelluloses [23]. However, in the case of fast pyrolysis, which is more like the welding process, cross-linking reactions are limited [22]. Additional condensation reactions involving furanic compounds originating from amorphous polysaccharides (furfural, hydroxylmethylfurfural) and lignin may occur [23]. The XPS results suggest that the higher tensile strengths obtained for maple wood than for birch wood samples [7] could be explained by: (1) increase of free phenolic functions in structures originating from lignin, which can support hydrogen bonding and closer association between lignin and the fibers; (2) cross-linking *via* methylenic bridges leading to new lignin based polymers with newly formed C–C bonds which are the basis of more condensed lignin structure which can also involve degradation products of hemicelluloses [24] as well as formation of a lignin–cellulose complex [25]; (3) a less altered welded material due to the lower polysaccharide decomposition.

3.3. DSC Analysis

DSC has been proven to provide valuable information about wood components modification or degradation [26–29]. DSC curves from the birch and maple wood samples are shown in Fig. 4(a and b), respectively, while peak temperatures are presented in Table 7. Two main combustion peaks were observed at ca. 340 and 515°C and small shoulders at around 295 and 440°C. Tsujiyama and Miyamori [26] assigned these transitions to the combustion of the amorphous polysaccharides and to mixtures of lignin and polysaccharides (LCC, lignin carbohydrate complex), respectively. DSC curves, shown in Fig. 4(a and b), indicate that the wood-to-wood welding affects differently the wood components, depending on the species. From Fig. 4(a) corresponding to birch wood we can note a decrease in the two peak temperatures as well as a slight shift of the LCC peak (2nd exotherm) for the welded material. This result indicates that the wood-to-wood welding altered the wood structure, mainly the amorphous polysaccharide part. This finding confirms the significant increase of C=O function in birch welded material as determined by XPS analysis which can be related to the formation of furfural derivatives. It seems that no condensation with lignin fragments occurred during welding of birch. Addition-

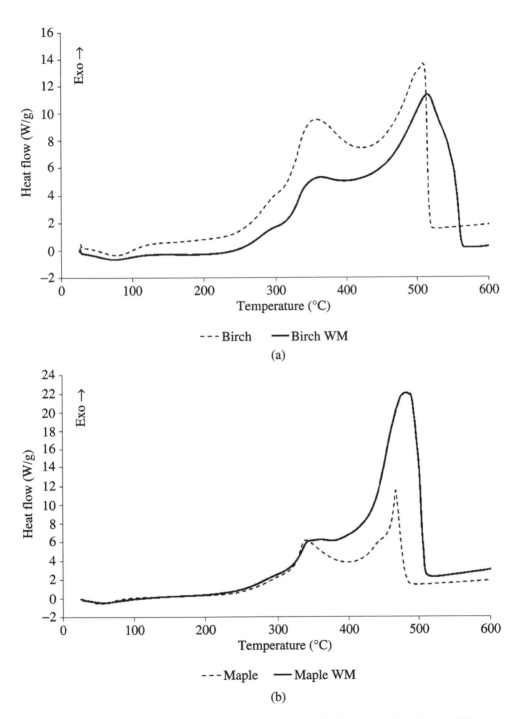

Figure 4. DSC curves of reference and welded materials for birch (a) and maple wood (b).

Table 7.

Peak temperatures of decomposition processes of wood samples
obtained from DSC curves

Sample	Amorphous polysaccharide peak (°C)	LCC peak (°C)
Birch Ref	357.3	508.4
Birch WM	364.1	514.5
Maple Ref	339.6	468.0
Maple WM	340.8	483.8

Ref: wood reference; WM: welded material; LCC: lignin–carbohydrate complex.

ally, the shift of LCC peak may indicate a light increase of lignin content in wood welded material. As for the welding of maple wood, the DSC curve presented in Fig. 4(b) demonstrates a significant increase and a shift to higher temperature of LCC part, as well as less distinctive peak of amorphous polysaccharide part (shoulder). It seems that during the friction, a new LCC based polymer is formed resulting from (1) the condensation reaction between lignin fragments and furfural derivatives released from hemicelluloses, and (2) the formation of a lignin–cellulose complex as also demonstrated previously by XPS. In addition, the significant increase of the LCC peak height may be due to the presence of highly oxygenated compounds which mainly constitute the interfacial welded material. This result seems to be in accord with the pyrolysis study results which revealed an increase of furfural content from the welded material which we interpreted by an enhanced mixing of the xylan polymer with the LCC complex (which can be regarded as a co-polymer of lignin and carbohydrates) with the increase of temperature in the welding zone (again, more important for maple than for birch). The differences determined by the Py-GC/MS, XPS and DSC between the two studied species may explain the higher tensile strengths determined for maple wood than for birch wood samples.

4. Conclusion

The results obtained in this research clearly demonstrate that the physico-chemical analyses performed for the first time on the welding of two Canadian wood species: sugar maple and yellow birch using Py-GC/MS, XPS and DSC are very useful to explain the differences in chemical changes involved in wood welding. The major difference in welding performance of the two species seems to be related to the difference in lignin structures in the reference wood samples. The somewhat more pronounced guaiacyl character of maple wood lignin seems to be responsible for the higher reactivity of this lignin which is more susceptible to condensation reactions during welding, both with formaldehyde and with carbohydrates furanic degradation products. It seems that for both wood species studied in this research, lignin is the most important wood constituent for the successful welding. An interesting finding of this research is the presence of LCC in the welded material, which could be related to the enhanced miscibility of lignin and hemicelluloses (xylan in the case of studied hardwoods) polymers in the presence of the common co-polymer (LCC), which can explain the high furfural release from the welded material in pyrolysis study. Different DSC patterns obtained for the welded material *versus* reference wood for the two studied species could be again related to the somewhat higher welding temperature determined for maple. Our study has also provided a very important finding that the chemical changes in lignin structure during welding process (and any other thermal or chemical treatment) cannot be solely based on the S/G ratio. The results of this study have, therefore, clearly demonstrated that the differences in the wood polymer composition influence the performance of wood in the welding process and, consequently, the mechanical properties of the welded materials.

Acknowledgements

The authors are very grateful to the Fond Québécois de Recherche sur la Nature et les Technologies (FQRNT) for the financial support to this project, and to Mr. Yves Bédard and Mr. Luc Germain for the technical support.

References

1. A. Pizzi, J.-M. Leban, F. Kanazawa, M. Properzi and F. Pichelin, *J. Adhesion Sci. Technol.* **18**, 1263–1278 (2004).
2. C. Ganne-Chedeville, A. Pizzi, A. Thomas, J.-M. Leban, J.-F. Bocquet, A. Despres and H. R. Mansouri, *J. Adhesion Sci. Technol.* **19**, 1157–1174 (2005).
3. F. Kanazawa, A. Pizzi, M. Properzi, L. Delmotte and F. Pichelin, *J. Adhesion Sci. Technol.* **19**, 1025–1038 (2005).
4. D. Fengel and G. Wegener, *Wood: Chemistry, Ultrastructure and Reactions*. Walter De Gruyter, Berlin (1989).
5. D.-Y. Kim, Y. Nishiyama, M. Wada, S. Kuga and T. Okano, *Holzforschung* **55**, 521–524 (2001).
6. M. M. Nassar and G. D. M. Mackay, *Wood Fiber Sci.* **16**, 441–453 (1984).
7. G. Rodriguez, P. N. Diouf, P. Blanchet and T. Stevanovic, *J. Adhesion Sci. Technol.*, in press.
8. J. C. del Rio, A. Gutierrez, M. Hernando, P. Landrin, J. Romero and A. T. Martinez, *J. Anal. Appl. Pyrolysis* **74**, 110–115 (2005).
9. M. F. Nonier, N. Vivas, N. de Gaulejac, C. Absalon, P. Soulie and E. Fouquet, *J. Anal. Appl. Pyrolysis* **75**, 181–193 (2006).
10. C. R. Pettersen, in: *The Chemistry of Solid Wood*, R. M. Rowell (Ed.), Adv. Chem. Ser. No. 207, pp. 57–126. American Chemical Society, Washington, DC (1984).
11. J. Ralph, G. Brunow and W. Boerjan, in: *Encyclopedia of Life Science*, F. Rose and K. Osborne (Eds), pp. 1–10. Wiley, Chichester (2007).

12. C. Assor, V. Placet, B. Chabbert, A. Habrant, C. Lapierre, B. Pollet and P. Perré, *J. Agric. Food Chem.* **57**, 6830–6837 (2009).

13. J. C. del Rio, A. Gutierrez, M. J. Martinez and A. T. Martinez, *J. Anal. Appl. Pyrolysis* **58**, 441–452 (2001).

14. R. M. Rowell, in: *Chemical Modification of Lignocellulosic Materials*, D. N.-S. Hon (Ed.), pp. 295–310. Marcel Dekker, New York, NY (1996).

15. J. Kjallstrand, O. Ramnas and G. Petersson, *J. Chromatography. A* **824**, 205–210 (1998).

16. F. Tjeerdsma, M. Boonstra, A. Pizzi, P. Tekely and H. Militz, *Holz Roh Werkst.* **56**, 149–153 (1998).

17. B. J. Tjeerdsma and H. Militz, *Holz Roh Werkst.* **63**, 102–111 (2005).

18. G. Henriksson, M. Lawoko, M. E. E. Martin and G. Gellerstedt, *Holzforschung* **61**, 668–674 (2007).

19. M. Shigematsu, M. Morita, I. Sakata and M. Tanahashi, *Macromol. Chem. Phys.* **195**, 2827–2837 (1994).

20. K. David, Y. Pu, M. Foston, J. Muzzy and A. Ragauska, *Energy Fuels* **23**, 498–501 (2009).

21. B. Stamm, E. Windeisen, J. Natterer and G. Wegener, *Wood Sci. Technol.* **40**, 615–627 (2006).

22. D. Mohan, C. U. Pittman and P. H. Steele, *Energy Fuels* **20**, 848–889 (2006).

23. M. Nuopponen, T. Vuorinen, S. Jamsä and P. Viitaniemi, *J. Wood Chem. Technol.* **24**, 13–26 (2004).

24. L. Delmotte, C. Ganne-Chedeville, J. M. Leban, A. Pizzi and F. Pichelin, *Polym. Degrad. Stabil.* **93**, 406–412 (2008).

25. B. Košíková, M. Hricovini and C. Cosetino, *Wood Sci. Technol.* **33**, 373–380 (1999).

26. S.-I. Tsujiyama and A. Miyamori, *Thermochim. Acta* **351**, 177–181 (2000).

27. H.-L. Lee, G. C. Chen and R. M. Rowell, *J. Appl. Polym. Sci.* **91**, 2465–2481 (2004).

28. E. Franceschi, I. Cascone and D. Nole, *J. Therm. Anal. Calorim.* **91**, 119–125 (2008).

29. M. Krzesinska, J. Zachariasz, J. Muszynski and S. Czajkowska, *Bioresource Technol.* **99**, 5110–5114 (2008).

Moisture Sensitivity of Scots Pine Joints Produced by Linear Frictional Welding

Mojgan Vaziri [a,*], **Owe Lindgren** [a], **Antonio Pizzi** [b] **and Hamid Reza Mansouri** [b]

[a] Department of Wood Science and Technology, Luleå University of Technology, Forskargatan 1, 931 87 Skellefteå, Sweden

[b] ENSTIB-LERMAB, Université Henri Poincaré — Nancy 1 27 Rue du Merle Blanc, BP 1041, 88051 EPINAL Cedex 9, France

Abstract

The industrial application range of welded wood so far has been limited to interior use because of its poor moisture resistance. Influences of some welding and wood parameters such as welding pressure, welding time, and heartwood/sapwood on water resistance of Scots pine (*Pinus sylvestris*) were investigated. An X-ray Computed Tomography scanner was used to monitor density change in weldlines during water absorption–desorption. Axial samples measuring 200 mm × 20 mm × 20 mm from Scots pine were welded and placed standing in 5-mm-deep tap water. Then they were taken out of the water one at a time and scanned at 10-min intervals until the first crack appeared in the weldline where the two parts of each specimen made connection.

Results showed that the X-ray Computed Tomography can be used as an effective tool to study welded wood. Welding pressure, welding time, and heartwood/sapwood showed significant effect on length and location of the crack in the welded zone. Data evaluation showed that combination of 1.3 MPa welding pressure, 1.5 s welding time and using heartwood led to highest moisture resistance, which produced only a very short crack in the beginning of the weldline.

Keywords

Linear welding, tomography, water resistance, welding conditions

1. Introduction

One disadvantage of welded wood produced by linear welding is its sensitivity to moisture. Therefore, its application is limited mainly for interior use. Moisture leads to splitting of the bondline and makes it unsuitable for structural use in spite of its high dry strength [1].

The initial objective of this study was to demonstrate how X-ray Computed Tomography (CT-) scanning and image processing could also be successfully used for

---

[*] To whom correspondence should be addressed. Tel.: +46910 58 57 04; Fax: +46910 58 53 99; e-mail: mojgan.vaziri@ltu.se

Wood Adhesives

wood welding research. Secondly to weld and study a new species (Scots pine) and thirdly to investigate the machine settings and material parameters that led to less water damage (shorter cracks) in the welded zone. The final objective was to see if welding factors had any significant effect on the location of crack in the weldline.

According to shear tensile tests on Norway spruce (*Picea abies*) and beech (*Fagus sylvatica*), an increase of normal pressure and welding frequency led to a reduction of the welding time and coefficient of friction [2]. Water resistance test on beech (*Fagus sylvatica*) immersed in cold water (15°C) showed that short displacement (2 mm), high vibration frequency (150 Hz), and short welding time (1.5 s) increased the resistance of the linear welded joints [1]. The investigation on welding of wood components showed that edge-to-edge welding of particleboard, oriented strand board (OSB), medium density fibreboard (MDF), and plywood gave better strength than face-to-face panel welding. In general, the edge-to-edge weldline was slightly weaker than the panel itself [3].

To date, investigations on water resistance of welded woods have mainly carried out using water immersion method. However, classical water immersion is not practical for investigation of changes that occur within the welded joint during absorption and desorption in outdoor conditions. Recently, there has been an increasing interest in non-destructive methods like X-ray Computer Tomography (CT-) scanning which can overcome such limitations. CT-scanning offers a non-contact, non-destructive, and 3-D measurement method. The measured X-ray absorption is very dependent on wood density and moisture changes with known time intervals are accurately measured [4]. The results of this research are based on investigations on test pieces of Scots pine (*Pinus sylvestris*) as one of the most used wood species for which welding process has not been investigated so far.

2. Materials and Methods

2.1. Wood Welding

The mechanical welding machine used was a Mecasonic linear vibration welding machine made in France, type LVW 2061, with vibration frequency up to 260 Hz, normally used to weld plastics. 40 samples of dimensions 200 mm × 20 mm × 20 mm were cut from heartwood and sapwood of Scots pine in longitudinal direction of wood grain. They were welded together two at a time to form a bonded joint of 200 mm × 20 mm × 40 mm by linear vibration. A certain range of welding machine's settings was used and some parameters such as welding pressure, welding time, and heartwood/sapwood were selected as design factors for evaluation in this study (Table 1).

Wood samples are inhomogeneous and to minimize the variability we randomly chose the samples and ignored sample to sample variability.

According to theoretical understanding and practical experience based on previous experiments, two pressure levels of 0.75 and 1.3 MPa and two welding times of 1.5 and 2.5 s were chosen for evaluation.

Table 1.
Parameters used for welding machine's setting

Parameter	Unit	Value	
Welding pressure (WP)	(MPa)	1.3	0.75
Welding time (WT)	(s)	1.5	2.5
Welding displacement (WD)	(mm)	2	
Frequency	(Hz)	150	
Holding pressure (HP)*	(MPa)	2.75	
Holding time (HT)**	(s)	50	
Equilibrium moisture content (EMC)	(%)	12	

*The clamping pressure exerted on the surface of the specimen after the welding vibration had stopped.

**The pressure holding time maintained after the welding vibration had stopped.

2.2. Water Absorption

Test objects were conditioned in the laboratory, at ambient conditions for 30 days before scanning. Moisture content of the specimens was determined by 10 additional samples. These samples were regarded as representative of the specimens as a whole. Therefore, not every specimen was tested individually with regard to moisture content. The average moisture content of the samples was 5%. Conditioned samples were scanned by a CT-scanner and one CT image was taken for each sample as shown in Fig. 1(a). Then the samples were placed with butt ends in a basin on bars of stainless steel in 5-mm-deep tap water for water absorption.

2.3. CT-Scanning and Image Processing

A SIEMENS Emotion Duo medical X-ray CT-scanner was used and scanning was carried out at ambient conditions, relative humidity (RH) 65% and temperature 22°C, according to scanner settings in Table 2. To create an environment with varying humidity like in exterior use, the test objects were transferred from water basin to CT-scanner for each scanning as shown in Fig. 2. Therefore, samples lost some water during scanning and transport. Each sample was scanned for equal intervals of 10-min according to a time schedule until the first crack appeared in the weldline as shown in Fig. 1(b).

With the assistance of image-processing software (Image J) from the National Institutes of Health (NIH) X-ray absorption was measured as a profile which is shown in Fig. 3 in an area along the weldline (Fig. 4).

3. Results

Data for length and location of first cracks in the welded zone, irrespective of when they appeared, were analyzed by means of the statistical software Minitab.

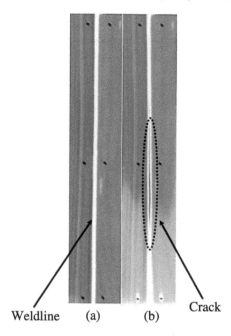

Weldline   (a)   (b)   Crack

Figure 1. One observation (sapwood welded by 0.75 MPa welding pressure and 1.5 s welding time) was chosen to show two examples of CT-images before and after water absorption. (a) CT-image before water absorption. (b) CT-image after water absorption. A crack appeared as a black streak in the weldline after water absorption for 7 h.

Table 2.
Parameters used for CT-scanner's setting

Parameter	Unit	Value
Voltage	kV	110
Current	mA	70
Scan time	s	2
Scan thickness	mm	5
Matrix	Pixels	512*512
Resolution	Pixels/mm	2.3

[*] Program and algorithm used in scanning were Body PCT Sequence and Algorithm, respectively.

All the three evaluated factors had significant effect on the length and location of the crack with 95% confidence interval. Graphs of the average responses for each factors combination are shown in Figs 5 and 6. The best combination of factors which led to only a very short crack in the beginning of the weldline is shown as a CT-image in Fig. 7 and a CT-number profile in Fig. 8. Low welding pressure (0.75 MPa), short welding time (1.5 s), and using sapwood showed the longest

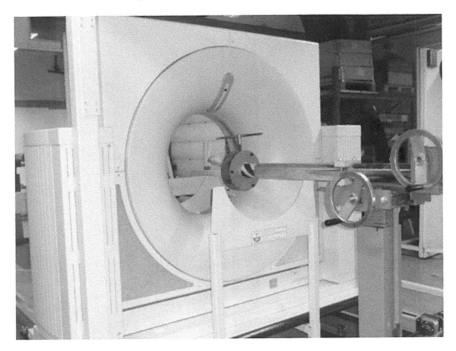

Figure 2. CT-scanner and scanning of a sample taken out of the water.

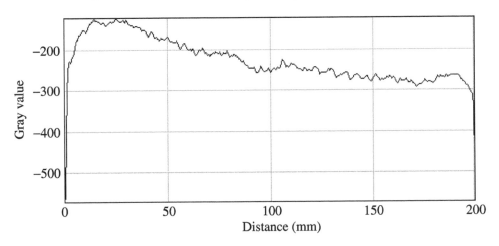

Figure 3. Gray value profile. CT data for each pixel in the longitudinal cross section of the weldline were measured in a gray-scale value between 0–255. Negative gray value indicates that the X-ray absorption is less than that of water.

crack in the middle of weldline as shown in Fig. 1(b). This can also be seen in dotted circle in Fig. 9.

This investigation involved several parameters and it was necessary to study the joint effect of parameters on response so a factorial design was considered. An experiment based on a 2^3 full factorial design and 5 replicates was performed with

A

Figure 4. (A) Area of 3 pixels wide and 202 pixels long corresponding to weldline.

three parameters (welding pressure, welding time, and heartwood/sapwood), each at only two levels of high and low. In a 2^k factorial design, it is easy and intuitive to express the results of the experiments in terms of a mathematical model called a regression model. There were two dependent variables or responses y (length and location of the crack) that depended on three independent variables (welding pressure, welding time, and sapwood or heartwood). The relationship between these variables was characterized by a linear regression model fitted to the data [5]. The analysis gave with 95% confidence interval three significant factors and linear regression models were found to be:

$$y_1 = 55 - 6.55x_1 - 6x_2 - 8.41x_3 + 9.28x_1x_2 - 10.11x_1x_3 + 16.81x_2x_3, \quad (1)$$

$$y_2 = 20.74 - 11x_1 - 0.38x_2 - 6.57x_3 + 6.97x_1x_2 + 2.92x_1x_3, \quad (2)$$

where y_1 = crack length (mm); y_2 = crack location (mm) — height of the crack along the weldline (distance between crack and beginning of the weldline where the sample is in contact with water); x_1 = welding pressure (MPa); x_2 = welding time (s); x_3 = type of wood (heartwood/sapwood). The prediction ability was calculated to 0.96 for crack length and 0. 94 for crack location and the model based on this data set could be regarded as satisfactory at predicting new responses as shown in Figs 10 and 11.

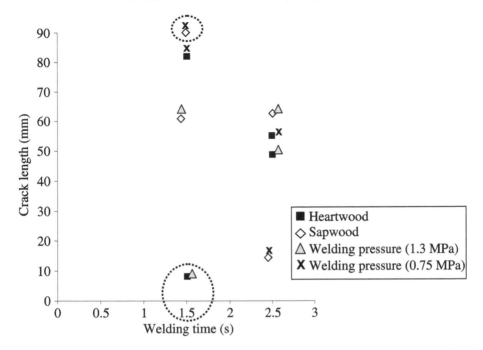

Figure 5. Influence of welding pressure, welding time, and heartwood/sapwood on length of crack in the weldline. 1.3 MPa welding pressure, 1.5 s welding time and using heartwood lead to a short crack in the weldline. However, 0.75 MPa welding pressure, 1.5 s welding time, and using sapwood produce a long crack in the weldline. The combinations of parameters which cause the shortest and longest cracks are shown in dotted circles.

4. Discussion

The linear regression equation and its coefficient of determination (R^2) values of 0.96 and 0.94 indicate that:

(a) CT-scan images can accurately estimate length and location of crack and can also be successfully applied in wood welding research.

(b) Welding parameters and wood properties have marked influence on the final properties of the welded connections such as crack length and location.

According to regression models by changing welding parameters from 0.75 MPa welding pressure, 1.5 s welding time, and sapwood to 1.3 MPa welding pressure, 1.5 s welding time, and heartwood the crack length can be decreased from 92 to 10 mm. Location of the crack can also be changed along the weldline from 0.5 to 48.58 mm from beginning of the welded zone.

Results showed that a combination of 1.3 MPa welding pressure, 1.5 s welding time, and using heartwood led to shortest crack in the beginning of weldline.

There are five reasons why high welding pressure and short welding time yield stronger bondline.

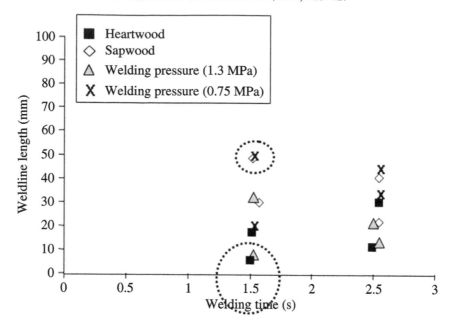

Figure 6. Influence of welding pressure, welding time, and heartwood/sapwood on location of crack at different heights of the weldline. 1.3 MPa welding pressure, 1.5 s welding time, and using heartwood lead to a crack in the beginning of weldline. However, 0.75 MPa welding pressure, 1.5 s welding time, and using sapwood produce a crack in the middle of weldline. The combinations of factors which cause crack in the beginning or middle of the weldline are shown in dotted circles.

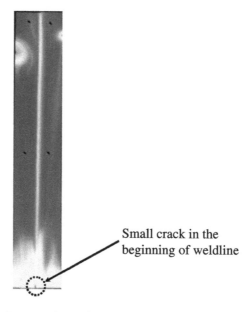

Figure 7. CT-image of a heartwood sample welded by 1.3 MPa welding pressure and 1.5 s welding time as an example for length and location of crack. This parameters combination produces only a short crack in the beginning of the weldline as shown in a dotted circle.

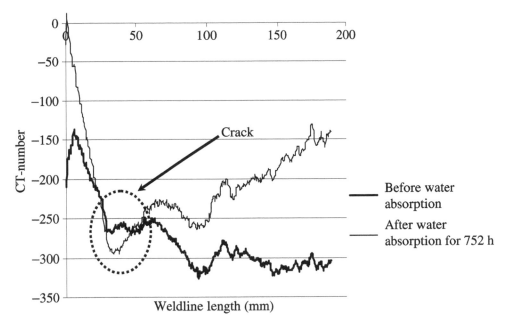

Figure 8. CT-number *versus* weldline length (mm) in a sample of heartwood, welded by 1.3 MPa welding pressure and 1.5 s welding time as an example. This factors combination produces only a very short crack in the beginning of the weldline which is shown in dotted circle. Minus sign indicates that the X-ray absorption is less than that of water.

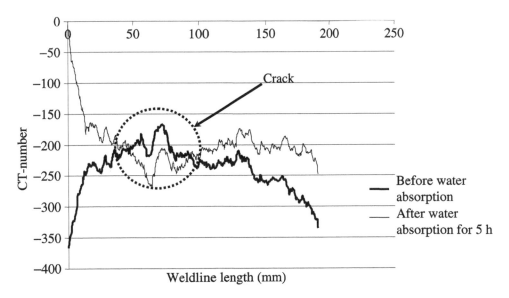

Figure 9. CT-number *versus* weldline length (mm) in a sample of sapwood welded by 0.75 MPa welding pressure and 1.5 s welding time as an example. This factors combination produces a long crack in the middle of the weldline which is shown in dotted circle. Minus sign indicates that the X-ray absorption is less than that of water.

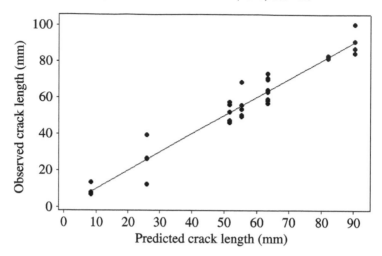

Figure 10. Scatter plot for observed crack length *versus* predicted crack length (mm). Crack lengths of 40 specimens were measured by image processing program (Image J). Predicted values were based on the regression model with $R^2 = 0.92$. Each group of multiple data points shows the measured crack lengths of five samples, welded with the same factors combination (replicates). Replication means an independent repeat of each factors combination which enables us to first obtain an estimate of experimental error, and second to obtain a more precise estimate of ($\bar{y}$) [5].

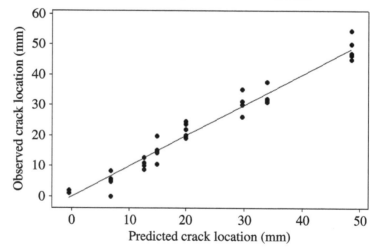

Figure 11. Scatter plot for observed crack location *versus* predicted crack location (mm). Crack locations of 40 specimens were measured by image processing program (Image J). Predicted values were based on the regression model with $R^2 = 0.94$. Each group of multiple data points shows the measured crack locations for five samples welded with the same factors combination (replicates). Replication means an independent repeat of each factors combination which enables us to first obtain an estimate of experimental error, and second to obtain a more precise estimate of ($\bar{y}$) [5].

Stamm *et al.* [2] showed in different investigations that:

1. Welding process is governed by the thermal energy generated at the interface and there is a relation between welding pressure and welding time. An in-

crease in welding pressure leads to an increase of the heat generation and a decrease of welding time consequently. Therefore, rate of welding pressure active during friction welding has a significant influence on the frictional energy generation [3].

2. The higher the pressure is, the more regular and homogeneous the welding process is and therefore, chemical and mechanical connection created during different phases of welding time is stronger [6].

3. During frictional welding softening and thermal decomposition of wood compounds makes a viscous layer of thermally-altered wood decomposition products (cellulose, hemicelluloses, and lignin) as a connector between two wooden pieces. Thermal degradation of the adjacent cell structure occurs only within a thin layer close to the contact zone. Increasing pressure leads to a decreasing coefficient of friction (viscosity) and decreasing thickness of the viscous layer. The thinner is this viscous layer the stronger is the connection between two pieces of wood [6].

Two other reasons are:

4. X-ray micro-densitometry tests on welded wood samples of beech (*Fagus sylvatica*) and spruce (*Picea abies*) showed that bondline of beech which was narrower than spruce joints had better mechanical performance [7].

5. Shorter welding time creates stronger welded joints. Viscous layer in contact zone contains a network of long wood cells and wood fibres in a matrix of molten material which then solidifies. During welding some of detached cells in the contact zone are pushed out of the joint as excess fibre. With shorter welding time, on the one hand, only a few fibres can be expelled and, on the other hand, no charring occurs in the welded zone. Therefore, weldline is not black. More remaining fibres and less charring in the contact zone ensure better connection [8].

Combination of low pressure (0.75 MPa), short welding time (1.5 s), and using sapwood showed the poorest water resistance. For wood as composite material no melting occurs during wood welding. Nevertheless, softening combined with thermal decomposition is observed during friction welding of wood, which occurs at a certain temperature limit. Massive decomposition starts if surface temperature reaches 350–380°C during the welding process. It is characterised by an incipient smoke generation, a significantly darker colour of the sample, and a rampant increase of the coefficient of friction. The chemical reactions taking place during this very short period are very important for forming the joint [6]. With regard to the chemical reactions and degradation behaviour taking place during wood welding, the heat generated is very important. Apparently, a certain pressure in combination with a certain welding frequency and welding time is required to generate the heat necessary for decomposition of wood and forming a joint. Combination of

0.75 MPa welding pressure and 1.5 s welding time cannot generate heat required to make a strong connection. At short welding time (1.5 s), higher welding pressure is required to achieve a satisfactory connection [3]. Heartwood of Scots pine showed higher water resistance than sapwood in this study. Two basic reasons for this are:

1. Heartwood of Scots pine is more durable than sapwood because it generally contains more extractives and absorbs less water. Heartwood is nearly always much less permeable to water than sapwood due to pit aspiration, obstruction by extractives, and vessel tyloses (outgrowth from an adjacent ray or axial parenchyma cell through a pit cavity in a vessel wall, partially or completely blocking the vessel lumen) [9].

Durability test on Scots pine exposure to weather in Sweden showed that sapwood had a higher moisture uptake and a higher mass loss compared with heartwood [10].

2. As the welded interface is exposed to heat generated due to frictional movement one can anticipate that its shrinkage and swelling have decreased as a consequence of the thermal treatment. In several heat treatment methods developed to modify the wood substrate using elevated temperatures (180–220°C), moisture resistance of wood had increased and its shrinkage and swelling had decreased.

Water absorption test on thermally modified wood showed that the differences between sapwood and heartwood of Scots pine were significantly larger than Spruce [11]. Heat treatment actually increased the water absorption of pine sapwood and decreased the water absorption of pine heartwood.

5. Conclusions and Future Work

- CT-scanning can accurately determine the length and location of the crack and can also be successfully used for wood welding studies.

- This study shows that the welding parameters and wood properties have significant influences on crack length and location. Water resistance of welded Scots pine can be increased by setting welding machine for welding pressure of 1.3 MPa, welding time of 1.5 s, and using heartwood. Cracks created in this parameters combination are smaller and are located at the beginning of the weldline.

- As these examinations were not extensive enough to result in a definite conclusion that welded Scots pine is suitable for outdoor use, the subject needs further research.

- This investigation was carried out only for Scots pine. Other species might show different behaviour.

Acknowledgements

This research was supported by the Luleå University of Technology (LTU) in Sweden and ENSTIB-LERMAB, Université Henri Poincare, Epinal, France.

References

1. H. R. Mansouri, P. Omrani and A. Pizzi, *J. Adhesion Sci. Technol.* **17**, 23–63 (2008).
2. B. Stamm, J. Natterer and P. Navy, *Holz Roh Werkst.* **63**, 313–320 (2005).
3. C. Ganne-Chedeville, M. Properzi, A. Pizzi, J.-M. Leban and F. Pichelin, *Holz Roh Werkst.* **65**, 83–85 (2006).
4. O. Lindgren, *Wood Sci. Technol.* **25**, 425–432 (1991).
5. D. C. Montgomery, *Design and Analysis of Experiments*. Wiley (2005).
6. B. Stamm, E. Windeisen, J. Natterer and G. Wegener, *Holz Roh Werkst.* **63**, 388–389 (2005).
7. J.-M. Leban, A. Pizzi, S. Wieland, M. Zanetti, M. Properzi and F. Pichelin, *J. Adhesion Sci. Technol.* **18**, 73–685 (2004).
8. B. Gfeller, A. Pizzi, M. Zanetti, M. Properzi, F. Pichelin, M. Lehmann and L. Delmotte, *Holzforschung* **58**, 45–52 (2004).
9. J. F. Siau, Wood: influence of moisture on physical properties, *PhD Thesis*, Virginia Polytechnic Institute and State University, Department of Wood Science and Forest Products, Blacksburg, VA, USA (1995).
10. Å. Rydell, M. Bergström and T. Elowson, *Holzforschung* **59**, 183–189 (2005).
11. S. Metsä-Kortelainen, in: *Proceedings of the Second European Conference on Wood Modification*, Göttingen, Germany, pp. 70–73 (2005).

High Density Panels Obtained by Welding of Wood Veneers without any Adhesives

H. R. Mansouri [a,b], J.-M. Leban [c] and A. Pizzi [a,*]

[a] LERMAB, ENSTIB, University Henri Poincare-Nancy 1, 27 rue du Merle Blanc, BP 1041,
88051 Epinal, France
[b] Department of Wood Science and Technology, Faculty of Natural Resources, University of Zabol,
P.O. Box 98615-538 Zabol, Iran
[c] LERFOB, INRA, Institute Nationale de la Recherche Agronomique, centre de recherche de
Champenoux, 54280 Champenoux, France

Abstract

High density panels obtained by pressing wood veneers together in plywood or laminated veneer lumber (LVL) assembly configurations, at temperatures of 225°C to 250°C, high pressure and long press times, as long as 60 min or longer, have been shown by tensile tests, X-ray microdensitometry and scanning electron microscopy to present interphases where wood has welded just by application of the press heat alone. Short press times under the same conditions of pressure and temperature yield poor welding, or no welding at all, in the veneers interphases.

Keywords

Wood veneer welding, plywood, laminated veneer lumber, hot press welding

1. Introduction

Mechanically-induced friction welding techniques have recently been applied to joining wood, without the use of any adhesives [1–6]. Several welding methods for solid wood have been published such as linear vibration welding [1, 2, 5], rotational dowel welding [3], orbital welding [4], and spindle welding and ultrasonic welding of wood [6]. All these methods rely on some friction of the wood surfaces to be welded to achieve the increase of temperature allowing intercellular material melting and thus interfacial welding. Once the main physical and chemical mechanisms of the effect have been understood [1–4, 7–9] other methods to reproduce the same effect with different techniques and for different wood products can be sought. Thus, this paper deals with a different welding technique, for a different

---

* To whom correspondence should be addressed. Tel.: (+33) 329296117; Fax: (+33) 329296138; e-mail: antonio.pizzi@enstib.uhp-nancy.fr

wood product, namely press heating wood veneers to achieve welding to obtain multilayer plywood panels without adhesives.

2. Experimental

2 mm thick rotary-peeled beech (*Fagus sylvatica*) wood veneers were pressed in a laboratory press at 200°C, 225°C and 250°C to form 5-ply panels. The pressures used were 20 bars, 30 bars and 40 bars, for 30, 60 or 75 min. Panel constructions in which the veneers were maintained cross-grained as in plywood, or all with the grain in the same direction, as in laminated veneer lumber (LVL) were studied. Assemblies in which the veneers were dry, as well as assemblies where dry and wet veneers were alternated in the panel construction were investigated. Assemblies in which ethylene glycol was added as a wood plasticizer on the surface of the veneers were also studied. The plywood panels were cut and tested for dry tensile strength according to European Norm EN 314. The results were quite variable, and they are shown in Table 1 for only a few of the best cases.

A second series of samples was pressed in the conditions reported by Cristescu [11], i.e., at 250°C, 40 bars and 4 and 6 min press times.

The samples were examined by variable pressure scanning electron microscopy (SEM) with a LEO 1450 VP microscope (Zeiss, Oberkochen, Germany).

The samples were also examined by X-ray microdensitometry according to the technique already reported [2]. The X-ray microdensitometry equipment used consisted of an X-ray tube producing 'soft rays' (low energy level) with long wavelength emitted through a beryllium window. These were used to produce an X-ray negative photograph of approx. 2 mm thick samples, conditioned at 12% moisture content, at a distance of 2.5 m from the tube. This distance is important to minimise blurring of the image on the film frame (18 × 24 cm) used. The usual exposure conditions were: 4 h, 7.5 kW and 12 mA. Two calibration samples were placed on each negative photograph in order to calculate wood density values. The specimens were

Table 1.

Tensile strength results on welded 5-layer panels. All veneers have parallel grain direction

Press temperature (°C)	Press time (min)	Pressure (bar)	Tensile strength (MPa)
250	75	20	3.19 ± 0.70
225	60	30	1.81 ± 0.59
200	45	30	0.78 ± 0.39
250	60	40	2.32 ± 1.09
250	4	40	0.0
250	6	40	0.0
EN 314 dry strength			$\geqslant 1.0$

tested in this manner on an equipment consisting of an electric generator (INEL XRG3000), a X-ray tube (SIEMENS FK60-04 Mo, 60 kV, 2.0 kW) and a KODAK negative film (Industrex type M100).

3. Results and Discussion

The material obtained after pressing, in the best case, was a high density panel much denser than normal plywood, in reality a different material altogether. Figures 1 and 2 show the X-ray microdensitometry maps of two of the panels produced. The panel in Fig. 1 has the appearance of a solid piece of wood and no interphases can be observed between the 5 veneers that were originally assembled to form the panel. The very high average panel density of 1200 kg/m^3 and densities as high as 1400–1500 kg/m^3 in some parts of the panel justify such a solid wood appearance. The specimen in Fig. 2 has much lower average panel density of 1050 kg/m^3 because it was pressed under less drastic conditions. The 4 interphases between the veneers are still visible and present maximum densities of about 1200 kg/m^3. Comparison of the two panels in Figs 1 and 2 indicates how the welding occurs: first the wood veneer surfaces weld by forming high density interphases only (Fig. 2). As the press

kg/m^3

■ 1450–1500
■ 1400–1450
■ 1350–1400
■ 1300–1350
■ 1250–1300
■ 1200–1250
1150–1200

Figure 1. Density profile of a high density plywood panel pressed at the higher temperature. The panel density is so high that the interphases are not visible as the rest of the panel has a density as high as the interphases. Average panel density = 1203 kg/m^3, 5 plies, veneers stacked with the wood grain in the same direction, 250°C, 60 min press time, veneers pressed dry.

kg/m^3

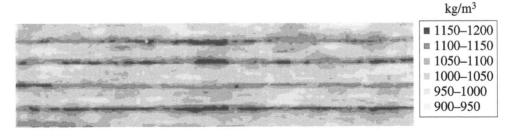

■ 1150–1200
■ 1100–1150
■ 1050–1100
■ 1000–1050
950–1000
900–950

Figure 2. Density profile of a high density plywood panel pressed at the lower temperature. The panel density is not as high as in Fig. 1 and the areas of higher density, the interphases, are visible. Average panel density = 1050 kg/m^3, 5 plless, veneers stacked with the wood grain in the same direction, 200°C, 60 min press time, veneers pressed dry.

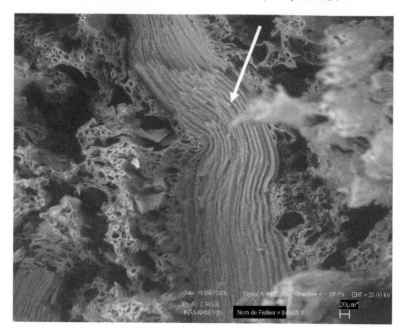

Figure 3. SEM micrograph showing the loss of wood cells morphology in the welded interphase (indicated by the arrow) and the conservation of wood cells, although damaged, in parts of the veneers unaffected by welding.

temperature increases it is the whole cellular morphology of the wood that changes, the density increasing throughout the panel to the point that the welded interphases are indistinguishable from the rest of the wood (Fig. 1). The SEM micrograph in Fig. 3 shows at the anatomical level what occurs to the cellular structure of the wood in the interphase. The figure shows that the cells in the interphase have been deformed by the high temperature and pressure used to form the welded interphase (indicated by an arrow) while the rest of the veneers conserve, although damaged, the cell structure characteristic of wood.

The initial results based on standard tensile strength tests for plywood indicated that the panel had strength satisfying well the European Norm EN 314 when the veneers were pressed at 225°C or 250°C for 60 to 75 min at a pressure of 30 or 40 bars. Plywood tensile tests are used to characterise how good the bond between veneers is. While the tensile strength results are good and satisfy the relevant European Norm EN 314, they are nonetheless of little industrial use as the press times used here are far too long. Plywood panels of the same thickness bonded with adhesives require press times of only 5–6 min at temperatures between 115° and 125°C. If one considers the panels obtained without adhesives as a new material altogether, thus not comparable to plywood, press times are still too long to be industrially and economically significant. Recent report [11] stating that similar panels were obtained using press times as short as 4 or 6 min at 250°C and the same maximum pressure used in this article needed checking. The work cited in [11] did not test for tensile strength but exclusively for bending strength, and while shear strength was mentioned its values were not reported. Bending strength in wood panels almost exclusively depends on panel assembly geometry and this test is not considered a measure of how good the bond is in the interphases between the wood veneers. The tensile test results on panels pressed under the same conditions as reported in [11], shown in Table 1, confirmed that at such short press times the panel obtained did not show good performance in this respect, namely that the interphases were not welded to a sufficient extent, if at all.

4. Conclusions

1. High density panels obtained by pressing wood veneers together at temperatures of 225°C to 250°C, high pressure and long press times, as long as 60 min or longer, have been shown to present interphases that are welded just by the application of heat alone.

2. Short press times under the same conditions of pressure and temperature as in (1) yield poor welding, or no welding at all, in the veneers interphases.

3. The claim in [11] that short press times are possible under the same conditions of pressure and temperature as in (1) has been proven incorrect. The tests in [11] rely only on bending strength values which do not correspond to the interphase and thus are not able to distinguish if welding has occurred or not.

Acknowledgements

The authors wish to thank the financial support of the CPER 2007–2013 'Structuration du Pôle de Compétitivité Fibres Grand'Est' (Competitiveness Fibre Cluster), through local (Conseil Général des Vosges), regional (Région Lorraine), national (DRRT and FNADT) and European (FEDER) funds.

References

1. B. Gfeller, M. Zanetti, M. Properzi, A. Pizzi, F. Pichelin, M. Lehmann and L. Delmotte, *J. Adhesion Sci. Technol.* **17**, 1425–1590 (2003).
2. J.-M. Leban, A. Pizzi, S. Wieland, M. Zanetti, M. Properzi and F. Pichelin, *J. Adhesion Sci. Technol.* **18**, 673–685 (2004).
3. A. Pizzi, J.-M. Leban, F. Kanazawa, M. Properzi and F. Pichelin, *J. Adhesion Sci. Technol.* **18**, 1263–1278 (2004).
4. B. Stamm, J. Natterer and P. Navi, *Holz Roh Werkstoff* **63**, 313–320 (2005).
5. H. R. Mansouri, P. Omrani and A. Pizzi, *J. Adhesion Sci. Technol.* **23**, 63–70 (2009).
6. G. Tondi, S. Andrews, A. Pizzi and J.-M. Leban, *J. Adhesion Sci. Technol.* **21**, 1633–1643 (2007).
7. P. Omrani, A. Pizzi, H. R. Mansouri, J.-M. Leban and L. Delmotte, *J. Adhesion Sci. Technol.* **23**, 827–837 (2009).
8. L. Delmotte, H. R. Mansouri, P. Omrani and A. Pizzi, *J. Adhesion Sci. Technol.* **23**, 1271–1279 (2009).

9. P. Omrani, H. R. Mansouri, G. Duchanois and A. Pizzi, *J. Adhesion Sci. Technol.* **23**, 2057–2072 (2009).
10. European Norm EN 314 — specifications for plywood.
11. C. Cristescu, Bonding veneers using only heat and pressure: focus on bending and shear strength, *Licentiate Thesis*, Luleå University of Technology, LTU Skellefteå, Sweden (2008).

Overview of European Standards for Adhesives Used in Wood-Based Products

Frédéric Simon and Guillaume Legrand *

FCBA, Institut Technologique Forêt Cellulose Bois-Construction Ameublement, Pôle Industries Bois-Construction, Allée de Boutaut, B.P. 227, 33028 Bordeaux, France

Abstract
This paper is an overview of European standards dealing with adhesives for wood-based products. No standard exists in Europe for the assessment of adhesives for wood-based panels except for the evaluation of the bonding quality of plywood panels. However, many standards exist and are described here for the assessment of adhesives for structural purposes such as glulam beams or non-structural purposes such as glued components for joinery.

Keywords
European standards, adhesives, wood, plywood, glulam, load-bearing timber structures, glued components

1. Introduction

The current European standards for wood-based panels deal with evaluation of properties of end-products such as internal bond strength or bending strength. In case of plywood, the bonding quality can be evaluated by tensile shear testing according to EN 314-1 (2005) [1]. These properties are fully linked to adhesives used and process parameters. Nevertheless, no standard exists in Europe for the assessment of adhesives for wood-based panels.

However, many European standards exist for the assessment of adhesives for structural purposes such as glulam beams for evident safety reasons. These standards allow to characterize in the most efficient way the adhesive properties such as, for example, shear strength according to EN 302-1 (2004) [2] or creep behaviour according to EN 15416-3 (2008) [3].

The same tendency can be observed for adhesives for non-structural purposes such as glued components for joinery. In this case, properties such as shear strength can be evaluated according to EN 205 (2003) [4].

---

* To whom correspondence should be addressed. Tel.: +33-55 64 36 448; e-mail: Guillaume.LEGRAND@fcba.fr

Wood Adhesives

Table 1.
Ageing treatments and service classes for plywood panels according to EN 314-2 (1993) [5]

	Ageing treatments			
	24 h in water at $(20 \pm 3)°C$	6 h in boiling water — cooling in water at $(20 \pm 3)°C$ for at least 1 h	4 h in boiling water — drying in oven for 16 to 20 h at $(60 \pm 3)°C$ — 4 h in boiling water — cooling in water at $(20 \pm 3)°C$ for at least 1 h	72 h in boiling water — cooling in water at $(20 \pm 3)°C$ for at least 1 h
Service class 1 dry use	X	Not required	Not required	Not required
Service class 2 wet use	X	X	Not required	Not required
Service class 3 exterior use	X	Not required	X	X

2. Adhesives for Plywood

Plywood panels produced in Europe for building purposes use urea-formaldehyde (UF), melamine-urea-formaldehyde (MUF) and phenol-formaldehyde (PF) adhesives. Standards requirements are not focused on direct properties of these adhesives but properties of the end-products (closely linked to adhesive properties and process parameters). Nevertheless, EN 314-1 (2005) [1] defines a test method for the evaluation of the bonding quality of plywood panels, but also blockboards, laminated boards and laminated veneer lumbers (LVL), by shear testing before and after ageing treatments. Relevant requirements are specified in EN 314-2 (1993) [5] which also defines the ageing treatments required for the intended service class of the tested panel as shown in Table 1.

Test samples are designed to determine the bonding quality of each ply tested in tensile shear, according to Fig. 1.

Requirements defined in EN 314-2 (1993) [5] allow to link shear strength and wood failure percentage, as shown in Table 2.

3. Adhesives for Load-Bearing Timber Structures

3.1. Phenolic and Aminoplastic Adhesives

Most of the glulam beams produced in Europe for load-bearing timber structures use melamine-urea-formaldehyde (MUF) and phenol-resorcinol-formaldehyde (PRF) adhesives. Nevertheless, PRF consumption has been decreasing continuously for several years because of the colour of the joints obtained, increased productivity requirements and environmental issues.

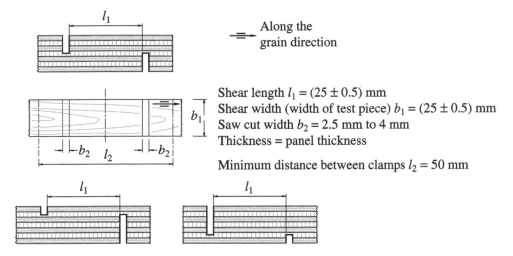

Figure 1. Test sample design for tensile shear testing according to EN 314-1 (2005) [1] for a 7 ply lay-up.

Table 2.
Performance requirements according to EN 314-2 (1993) [5] after tensile shear testing according to EN 314-1 (2005) [1]

Mean shear strength f_v (N/mm^2)	Wood failure (%)
$0.2 \leqslant f_v < 0.4$	$0.2 \leqslant f_v < 0.4$
$0.4 \leqslant f_v < 0.6$	$\geqslant 80$
$0.6 \leqslant f_v < 1.0$	$\geqslant 60$
$1.0 \leqslant f_v$	No requirement

These adhesives must meet the performance requirements specified in EN 301 (2006) [6] standard when tested in accordance with the following test methods:

- Tensile shear test according to EN 302-1 (2004) [2] on bonded test samples made from beech (*Fagus Sylvatica* L.);

- Delamination test according to EN 302-2 (2004) [7] on bonded test samples made from spruce (*Picea Abies* L.);

- Fibre damage test according to EN 302-3 (2004) and EN 302-3/A1 (2006) [8] on bonded test samples made from spruce (*Picea Abies* L.);

- Shrinkage stress test according to EN 302-4 (2004) [9] on bonded test samples made from spruce (*Picea Abies* L.).

All these tests have to be carried out with ready to use glue mixes, i.e., adhesive and hardener mixed before application. In case of separate spreading of adhesive and hardener, the delamination test according to EN 302-2 (2004) [7] and the fibre damage test according to EN 302-3 (2004) [8] must also be performed in this regard.

Two types of adhesives are defined in EN 301 (2006) [6] for use in different climatic conditions which are designated as Type I and Type II:

– Type I: prolonged exposure to high temperature or weather;

– Type II: heated and ventilated building; exterior protected from weather with short periods of exposure to weather is also accepted.

All these standards prepared under the direction of the CEN TC193/SC1 Committee are currently under revision. Revisions under discussion focus on test devices and parameters but the main requirements would stay the same. Nevertheless, it is interesting to underline that the new version of EN 301 [6] will probably include more types of adhesives in order to better correspond to the service classes of the glued products.

3.1.1. Tensile Shear Test According to EN 302-1 (2004) [2]

Single lap joint test samples with thin (0.1 mm) and thick (1 mm) gluelines are subjected to ageing treatments before longitudinal tensile loading at a constant loading rate of (2.0 ± 0.5 kN/min) up to failure (Fig. 2).

Bonded assemblies are prepared from straight-grained beech (*Fagus Sylvatica* L.) with a density of (700 ± 50) kg/m^3 at (12 ± 1)% moisture content and annual growth rings oriented at 30° to 90° to the surface to be bonded.

Before testing, test samples must be subjected to the ageing treatments listed in Table 3. After each ageing treatment, the mean results must fulfill the EN 301 (2006) [6] requirements shown in Table 4.

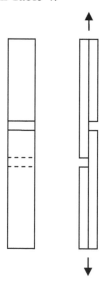

Figure 2. Test sample design (front view and edge view) for tensile shear testing according to EN 302-1 (2004) [2].

Table 3.
Ageing treatments prior to tensile shear testing according to EN 302-1 (2004) [2]

Treatment	Description
A1	Test immediately after obligatory 7 days in standard atmosphere at 20°C and 65% relative humidity
A2	7 days in standard atmosphere 4 days soaking in water at $(20 \pm 5)°C$ Samples tested in the wet state
A3	7 days in standard atmosphere 4 days soaking in water at $(20 \pm 5)°C$ Recondition in standard atmosphere to original mass Samples tested in the dry state
A4	7 days in standard atmosphere 6 h soaking in boiling water 2 h soaking in water at $(20 \pm 5)°C$ Samples tested in the wet state
A5	7 days in standard atmosphere 6 h soaking in boiling water 2 h soaking in water at $(20 \pm 5)°C$ Recondition in standard atmosphere to original mass Samples tested in the wet state

Table 4.
Performance requirements according to EN 301 (2006) [6] after tensile shear testing according to EN 302-1 (2004) [2]

Treatment	Minimum mean shear strength in N/mm^2			
	0.1 mm thin joint		1.0 mm thick joint	
	Type I	Type II	Type I	Type II
A1	10.0	10.0	8.0	8.0
A2	6.0	6.0	4.0	4.0
A3	8.0	8.0	6.4	6.4
A4	6.0	Not required	4.0	Not required
A5	8.0	Not required	6.4	Not required

3.1.2. Delamination Test According to EN 302-2 (2004) [7]

Glulam specimens are subjected to an impregnation/drying procedure. Then, the extent of open gluelines, delamination, as a result of this procedure is measured and compared with the total length of gluelines on both end-grain faces of the specimens.

Table 5.

Water impregnation/drying procedure parameters defined in EN 302-2 (2004) [7]

Treatment	Parameters	Units	High temperature procedure for Type I adhesive	Low temperature procedure for Type II adhesive
Water impregnation	Water temperature	°C	10–25	10–25
	Pressure	kPa	25 ± 5	25 ± 5
	Duration	min	15	15
	Pressure	kPa	600 ± 25	600 ± 25
	Duration	h	1	1
	Number of impregnation cycles	–	2	2
Drying	Air temperature	°C	65 ± 3	27.5 ± 2.5
	Air humidity	%	12.5 ± 2.5	30 ± 5
	Air speed	m/s	2.25 ± 0.25	2.25 ± 0.25
	Duration	h	20	90
Number of cycles	–	–	3	2

Bonded assemblies composed of 6 lamellae are prepared from straight-grained spruce with a density of (425 ± 25) kg/m^3 at $(12 \pm 1)\%$ moisture content. They must be bonded with the shortest and the longest closed assembly times given in the technical data sheet of the product to be tested. The open assembly time is set to be shorter than 5 min and the pressure applied for bonding is set to 0.6 N/mm^2.

Water impregnation/drying procedure parameters are given in Table 5.

The delamination measurement must take place within one hour after the final drying step (after each drying step, a mass control of each sample is performed to obtain the original weight of the samples). The total delamination should consider:

– Cohesive crack in the adhesive layer;

– Failure of the glueline between the adhesive layer and the wood lamella. No wood fibers are left attached to the adhesive layer;

– Failure of the glueline within the first layers of the wood cells beyond the adhesive layer, in which the failure path is not influenced by the grain angle and the growth ring structure.

At the end of the last drying cycle, delamination results must fulfill the EN 301 (2006) [6] requirements shown in Table 6.

It is important to point out that if the adhesive is claimed to be suitable for use with hardwood species and/or treated wood, then the adhesive must also be tested on bonded assemblies made from that species or wood treated in that way. Nev-

Table 6.
Performance requirements according to EN 301 (2006) [6] after delamination testing according to EN 302-2 (2004) [7]

Conditioning treatment	Adhesive type	Max. delamination in any sample (%)
High temperature procedure	I	5.0
Low temperature procedure	II	10.0

ertheless, testing parameters according to EN 302-2 (2004) [7] and performance requirements according to EN 301 (2006) [6] were established for softwoods and are not really adapted to hardwood species or treated woods. To solve the problem, the CEN TC193/SC1 Committee started in 2009 a new Work Item to define test methods and requirements for the assessment of bonding of such materials. First drafts are expected in 2010.

3.1.3. Determination of the Effect of Acid Damage to Wood Fibres on the Transverse Tensile Strength According to EN 302-3 (2004) and EN 302-3/A1 (2006) [8]

A simple joint test sample is subjected to temperature and humidity cycles before transverse tensile loading at a constant loading rate of (10 ± 1) kN/min up to failure (Fig. 3). The objective is to evaluate the effect on bond strength of the damage to wood fibres caused by the action of acids from the adhesive during the climatic cycle. Thus, it must be achieved only if either the adhesive mixture or one of the adhesive components when applied separately, show an acid pH value lower than 4.0.

Bonded assemblies are made from straight-grained spruce with a density of (425 ± 25) kg/m^3 at $(12 \pm 1)\%$ moisture content, free from knots and with growth rings no wider than 2 mm oriented at 30–60° to the surfaces to be bonded.

When adhesive and hardener are mixed before application, the glueline thickness must be set to 0.5 mm. In case of separate spreading of adhesive and hardener, the glueline thickness must be set to 0.1 mm. In all cases, pressure applied for bonding must be set to 0.6 N/mm^2 for 24 h and bonded assemblies must be conditioned for 7 to 14 days before cutting test samples and testing.

Test samples must be subjected to a climatic cycle composed of 3 parts described in Table 7.

After the climatic cycle, test samples are stored in standard atmosphere at 20°C and 65% relative humidity until constant mass before transverse tensile testing taking into account the failure types as follows:

– A: solid wood failure;

– B/C: cohesive or interfacial failure along the glueline (B), with or without a fine layer of fibres visible in the failure zone. The extent of fibres (C) visible within the failure area of type B must be given in % rounded to the nearest 10%.

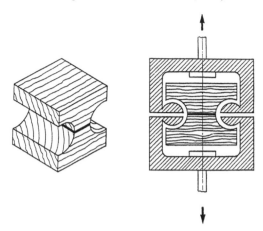

Figure 3. Test sample design (3/4 view and front view with test device) for transverse tensile testing according to EN 302-3 (2004) [8].

Table 7.
Climatic cycle according to EN 302-3/A1 (2006) [8]

Cycle parts designation	Duration (h)	Temperature (°C)	Relative humidity (%)
A	24	50 ± 2	87.5 ± 2.5
B	8	10 ± 2	87.5 ± 2.5
C	16	50 ± 2	<20

Number of repetition of the complete climatic cycle: 4.

According to EN 301 (2006) [6], the mean transverse tensile strength of untreated control test samples must not be lower than 2 N/mm^2 and the mean transverse tensile strength of the treated test samples must not be lower than 80% of the control.

3.1.4. Determination of the Effect of Wood Shrinkage on the Shear Strength According to EN 302-4 (2004) [9]

Crosswise double joint test samples with 0.5 mm thick gluelines are subjected to a drying cycle before compressive shear loading at a constant loading rate of (20 ± 5) kN/min up to failure.

Bonded assemblies are made from straight-grained spruce with a density of (425 ± 25) kg/m^3 at (12 ± 1)% moisture content, free from knots. Core and cover pieces must have a moisture content of (17 ± 0.5)% before bonding. Aluminium frames must be used to set the joint thickness to 0.5 mm and limit the bonded area to (100 ± 0.1) × (100 ± 0.1) mm^2. Pressure applied for bonding must be set to 0.8 N/mm^2 for 24 h and bonded assemblies must be conditioned for 7 days before testing.

Test samples must be subjected to drying at (40 ± 2)°C, (30 ± 2)% relative humidity under an air speed of (0.7 ± 0.1) m/s until a decrease of their moisture

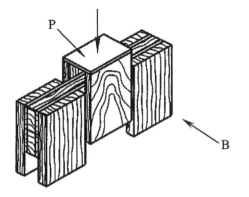

P: Metallic plate to apply the
compressive force
B: Supplementary boards to
maintain the test sample

Figure 4. Test sample design for compressive shear testing according to EN 302-4 (2004) [9].

content of 9 points of humidity. Then, they are stored in standard atmosphere at 20°C and 65% relative humidity for 2 weeks before compressive shear testing. As shown in Fig. 4, four supplementary boards (B) are screwed on the core piece in order to maintain the specimen during the test and a metallic plate (P) is used to apply the compressive force on the cover pieces.

According to EN 301 (2006) [6], the mean compressive shear strength must not be lower than 1.5 N/mm^2.

3.1.5. Determination of Conditions for Use According to prEN 302-5 (2007) [10], EN 302-6 (2004) [11] and EN 302-7 (2004) [12]

If required, conditions for use of adhesives can be determined by the following test methods:

prEN 302-5 (2007) [10] deals with determination of the maximum assembly time. Glulam beam test samples prepared with increasing assembly times are tested according to EN 302-2 [7] until a delamination exceeding 5% is found.

EN 302-6 (2004) [11] deals with determination of the conventional pressing time. Single lap test samples prepared with various pressing times at different temperature levels are tested according to EN 302-1 [2] until a failure strength exceeding 4 N/mm^2 is found.

It is important to point out that these methods have shown weakness in repeatability and reproducibility as adhesives are tested to their limits of use. Thus, results obtained should be considered as indication only.

EN 302-7 (2004) [12] deals with the determination of the conventional adhesive duration of use. The viscosity of a specified volume of adhesive at 20°C is monitored using a Brookfield type viscometer until a viscosity exceeding 25 000 mPa s is found.

3.2. One-Component Polyurethane Adhesives

For the last 10 years, an increasing portion of glulam beams produced in Europe for load-bearing timber structures has used one-component polyurethane (PUR) adhesives because of productivity requirements and environmental issues. These adhesives must meet the performance requirements specified in EN 15425 (2008) [13] standard when tested in accordance with the same test methods as amino-plastic and phenolic adhesives taking into account specific characteristics of such adhesives.

3.2.1. Tensile Shear Test According to EN 302-1 (2004) [2]

Test must be performed in accordance with EN 302-1 [2] with thin (0.1 mm) and thick (0.5 mm instead of 1.0 mm) gluelines. Added ageing treatments A6 and A7 must be performed according to EN 15425 (2008) [13] requirements described in Table 8.

After each ageing treatment, the mean results must fulfill the EN 15425 (2008) [13] requirements shown in Table 9.

3.2.2. Determination of Resistance to Delamination According to EN 302-2 (2004) [7]

Tests must be carried out in accordance with EN 302-2 (2004) [7] and must fulfill the same requirements as aminoplastic and phenolic adhesives.

Table 8.
Added ageing treatments according to EN 15425 (2008) [13] prior to tensile shear testing of PUR adhesives according to EN 302-1 (2004) [2]

Treatment	Description
A6	7 days in standard atmosphere at 20°C and 65% relative humidity 72 h wrapped in an aluminium foil at (50 ± 2)°C Samples tested hot in a temperature controlled test cabinet at (50 ± 2)°C
A7	7 days in standard atmosphere at 20°C and 65% relative humidity 72 h wrapped in an aluminium foil at (70 ± 2)°C Samples tested hot in a temperature controlled test cabinet at (70 ± 2)°C

Table 9.
Performance requirements according to EN 15425 (2008) [13] after tensile shear testing according to EN 302-1 (2004) [2]

Treatment	Minimum mean shear strength in N/mm^2			
	0.1 mm thin joint		0.5 mm thick joint	
	Type I	Type II	Type I	Type II
A1	10.0	10.0	9.0	9.0
A2	6.0	6.0	5.0	5.0
A3	8.0	8.0	7.2	7.2
A4	6.0	Not required	5.0	Not required
A5	8.0	Not required	7.2	Not required
A6	Not required	9.5	Not required	7.2
A7	8.0	Not required	6.5	Not required

3.2.3. Determination of the Effect of Acid Damage to Wood Fibres on the Transverse Tensile Strength According to EN 302-3 (2004) and EN 302-3/A1 (2006) [8]

Tests must be carried out in accordance with EN 302-3 (2004) and EN 302-3/A1 (2006) [8] replacing spruce by beech. According to EN 15425 (2008) [13], the mean transverse tensile strength of untreated control test samples must not be lower than 5 N/mm^2. The test is mandatory for all adhesives irrespective of pH-value.

3.2.4. Determination of the Effect of Wood Shrinkage on the Shear Strength According to EN 302-4 (2004) [9]

Tests must be carried out in accordance to EN 302-4 (2004) [9] and must fulfill the same requirements as aminoplastic and phenolic adhesives.

3.2.5. Determination of Resistance to Static Loading in Compression Shear According to EN 15416-2 (2008) [14]

The test is based on ASTM D 3535-90. Multiple bondline specimens are subjected to a constant load equal to 3870 N (inducing shear stresses on the bondline of 3 N/mm^2) under a climatic cycle composed of 3 parts described in Table 10.

Bonded assemblies are prepared from straight-grained beech (*Fagus Sylvatica* L.) with a density of (700 ± 50) kg/m^3 at (12 ± 1)% moisture content.

If at least 5 test samples are still intact after 336 h, test samples are unloaded and the creep deformation is measured to the nearest 0.01 mm for all gluelines on both sides. According to EN 15425 (2008) [13] requirements, the mean creep deformation of the gluelines must not exceed 0.05 mm.

3.2.6. Determination of Resistance to Creep in Bending Shear According to EN 15416-3 (2008) [3]

Simple joint test samples are subjected to 4-point constant load equal to 2000 N under climatic cycles (Fig. 5).

Bonded assemblies are made from straight-grained spruce with a density of (425 ± 25) kg/m^3 at (12 ± 1)% moisture content, free from knots. 4 lamellae are prepared from one board and matched according to Fig. 6(a) in order to obtain 2 specimens, one glued with the PUR adhesive to be tested and the other with a PRF adhesive already approved according to EN 301 (Fig. 6(b)). A total of 5 pairs of

Table 10.
Climatic cycle according to EN 15416-2 (2008) [14]

Cycle parts designation	Temperature (°C)	Relative humidity (%)	Equilibrium moisture content (%)	Duration (h)
1	70 ± 2	Ambient (i.e., 5–10)	1–1.5	336
2	20 ± 2	85 ± 5	18.5	336
3	50 ± 2	75 ± 5	13	336

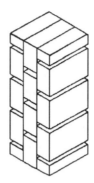

Figure 5. Test sample design for static loading in compression shear according to EN 15416-2 (2008) [14].

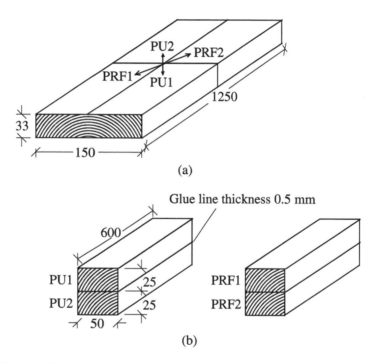

Figure 6. Wood sampling for preparation of test samples for bending shear according to EN 15416-3 (2008) [3] (numbers in mm).

matching specimens must be prepared. Adhesives have to be used as recommended by the adhesive manufacturer.

After pressing and conditioning for at least 7 days in standard climate at 20°C and 65% relative humidity, final test samples are prepared and subjected to bending shear as shown in Fig. 7 under a climatic cycle composed of 2 parts described in Table 11.

The mid-span deflection (w) must be recorded to the nearest 0.01 mm 1 min after loading ($w(0)$) and subsequently at least once a week ($w(t)$) for 26 weeks.

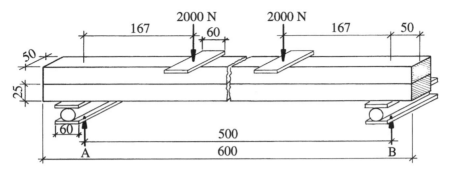

Figure 7. Bending test according to EN 15416-3 (2008) [3] (numbers in mm).

Table 11.
Climate cycle according to EN 15416-3 (2008) [3]

Cycle parts designation	Temperature (°C)	Relative humidity (%)	Duration (h)
1	20 ± 2	85 ± 5	168
2	45 ± 2	40 ± 5	168

For each specimen, the relative creep deformation value $k_{\text{def}}(t)$ must be calculated according to:

$$k_{\text{def}}(t) = (w(t)/w(0)) - 1.$$

For each pair of matching specimens, the relative creep deformation ratio $RC_i(t)$ must be calculated according to:

$$RC_i(t) = k_{\text{def}}(t), \text{PU},i / k_{\text{def}}(t), \text{PRF},i, \quad i = 1, \ldots, 5.$$

For all 5 pairs of matching specimens, the mean relative creep deformation ratio after 26 weeks must be calculated according to:

$$RC_{\text{mean}} \text{ (26 weeks)} = \frac{1}{5} \sum_{1}^{5} RC_i \text{ (26 weeks)}.$$

According to EN 15425 (2008) [13] requirements, if CR_{mean} (26 weeks) is lower than 1.12, the test is completed. If not, the test must be continued for a second period of 26 weeks and CR_{mean} (52 weeks) must be lower than 1.15.

3.3. EPI Adhesives

Emulsion Polymer Isocyanate (EPI) adhesives appeared 2 years ago in Europe for the manufacture of glulam beams because of environmental issues. These adhesives are still not covered by a European standard but must meet the performance requirements specified by the CEN TC193/SC1 Committee in a draft that will be published in 2010. Associated test methods are the same as for PUR adhesives, taking into account specificities of EPI adhesives such as the gap filling capacity. In this way, testing according to EN 302-1 [2], EN 302-3 [8], EN 302-4 [9] and EN 15416-3 [3] should be performed with maximum glueline thickness of 0.3 mm. This explains that the use of EPI adhesives will be limited to straight beams of maximum 150×300 mm^2 cross section and 12 m length made with softwoods for which it is possible to guarantee that glueline thickness will not exceed 0.2 mm.

3.4. Fire Behaviour of Adhesives, Rehabilitation of Timber Structures and Adhesives for Glued-in Rods Used for the Assembly or the Reinforcement of Glulam Beams

These topics are of great interest because of lack of knowledge or fast development of new building technologies. In this regard, work items and draft standards are actually under discussion and development by the CEN TC193/SC1 Committee. First drafts would probably be available in 2010.

4. Adhesives for Non-structural Applications

Most of the glued components produced in Europe for non-structural purposes such as joinery use urea-formaldehyde (UF), poly(vinyl acetate) (PVAc) or polyurethane (PUR) adhesives. These adhesives must meet the performance requirements speci-fied in EN 204 (2002) [15] for thermoplastic adhesives or EN 12765 (2002) [16] for thermosetting adhesives when tested in accordance with the following test methods:

- Tensile shear test according to EN 205 (2003) [4] on test samples made from beech (*Fagus Sylvatica* L.).

4 types of adhesives are defined in these standards for use in different climatic conditions denoted hereafter as:

- D1 or C1 class: inside;

- D2 or C2 class: inside, rare contact with water;

- D3 or C3 class: protected outside;

- D4 or C4 class: outside, exposed to the weather.

4.1. Tensile Shear Test According to EN 205 (2003) [4]

This test method is very close to the one described in EN 302-1 (2004) [2]. The main difference is the loading rate equal to 50 mm/min for thermoplastic adhesives and 6 to 9 mm/min for thermosetting adhesives.

Before testing, test samples must be exposed to the ageing treatments defined in EN 204 (2002) [15] and EN 12765 (2002) [16] as shown in Tables 12 and 13.

Table 12.
Ageing treatments according to EN 204 (2002) [15] prior to tensile shear testing according to EN 205 (2003) [4] (NR = not required)

Treatment	Description	Minimum mean shear strength in N/mm^2 for each class			
		D1	D2	D3	D4
1	Test immediately after 7 days in standard atmosphere at 20°C and 65% relative humidity	10	10	10	10
2	7 days in standard atmosphere 3 h soaking in water at $(20 \pm 5)°C$ Recondition 7 days in standard atmosphere Samples tested in the dry state	NR	8	NR	NR
3	7 days in standard atmosphere 4 days soaking in water at $(20 \pm 5)°C$ Samples tested in the wet state	NR	NR	2	4
4	7 days in standard atmosphere 4 days soaking in water at $(20 \pm 5)°C$ Recondition 7 days in standard atmosphere Samples tested in the dry state	NR	NR	8	NR
5	7 days in standard atmosphere [20/65] 6 h soaking in boiling water 2 h soaking in water at $(20 \pm 5)°C$ Samples tested in the wet state	NR	NR	NR	4

5. Conclusion

The aim of this paper was to present European standards for the assessment of adhesives for wood-based products.

For wood-based panels, European standards are focused on general properties of end-products. The only standard really linked to adhesion is EN 314-1 (2005) [1] which defines a test method for evaluation of the bonding quality of plywood panels.

For glulam beams and load bearing timber structures, standards are numerous and stringent. A significant distinction is made between adhesives types, taking into account their chemical and mechanical properties. In this way, it is necessary to consider EN 301 (2006) [6] for aminoplastic and phenolic adhesive and EN 15425 (2008) [13] for polyurethane adhesives. Associated test methods are described to assess the critical properties of these adhesives and to evaluate the differences between them in more or less the same conditions.

For glued components for non-structural applications, the same tendency can be seen and one must consider EN 204 (2002) [15] for thermoplastic adhesives and EN 12765 (2002) [16] for thermosetting adhesives.

Table 13.
Ageing treatments according to EN 12765 (2002) [16] prior to tensile shear testing according to EN 205 (2003) [4] (NR = not required)

Treatment	Description	Minimum mean shear strength in N/mm^2 for each class			
		C1	C2	C3	C4
1	Test immediately after 7 days in standard atmosphere at 20°C and 65% relative humidity	10	10	10	10
2	7 days in standard atmosphere 1 day soaking in water at (20 ±5)°C Samples tested in the wet state	NR	7	7	7
3	7 days in standard atmosphere 3 h soaking in water at (67 ± 2)°C 2 h soaking in water at (20 ± 5)°C Samples tested in the wet state	NR	NR	4	NR
4	7 days in standard atmosphere [20/65] 6 h soaking in boiling water 2 h soaking in water at (20 ± 5)°C Samples tested in the wet state	NR	NR	NR	4

Last, but not least, a serious interest is to reduce formaldehyde emissions coming from all types of glued products. This is the reason for development of new formaldehyde-free products such as polyurethane adhesives and emulsion polymer isocyanate (EPI) adhesives for load bearing structures.

References

1. EN 314-1:2005, Plywood, Bonding quality — Part 1: Test methods.
2. EN 302-1:2004, Adhesives for load bearing structures — Test methods — Part 1: Determination of bond strength in longitudinal tensile shear.
3. EN 15416-3:2008, Adhesives for load bearing timber structures — Test methods — Part 3: Creep deformation test at cyclic climate conditions with specimens loaded in bending shear.
4. EN 205:2003, Test methods for wood adhesives for non-structural applications — Determination of tensile shear strength of lap joints.
5. EN 314-2:1993, Plywood, Bonding quality — Part 2: Performance requirements.
6. EN 301:2006, Adhesives, phenolic and aminoplastic, for load bearing structures — Classification and performance requirements.
7. EN 302-2:2004, Adhesives for load bearing structures — Test methods — Part 2: Determination of resistance to delamination.
8. EN 302-3:2004 and EN 302-3/A1:2006, Adhesives for load bearing structures — Test methods — Part 3: Determination of the effect of acid damage to wood fibres by temperature and humidity cycling on the transverse tensile strength.

9. EN 302-4:2004, Adhesives for load bearing structures — Test methods — Part 4: Determination of the effect of wood shrinkage on the shear strength.

10. prEN 302-5:2007, Adhesives for load bearing structures — Test methods — Part 5: Determination of the maximum open assembly time.

11. EN 302-6:2004, Adhesives for load bearing structures — Test methods — Part 6: Determination of the conventional pressing time.

12. EN 302-7:2004, Adhesives for load bearing structures — Test methods — Part 7: Determination of the conventional working life.

13. EN 15425:2008, Adhesives, one-component polyurethane, for load bearing timber structures — Classification and performance requirements.

14. EN 15416-2:2008, Adhesives for load bearing timber structures — Test methods — Part 2: Static load test of multiple bondline specimens in compression shear.

15. EN 204:2002, Classification of thermoplastic wood adhesives for non-structural applications.

16. EN 12765:2002, Classification of thermosetting wood adhesives for non-structural applications.